AF356592

The Science and Technology of 3D Printing

The Science and Technology of 3D Printing

Editor

Tuhin Mukherjee

MDPI • Basel • Beijing • Wuhan • Barcelona • Belgrade • Manchester • Tokyo • Cluj • Tianjin

Editor
Tuhin Mukherjee
Department of Materials Science and Engineering
The Pennsylvania State University
United States

Editorial Office
MDPI
St. Alban-Anlage 66
4052 Basel, Switzerland

This is a reprint of articles from the Special Issue published online in the open access journal *Materials* (ISSN 1996-1944) (available at: www.mdpi.com/journal/materials/special_issues/tdp20).

For citation purposes, cite each article independently as indicated on the article page online and as indicated below:

LastName, A.A.; LastName, B.B.; LastName, C.C. Article Title. *Journal Name* **Year**, *Volume Number*, Page Range.

ISBN 978-3-0365-2585-3 (Hbk)
ISBN 978-3-0365-2584-6 (PDF)

Contents

About the Editor . . . . **vii**

Tuhin Mukherjee
Special Issue: The Science and Technology of 3D Printing
Reprinted from: *Materials* **2021**, *14*, 6261, doi:10.3390/ma14216261 . . . . **1**

Yitao Chen, Xinchang Zhang, Mohammad Masud Parvez and Frank Liou
A Review on Metallic Alloys Fabrication Using Elemental Powder Blends by Laser Powder
Directed Energy Deposition Process
Reprinted from: *Materials* **2020**, *13*, 3562, doi:10.3390/ma13163562 . . . . **5**

Jaime Varela, Edel Arrieta, Muktesh Paliwal, Mike Marucci, Jose H. Sandoval, Jose A. Gonzalez, Brandon McWilliams, Lawrence E. Murr, Ryan B. Wicker and Francisco Medina
Investigation of Microstructure and Mechanical Properties for Ti-6Al-4V Alloy Parts Produced
Using Non-Spherical Precursor Powder by Laser Powder Bed Fusion
Reprinted from: *Materials* **2021**, *14*, 3028, doi:10.3390/ma14113028 . . . . **25**

Lei-Lei Xing, Wen-Jing Zhang, Cong-Cong Zhao, Wen-Qiang Gao, Zhi-Jian Shen and Wei Liu
Influence of Powder Bed Temperature on the Microstructure and Mechanical Properties of
Ti-6Al-4V Alloy Fabricated via Laser Powder Bed Fusion
Reprinted from: *Materials* **2021**, *14*, 2278, doi:10.3390/ma14092278 . . . . **41**

Peter Pokorný, Štefan Václav, Jana Petru and Michaela Kritikos
Porosity Analysis of Additive Manufactured Parts Using CAQ Technology
Reprinted from: *Materials* **2021**, *14*, 1142, doi:10.3390/ma14051142 . . . . **55**

S. Mohammad H. Hojjatzadeh, Qilin Guo, Niranjan D. Parab, Minglei Qu, Luis I. Escano, Kamel Fezzaa, Wes Everhart, Tao Sun and Lianyi Chen
In-Situ Characterization of Pore Formation Dynamics in Pulsed Wave Laser Powder Bed Fusion
Reprinted from: *Materials* **2021**, *14*, 2936, doi:10.3390/ma14112936 . . . . **69**

Alexander E. Wilson-Heid, Erik T. Furton and Allison M. Beese
Contrasting the Role of Pores on the Stress State Dependent Fracture Behavior of Additively
Manufactured Low and High Ductility Metals
Reprinted from: *Materials* **2021**, *14*, 3657, doi:10.3390/ma14133657 . . . . **81**

Rhys Jones, Calvin Rans, Athanasios P. Iliopoulos, John G. Michopoulos, Nam Phan and Daren Peng
Modelling the Variability and the Anisotropic Behaviour of Crack Growth in SLM Ti-6Al-4V
Reprinted from: *Materials* **2021**, *14*, 1400, doi:10.3390/ma14061400 . . . . **97**

Praveen S. Vulimiri, Hao Deng, Florian Dugast, Xiaoli Zhang and Albert C. To
Integrating Geometric Data into Topology Optimization via Neural Style Transfer
Reprinted from: *Materials* **2021**, *14*, 4551, doi:10.3390/ma14164551 . . . . **107**

Jaime Eduardo Regis, Anabel Renteria, Samuel Ernesto Hall, Md Sahid Hassan, Cory Marquez and Yirong Lin
Recent Trends and Innovation in Additive Manufacturing of Soft Functional Materials
Reprinted from: *Materials* **2021**, *14*, 4521, doi:10.3390/ma14164521 . . . . **123**

Salman Pervaiz, Taimur Ali Qureshi, Ghanim Kashwani and Sathish Kannan
3D Printing of Fiber-Reinforced Plastic Composites Using Fused Deposition Modeling: A Status Review
Reprinted from: *Materials* **2021**, *14*, 4520, doi:10.3390/ma14164520 . . . . **159**

Daniel Günther, Patricia Erhard, Simon Schwab and Iman Taha
3D Printed Sand Tools for Thermoforming Applications of Carbon Fiber Reinforced Composites—A Perspective
Reprinted from: *Materials* **2021**, *14*, 4639, doi:10.3390/ma14164639 . . . . **187**

About the Editor

Tuhin Mukherjee

Tuhin Mukherjee received a Bachelor of Technology degree in Mechanical Engineering from the West Bengal University of Technology, West Bengal, India, in 2012, and a Master of Technology degree in Mechanical Engineering (Manufacturing Engineering) from the Indian Institute of Technology, Bombay, Mumbai, India, in 2014. He finished his Ph.D. in Materials Science and Engineering from the Pennsylvania State University, United States, in 2019. Currently, he is a postdoctoral scholar at the Pennsylvania State University. His research interests include additive manufacturing, welding, numerical modeling, heat and mass transfer, thermal distortion and residual stress, composition change, and machine learning. He has been published in several international journals including Nature Reviews Materials, Nature Materials, Progress in Materials Science, Acta Materialia, Nature Group Scientific Report, and npj Computational Materials. He was awarded the American Welding Society Graduate Research fellowship in 2016, and was awarded the Robert E. Newnham Research Excellence Award by Penn State University in 2017.

 **materials**

Editorial

Special Issue: The Science and Technology of 3D Printing

Tuhin Mukherjee 

Department of Materials Science and Engineering, The Pennsylvania State University, University Park, PA 16802, USA; tuhin@psu.edu

Citation: Mukherjee, T. Special Issue: The Science and Technology of 3D Printing. *Materials* **2021**, *14*, 6261. https://doi.org/10.3390/ma14216261

Received: 9 October 2021
Accepted: 18 October 2021
Published: 21 October 2021

Publisher's Note: MDPI stays neutral with regard to jurisdictional claims in published maps and institutional affiliations.

1. Introduction

Additive manufacturing, commonly known as three-dimensional printing (3D printing), is becoming an increasingly popular method for making components that are difficult to fabricate using traditional manufacturing processes. It enables a one-step fabrication of complex parts directly from a 3D design. 3D printed parts are now regularly used in medical, aerospace, automotive, energy, marine, and consumer product industries [1]. Examples of printed parts include patient-specific, customized medical implants; aero-engine components; parts with complex, intricate features and internal channels; lattice structures; and materials with site-specific chemical compositions, microstructures, and properties [2]. These parts are printed using metallic alloys, polymers, ceramics, and composites. However, the printing of metals and metallic alloys is the fastest developing field because of its applications, demand, and ability to print unique, functional parts. Depending on the material, geometry, and complexity of the part, several 3D printing processes can be employed [2]. For example, for printing metallic parts, powder bed fusion and directed energy deposition processes are commonly used. Thin layers of the powder of wire feedstocks are melted using a high-energy laser, electron beam, or electric arc, which form the part after solidification. Similarly, several processes are used in the industry to print parts with polymers, ceramic, and composites.

Several scientific and technological aspects of 3D printing processes are poorly understood [1]. For example, metal printing involves rapid melting, heat transfer, the convective flow of liquid metal, solidification, and cooling, all of which affect the part's geometry, microstructure, and properties [2]. Depending on the printing process, materials, and processing conditions, the cooling rates, temperature gradient, and solidification growth rates may vary significantly, which can produce a wide variety of grain structures, morphologies, and textures. Printed parts often suffer from defects such as porosity and cracking that degrade the mechanical properties, quality, and serviceability of the components. In addition, process planning and control to increase productivity without affecting the part quality is a challenging task. All of the scientific and technological issues of 3D printing, as discussed, affect the cost and market penetration of printed parts.

Research and development projects are being performed worldwide to provide a better understanding of the science and technology of 3D printing to make high-quality parts in a cost-effective and time-efficient manner. This Special Issue includes contemporary, unique, and impactful research on 3D printing from leading organizations worldwide.

2. Contributions to This Special Issue

This Special Issue contains eleven articles, including three reviews [3–5], one perspective article [6], and seven research articles [7–13] from leading institutes in the United States, China, Australia, Germany, Sweden, the Netherlands, Slovakia, the Czech Republic, Egypt, and United Arab Emirates. These articles cover the 3D printing of diverse materials such as metallic materials [5,7–13], composites [4,6], and soft materials [3].

The articles in this Special Issue cover a wide variety of experimental [8–11,13], theoretical [12], and data-science [7]-based research on the science and technology of 3D printing. For example, experimental investigations were performed to identify the most important factors that affect the microstructure and properties of Ti-6Al-4V parts printed using powder bed fusion [9,11]. In particular, the effects of powder morphology, preheating temperature, and post-process heat treatment on the size of α-grains, the hardness, and the tensile properties were studied. In addition, formation mechanisms of defects such as pores and cracks, and their harmful effects on the mechanical properties of printed parts were investigated [8,10,12,13]. A high-speed synchrotron X-ray imaging technique [10] was used to reveal the mechanisms of the evolution of pores during powder bed fusion. A new computer-aided quality (CAQ) technology [13] was proposed and used to determine the amount of porosities in the parts printed using powder bed fusion. The mechanical properties of stainless steel and titanium alloy parts were shown to be affected by the presence of pores [8]. A theoretical model was proposed and used to identify the conditions of crack growth and to discuss the detrimental effects of cracking on the mechanical properties of printed parts [12].

The three reviews and the perspective article indicate the progress made, the existing challenges, and the research needs in the contemporary fields of study. Recent trends and innovations in the printing of soft materials for wearable devices, soft robotics, and tissue engineering were reviewed by Regis et al. [3]. Pervaiz et al. [4] reviewed the 3D printing of fiber-reinforced plastic composites using fused deposition modeling. In fiber-reinforced plastic composites, fibers are mixed in a polymeric matrix. These composite materials are printed to make parts for the defense, automotive, aerospace, and sports equipment industries. The review also emphasized several challenges in the printing of these composites that need to be solved to increase market penetration. The printing of metallic materials allows for the fabrication of parts with unique chemical compositions using elemental powder blends. Chen et al. [5] critically reviewed the progress made in this area and emphasized several critical technical challenges that require further research. Finally, a perspective on the printing of tools for carbon fiber-reinforced composite applications using the binder jet process was provided [6].

3. Summary and Outlook

The eleven articles published in this Special Issue focus on the recent advancements in the 3D printing of metallic materials, composites, and soft materials. The achievements reported in this Special Issue are unique, high-quality, and impactful. Due to the wide variety of published work, including experimental, theoretical, and data science-based research, this Special Issue would be of significant and immediate interest to a diverse materials community.

Apart from the exciting advancements, the articles of this Special Issue also identified several scientific, technological, and economic challenges that need immediate attention. The articles pointed towards future research and development that are necessary to print high-quality parts in a cost-effective manner. Clearly, the work in these important areas of 3D printing is just beginning and several decades of research are needed to make 3D printing commercially viable for small and medium-sized companies worldwide.

Funding: This research received no external funding.

Conflicts of Interest: The author declares no conflict of interest.

References

1. DebRoy, T.; Mukherjee, T.; Milewski, J.O.; Elmer, J.W.; Ribic, B.; Blecher, J.J.; Zhang, W. Scientific, technological and economic issues in metal printing and their solutions. *Nat. Mater.* **2019**, *18*, 1026–1032. [CrossRef]
2. DebRoy, T.; Mukherjee, T.; Wei, H.L.; Elmer, J.W.; Milewski, J.O. Metallurgy, mechanistic models and machine learning in metal printing. *Nat. Rev. Mater.* **2021**, *6*, 48–68. [CrossRef]
3. Regis, J.E.; Renteria, A.; Hall, S.E.; Hassan, M.S.; Marquez, C.; Lin, Y. Recent Trends and Innovation in Additive Manufacturing of Soft Functional Materials. *Materials* **2021**, *14*, 4521. [CrossRef] [PubMed]
4. Pervaiz, S.; Qureshi, T.A.; Kashwani, G.; Kannan, S. 3D Printing of Fiber-Reinforced Plastic Composites Using Fused Deposition Modeling: A Status Review. *Materials* **2021**, *14*, 4520. [CrossRef] [PubMed]

5.   Chen, Y.; Zhang, X.; Parvez, M.M.; Liou, F. A Review on Metallic Alloys Fabrication Using Elemental Powder Blends by Laser Powder Directed Energy Deposition Process. *Materials* **2020**, *13*, 3562. [CrossRef] [PubMed]
6.   Günther, D.; Erhard, P.; Schwab, S.; Taha, I. 3D Printed Sand Tools for Thermoforming Applications of Carbon Fiber Reinforced Composites—A Perspective. *Materials* **2021**, *14*, 4639. [CrossRef] [PubMed]
7.   Vulimiri, P.S.; Deng, H.; Dugast, F.; Zhang, X.; To, A.C. Integrating Geometric Data into Topology Optimization via Neural Style Transfer. *Materials* **2021**, *14*, 4551. [CrossRef] [PubMed]
8.   Wilson-Heid, A.E.; Furton, E.T.; Beese, A.M. Contrasting the Role of Pores on the Stress State Dependent Fracture Behavior of Additively Manufactured Low and High Ductility Metals. *Materials* **2021**, *14*, 3657. [CrossRef] [PubMed]
9.   Varela, J.; Arrieta, E.; Paliwal, M.; Marucci, M.; Sandoval, J.H.; Gonzalez, J.A.; McWilliams, B.; Murr, L.E.; Wicker, R.B.; Medina, F. Investigation of Microstructure and Mechanical Properties for Ti-6Al-4V Alloy Parts Produced Using Non-Spherical Precursor Powder by Laser Powder Bed Fusion. *Materials* **2021**, *14*, 3028. [CrossRef] [PubMed]
10. Hojjatzadeh, S.M.H.; Guo, Q.; Parab, N.D.; Qu, M.; Escano, L.I.; Fezzaa, K.; Everhart, W.; Sun, T.; Chen, L. In-Situ Characterization of Pore Formation Dynamics in Pulsed Wave Laser Powder Bed Fusion. *Materials* **2021**, *14*, 2936. [CrossRef]
11. Xing, L.-L.; Zhang, W.-J.; Zhao, C.-C.; Gao, W.-Q.; Shen, Z.-J.; Liu, W. Influence of Powder Bed Temperature on the Microstructure and Mechanical Properties of Ti-6Al-4V Alloy Fabricated via Laser Powder Bed Fusion. *Materials* **2021**, *14*, 2278. [CrossRef]
12. Jones, R.; Rans, C.; Iliopoulos, A.P.; Michopoulos, J.G.; Phan, N.; Peng, D. Modelling the Variability and the Anisotropic Behaviour of Crack Growth in SLM Ti-6Al-4V. *Materials* **2021**, *14*, 1400. [CrossRef] [PubMed]
13. Pokorný, P.; Václav, Š.; Petru, J.; Kritikos, M. Porosity Analysis of Additive Manufactured Parts Using CAQ Technology. *Materials* **2021**, *14*, 1142. [CrossRef] [PubMed]

 *materials*

Review

A Review on Metallic Alloys Fabrication Using Elemental Powder Blends by Laser Powder Directed Energy Deposition Process

Yitao Chen, Xinchang Zhang * , Mohammad Masud Parvez and Frank Liou

Department of Mechanical and Aerospace Engineering, Missouri University of Science and Technology, Rolla, MO 65401, USA; yc4gc@mst.edu (Y.C.); mphf2@mst.edu (M.M.P.); liou@mst.edu (F.L.)

* Correspondence: xz25c@mst.edu

Received: 25 July 2020; Accepted: 10 August 2020; Published: 12 August 2020

Abstract: The laser powder directed energy deposition process is a metal additive manufacturing technique, which can fabricate metal parts with high geometric and material flexibility. The unique feature of in-situ powder feeding makes it possible to customize the elemental composition using elemental powder mixture during the fabrication process. Thus, it can be potentially applied to synthesize industrial alloys with low cost, modify alloys with different powder mixtures, and design novel alloys with location-dependent properties using elemental powder blends as feedstocks. This paper provides an overview of using a laser powder directed energy deposition method to fabricate various types of alloys by feeding elemental powder blends. At first, the advantage of laser powder directed energy deposition in manufacturing metal alloys is described in detail. Then, the state-of-the-art research and development in alloys fabricated by laser powder directed energy deposition through a mix of elemental powders in multiple categories is reviewed. Finally, critical technical challenges, mainly in composition control are discussed for future development.

Keywords: metal additive manufacturing; directed energy deposition; alloy design; elemental powder mixture; advanced materials; composition control

---

1. Introduction

Additive manufacturing (AM) is a novel manufacturing technique that can fabricate a wide range of materials and complex structures. The definition given in the American Society for Testing and Materials (ASTM) states that: AM is "The process of joining materials to make objects from 3D model data, usually layer upon layer, as opposed to subtractive manufacturing methodologies [1]". Due to its layer-based additive nature that is different from subtractive manufacturing, AM created a paradigm shift in the manufacturing industry [2]. AM has lots of advantages compared to traditional subtractive manufacturing. For example, AM can directly produce complex 3D parts without much tooling and assembly. It is also much more material-saving than conventional manufacturing since conventional manufacturing mainly uses the subtractive method to remove materials to reach the desired geometry [3]. Thus, AM has become more essential in the manufacturing industry. As metals and their alloys are of great importance in human life, efforts have been paid on the research and development of AM of metals and alloys [4]. Based on the mechanism and material, the AM process has been classified into seven categories [5]. There are four major categories associated with metal additive manufacturing, which are powder bed fusion (PBF), directed energy deposition (DED), binder jetting (BJ), and laminated object manufacturing (LOM). Among them, PBF and DED are more commercialized [5]. According to ASTM F3187-16 standard guide for DED technique [6], the DED process applies an energy source to fuse feedstock metallic materials by melting during deposition.

Metallic materials in powder or wire form are fed into the melt pool and solidify into a 2D solid layer. The tool path is guided by path planning to fill every layer, and the successive layers will be built until a 3D part is achieved [6]. The laser powder DED process applies laser as the energy source and metal powder as the raw material feedstock. In this paper, we focus on the laser powder DED, and for convenience, here we use DED to represent the laser powder DED process. Using computer-aided design (CAD) tools, a 3D model of a part can be created, and the slicing algorithm can be used to slice the 3D model into many 2D layers. Figure 1 shows the schematic diagram of the DED process using powder as feedstock. During the DED process, a laser beam is applied to create a melt pool, and metal powders are carried and blown into the melt pool by the inert gas. After the laser moves away, melted powders will join and cool down to form a solid layer. The laser travels according to the toolpath for each layer. By repeating this procedure for each layer, a 3D part can be constructed [7].

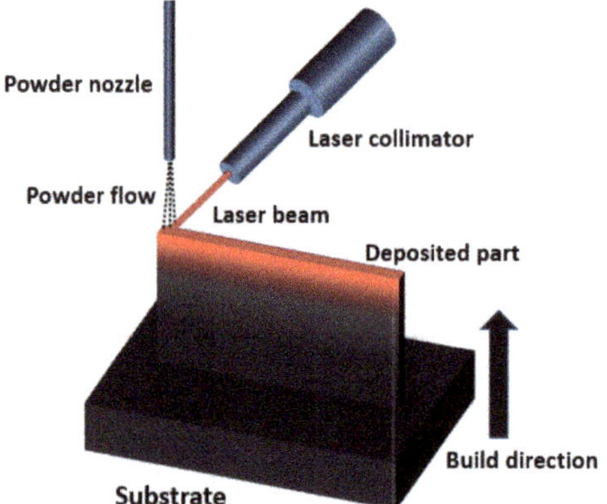

Figure 1. Schematic illustration of DED to build a 3D part.

During the past decade, typical metal alloys fabricated by DED has been studied. These alloys include austenitic stainless steels (304/304L [8,9] and 316/316L [10,11]), precipitation hardening stainless steels (17-4 PH [12]), nickel-based superalloys (Inconel 718 [13,14] and Inconel 625 [15,16]), titanium alloys (Ti-6Al-4V [17–20]), etc. DED and PBF have their own advantages according to their special features. For DED, it is able to build parts on a non-flat surface while PBF usually needs a horizontal area for powder spreading. Thus, modification of the curved surface and repairing of damaged parts are also possible using the DED process [21–23]. The part remanufacturing can be collaborated with reverse engineering to repair damaged parts by constructing a damaged profile and determine the laser scanning strategy, which will be of great significance in saving cost on tooling [24,25]. As an in-situ powder feeding process, it has the potential to fabricate parts without an enclosed chamber [26]. Therefore, the volume of building parts can be much larger than the PBF process. One more important feature is because of the in-situ powder feeding process, DED can flexibly create different material compositions from layer to layer by mixing different powders; however, this is difficult to be realized in PBF. Therefore, much more metallic alloys with various compositions can be potentially directly created by taking this advantage of DED. This novel aspect of DED will be the main topic to review in this article.

Metal powders are the commonly used raw materials and feedstocks in the DED process. Most of the powders used in DED mentioned in the former paragraphs are pre-alloyed powders, which indicate that each powder particle is designed with the prescribed composition. Since the composition is identical in each particle, the composition of the as-fabricated 3D part made by DED is usually close to constant. However, the cost of producing pre-alloyed powders are high. Therefore, similar to using cost-effective elemental powders in powder metallurgy [27–29], the idea of mixing elemental powders into the desired composition to synthesize alloys is also arisen in the area of AM, especially for DED. The cost of alloy powder manufacturing can be reduced by using elemental powders. Also, as each pre-alloyed powder particle has a constant elemental composition, the possibility of fabricating various types of alloys using DED is limited. If the pre-alloyed powders are replaced by elemental powder blends, it is possible to fabricate more alloys with a pre-designed elemental weight percentage or atomic percentage. This replacement can potentially make a great contribution to the development of novel alloys through the thorough investigation of different alloy systems. In addition, with the evolvement of highly automated DED manufacturing systems, the weight composition of multiple elemental powders can be changed during the manufacturing process by in-situ control of the feeding rate. Then, a variety of elemental compositions can coexist within a single part, which can be more functional than homogeneous alloys.

Although a few works have been done using DED to fabricate various alloys, as a relatively new method in DED, a comprehensive overview of the research progress of DED using elemental powder blends has not been done. Thus, an overview of the elemental powder-based DED process can provide new knowledge systems for the metal AM area and potentially develop new alloys, which can significantly widen the application of metal AM in the next generation of manufacturing fields. This review paper will summarize the current research progress in different types of applications via DED and discuss the major technical challenges and issues that remained in this area in order to provide guidelines for future studies.

2. Current Status of DED Using Elemental Powder

The typically reported research works in DED using elemental powders can be generally classified into two categories, which are listed and elaborated in Sections 2.1 and 2.2, respectively. Many types of industrial alloys can be potentially fabricated by mixing the specified composition of elemental powders. Also, conventional alloys can be flexibly modified by mixing elemental powders with other compositions to get an in-depth understanding of how a certain element affects the final properties. The study of the effect of specific elements on the as-fabricated parts will be more direct. With the flexible change in compositions, various types of alloys with a gradual shift in alloy compositions can be joined and form functionally graded materials (FGMs). Using DED and elemental powder blend method can investigate FGMs that are difficult to realize using conventional manufacturing. Novel alloys, especially high entropy alloys (HEAs), which need more types of elements, can be fabricated and designed flexibly by the DED process using elemental powders. Section 3 mainly covers the controlling of the deposition in multiple aspects. The outlook is discussed in Section 4, while the conclusion is summarized in Section 5.

2.1. Industrial Alloys and Intermetallics

There is a wide variety of alloys that are prevalent in the industry due to their excellent mechanical properties. However, due to the high manufacturing cost, they are mostly seen in specific areas. For instance, Ti-6Al-4V is an excellent industrial alloy with high specific strength and corrosion resistance [30]. However, the cost is high for certain industrial areas such as automobile and transportation [31]. Thus, reducing the manufacturing cost of Ti-6Al-4V is essential to expand the applications. Near net shape processing, such as powder metallurgy, was reported as a cost-effective approach to develop and expand the use of titanium alloys [32]. In the area of powder metallurgy, the elemental powder blend is applied to form titanium alloys and avoid the high cost of pre-alloyed powders [32]. Based on the elemental powder method applied in powder metallurgy, blending elemental powder to fabricate alloys has also become a potential method in powder-based metal AM, such as DED, to reduce the manufacturing cost and open new perspective research and industrial fields. Until now, several attempts were made in synthesizing Ti-6Al-4V by a mixture of Ti, Al, and V pure powders via DED. Differences were found between Ti-6Al-4V fabricated by DED using elemental powder blends and conventional Ti-6Al-4V. Manufacturing issues were also identified. More investigation of properties and performances of DED-processed Ti-6Al-4V using elemental powder blends are needed.

Hua et al., Yan et al., and Chen et al. [33–36] mixed Ti, Al, and V elemental powders to fabricate Ti-6Al-4V, which proved the feasibility of making industrial alloys like Ti-6Al-4V using elemental powders in a cost-effective approach. Apart from Ti-6Al-4V, Clayton [37] applied pure Fe, Cr, and Ni powders to make Fe-based alloys similar to stainless steels such as SS316 and SS430. It was found that the Fe-based alloy Fe-17Cr-12Ni made by elemental powder mixture got similar microstructure and mechanical properties with SS316. On the other hand, properties of the alloy Fe-17Ni fabricated by elemental powder mixture was not comparable with the conventional pre-alloyed SS430. More experimental investigations are needed to reveal the attributes of the difference in properties. Similarly, a number of other Fe-, Ti-, and Ni-based alloys in Fe-Cr-Ni and Ti-Al-V alloy systems can be potentially fabricated by mixing elemental powders with certain compositions. It is significantly beneficial to the alloy manufacturing industry that using only a small stock of Fe, Cr, Ni, Ti, Al, and V powders can generate a large number of alloy combinations.

Some types of intermetallic compounds possess excellent mechanical properties that could be widely used in various industries. Thanks to the manufacturing flexibility of the DED process, many hard-to-machine intermetallic compounds can now be fabricated by new methods. As a wide variety of intermetallic compounds only consist of two metal elements, mixing elemental powder can be a convenient way to synthesize those compounds. A couple of intermetallics are attractive for their high wear and corrosion resistance. For instance, Fe-Al intermetallics possess excellent wear, corrosion, and oxidation resistance. It can also function in a high-temperature environment. Conventionally, Fe-Al intermetallics are manufactured by sintering blended elemental powders. However, this process causes high energy consumption and cost [38]. Therefore, the DED process was also introduced for the fabrication of Fe-Al intermetallic [38]. Pęska et al. [39] applied elemental Fe and Al powder to synthesize Fe-Al intermetallics by DED. The hardness of the as-fabricated samples was very similar to the classically built Fe-Al intermetallics.

NiTi is a special intermetallic compound with unique shape memory effects and superelasticity. It is also biocompatible and corrosion resistive [40]. Thus, it is widely used in structures with shape-changing effect and biomedical implants. Machining is difficult to process NiTi. To reduce the machining procedure, attentions are paid on near-net shaping processes, especially AM. Attempts were made by using the DED process to fabricate NiTi and pre-alloyed powders were applied in [41–43]. As pre-alloyed NiTi powder is expensive, blending Ni and Ti elemental powder is an alternative way to in-situ synthesize Ni-Ti alloys and with a variety of composition design. Halani et al. [44] applied the DED process to fabricate NiTi using elemental powder. Different compositions such as Ni55Ti45 and Ni50Ti50 in atomic percentage were attempted for the deposition process. Similarly, Shiva et al. [45] studied the difference among premixed compositions of Ni45Ti55, Ni50Ti50, and Ni55Ti45. Bimber et al. [46–48] conducted works in fabricating NiTi by DED with elemental powder blends. They built large volume structures to find out the difference in the secondary phase, compressive properties, and martensitic transformation temperature regarding the spatial locations. Other works include comparative studies among DED, SLM, and EBM by Wang et al. [49]. Figure 2 shows the microstructure of NiTi alloy fabricated by DED in [49]. Different issues were identified in three metal AM processing methods, which will be discussed in the later sections.

Other types of intermetallics for surface strengthening using elemental powder mixture for DED were also studied. Most of the works focus on the surface strengthening of steel and titanium alloy to improve wear and corrosion resistance by synthesizing intermetallics on the surface. Yu et al. [50] applied elemental Al and Ni powder to synthesize pure Ni_3Al intermetallic compound coating on 1Cr18Ni9Ti stainless steel. Effects of laser energy density on tribological behavior and crystallographic orientation were studied. Wang et al. [51–54] have fabricated various types of intermetallics for surface coating applications using laser cladding. Those works include coating $TiCo/Ti_2Co$ on titanium alloy [51], coating $Ti_5Si_3/NiTi_2$ on titanium alloy [52], and coating Cr_3Si on austenitic stainless steel

1Cr18Ni9Ti by using Cr-Si-Ni elemental powder as the precursor material [53]. Si was also blended with metal powders to provide intermetallics and $Ni_2Si/NiSi$ on 0.2% C carbon steel [54]. Zhong et al. [55] applied the DED process to clad WC/Ni hardfacing alloy on 40Cr steel by using W, C, and Ni elemental powder. The in-situ reaction produced WC hard phase in the Ni matrix for surface wear resistance improvement. Different dissolution behaviors were observed in W and C/Ni powder within the melt pool at different locations in the melt pool. The dissolution situation of W, C, and N depends highly on the local temperature distribution and reheating, which is related to laser deposition parameters and deposition strategies.

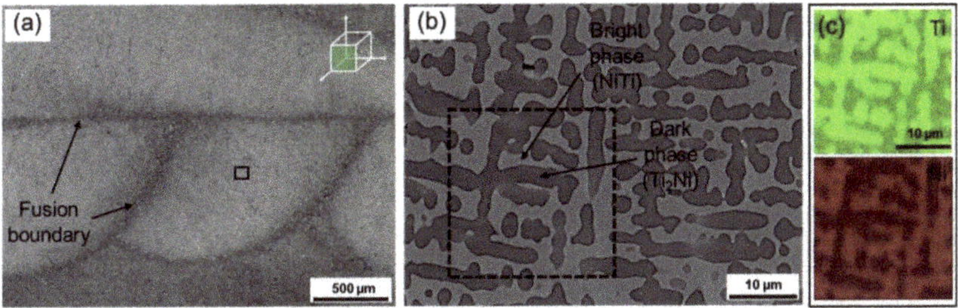

Figure 2. (**a**) Microstructure and fusion boundary of NiTi alloy fabricated by DED using elemental powder blends. (**b**) SEM image of the NiTi alloy with NiTi phase and Ti_2Ni phase. (**c**) EDS mapping of Ti and Ni [49]. (Reproduced with permission from Elsevier).

2.2. Develop Advanced Alloys

Due to the advantage of customization and small-batch manufacturing, the DED process with blended elemental powders is also a powerful tool for developing novel alloys and inventing innovative materials [33]. Changes in structure and property of adding different types and quantities of elements in an alloy system can be quickly observed by a small volume deposition, rather than making a large structure using conventional methods. By adding more alloying elements, the fabrication of advanced alloys such as FGMs, HEAs, and metallic glass can also be realized. Therefore, using elemental powder will then create many more probabilities in the field of alloy design in a cost-effective way.

2.2.1. Alloy Modification

Modifying alloys can be flexible by using elemental powders in DED, as the composition can be customized by varying parts of the elements or all elements. This method may either solve the difficulty in processing certain alloys or study the element effects in alloy systems using DED. This advantage can benefit the development of numerous binary and ternary alloy systems, including, but not limited to Cu-Ni, Ti-Nb, Fe-W, Ti-Al-Mo, etc. [56–59] and the development of new alloys. Cu and Ni are completely soluble, which attracts industrial interests in making Cu-Ni alloys with both high thermal conductivity from Cu and high mechanical properties from Ni. Karnati et al. [56] mixed elemental Cu and near-pure Ni powders in different composition levels, and they all produced solid solutions of Cu-Ni alloys. Li et al. [57] deposited 80W-20Fe using elemental W and Fe powders, which indicates DED is an effective and novel method to process W based alloys. Fallah et al. [58] deposited 55 wt.% Ti/45 wt.% Nb in elemental powder blend on the Ti-6Al-4V substrate to create a compositionally modified surface layer. Then the biocompatibility of Ti-Nb alloy at the surface can be utilized, and the cost can be reduced by avoiding manufacturing the entire part using Ti-Nb alloy. Zhang et al. [59] deposited a series of Ti-2Al-yMo (y = 2, 5, 7, 9, 12) to study the Ti-Al-X system and develop innovative alloys via DED.

2.2.2. FGMs

The elemental powder mixture can be used to modify alloys using different compositions of elemental powders. The DED process is also a flexible layer-based AM technique, which can produce different compositions at different locations. Then, within a certain binary or ternary alloy system, different compositions can be joined together by taking a proper usage of the DED process in a layer-wise fashion. More advanced metallic alloys can be designed for multifunctionality, and FGM is a good example. The concept of FGM originated from applying a graded composition between two materials with different properties to avoid delamination under extreme loading conditions at the interface [60]. Compared with selective laser melting (SLM), which is a laser-based metal additive manufacturing in the category of PBF, one important advantage for DED is the flexibility of in-situ control of the location-dependent chemical composition. The flexibility in composition control makes DED an excellent processing technique for fabricating FGMs [60]. DED can be fully utilized to deposit different compositions of elemental powder mixtures layer by layer without using complicated assembly processes in traditional manufacturing. One purpose of fabricating FGMs is to join two dissimilar pure alloys without a sharp interface [61–64]. This could be solved by adding compositionally graded interlayers between pure alloy A and pure alloy B. Figure 3 illustrates the concept of joining based FGM, and a real deposited sample from the literature [61] is shown in Figure 4, which joins Inconel 625 and 304L stainless steel via a compositional gradient.

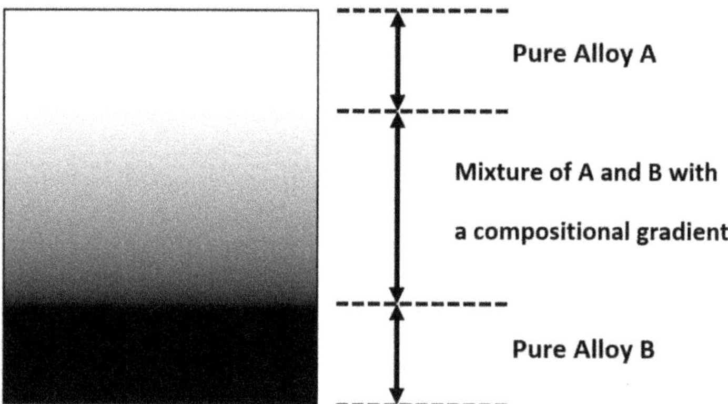

Figure 3. FGM joined by pure alloy A and pure alloy B with a compositional gradient.

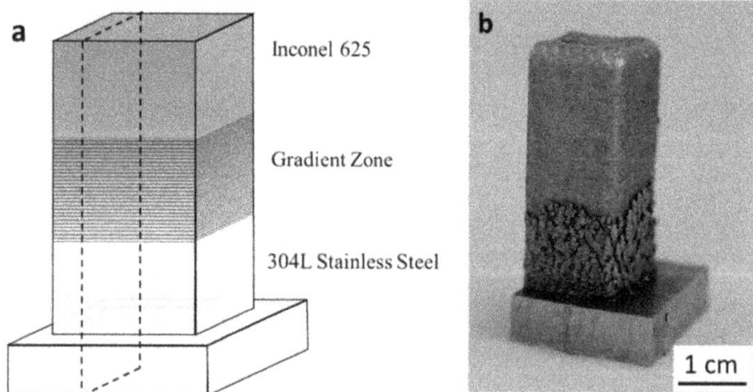

Figure 4. (**a**) Schematic of Inconel 625/304L stainless steel FGM. (**b**) Image of an Inconel 625/304L stainless steel FGM sample fabricated by DED [61]. (Reproduced with permission from Elsevier).

In addition, FGMs can be intentionally designed by combining different chemical compositions at different locations to fabricate multifunctional parts. This type of FGM has been done in using the DED method to fabricate binary or ternary alloy systems with a compositional gradient via elemental powder mixture. Banerjee et al. have made a decent amount of Ti-based binary alloys into FGMs using elemental powders, including Ti-V [65,66], Ti-Mo [66], and Ti-Ta [67]. Titanium can form α/β alloy systems with many other metal elements. As seen in Figure 5, a Ti/V FGM was fabricated by mixing graded Ti/V composition and the resultant composition gradient varies from 100% Ti on the left to 75 at.% Ti/25 at.% V on the right side. It is of great interest to apply elemental powder to study graded titanium alloys and identify their process-structure-property relations since many Ti-based binary alloys have not been widely considered in the AM category. Elemental powder mixture will be convenient to check the effect of alloying elements on microstructure, grain size, and mechanical properties. Those works focused on the relationship among composition, microstructure, and mechanical properties of Ti-based binary alloys with a composition change along the graded direction. Here the research in DED-processed FGMs also helps to establish the process-structure-property relationship of these promising alloy systems. Ti-Cr [68], Ti-Al [69], Ti-W [70], and Ti-Mo [71] systems were also investigated. Details are tabulated in Table 1. W element is good for grain refinement, while Mo does not have a significant refining effect. They also studied the Ti-Al-V system by varying the composition of V using the elemental powder mixture of Ti-8Al-xV [72]. Thus, graded ternary alloys can also be fabricated using DED to systematically study the effect of variation of alloying elements.

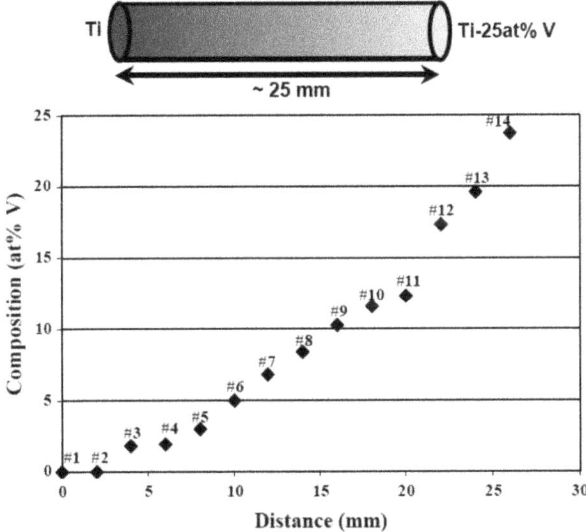

Figure 5. Schematic diagram of Ti-25 at.%V FGM alloy fabricated via elemental powder mixture and the compositional variation vs. distance [65]. (Reproduced with permission from Elsevier).

Table 1. Summary of Ti-based FGM alloys fabricated by elemental powder-based DED.

Alloy System	Ref.	Composition	Findings
Ti-Mo	[66]	Ti-25 at.% Mo	Hardness first increased and then decreased, a combination of grain size and alloy content.
Ti-Ta	[67]	Ti-50 wt.% Ta	The microhardness initially decreases, then increases, and finally decreases again.
Ti-Cr	[68]	Ti-60 at.% Cr	Hardness and modulus increase with Cr composition.
Ti-W	[70]	Ti-23 wt.% W	W has a significant effect of grain refinement across the gradient.

Karnati et al. [73] mixed elemental Cu and Ni powders to create Cu-Ni FGMs based on the complete solubility in Cu-Ni binary system. After the previous investigation of mixing Cu and Ni in different compositions, the different compositions were then combined and fabricated into Cu/Ni FGM. Li et al. [74,75] fabricated a new graded Fe-Cr-Ni alloy using Fe, Cr, and Ni elemental powders with a gradient in Cr and Ni composition, as shown in Figure 6. Thus, on the Cr-rich side, the graded system can exhibit excellent behaviors in corrosion resistance. On the Ni-rich side, the system possesses high plasticity owing to the large composition of austenite.

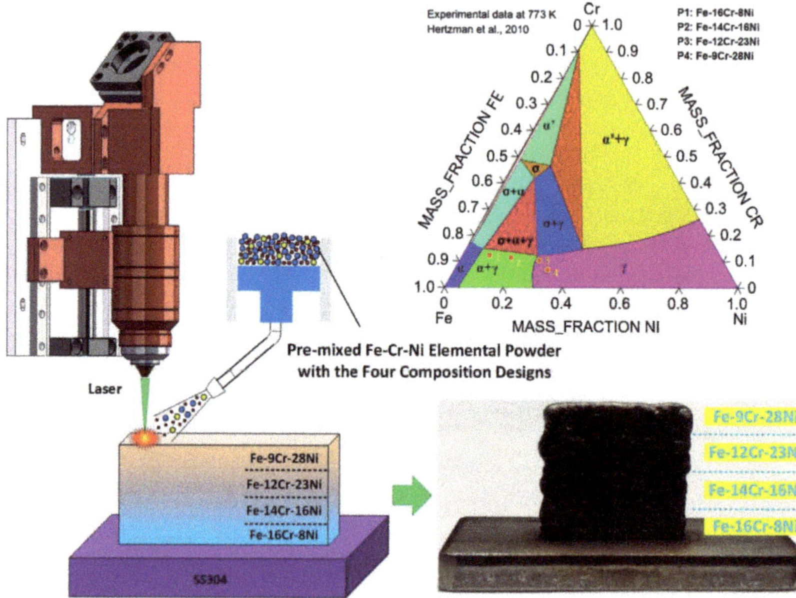

Figure 6. Fabricating Fe-Cr-Ni FGM by DED process using elemental Fe, Cr, and Ni powders [74]. (Reproduced with permission from Elsevier).

2.2.3. Magnetic Materials and Metallic Glass

Apart from common structural alloys, new types of alloy systems with special functions were also studied using elemental powder blend. Conteri et al. [76] studied a novel magnetic alloy Fe73.5Si13.5B9Nb3Cu1. Based on this study, Borkar et al. [77] synthesized a more complex magnetic alloy with a gradually changing Si/B ratio. Thus, this new design is also known as a functionally graded Fe-Si-B-Cu-Nb alloy with magnetic properties. Amorphous (or glassy) metals can be fabricated using this technique. Manna et al. [78] deposited 94Fe4B2C, 75Fe15B10Si, and 78Fe10BC9Si2Al1C by mixing glass-forming elemental powders on a substrate made by carbon steel. While in Hou et al.'s work [79], the Fe-based Fe-Cr-Mo-Co-C-B amorphous alloy was produced according to the weight percentage of Fe45.8Mo24.2Cr14.7Co7.8C3.2B4.3. The amorphous phase occupied 52.8% of the entire volume. The resulted hardness of the deposition has a maximum of 1200 HV0.5, which shows a significant improvement compared to the substrate that is 200 HV0.5.

2.2.4. HEAs

HEA was also proved to be feasible to be fabricated by DED using elemental powders. HEAs are known to possess high hardness, excellent high-temperature strength, corrosion resistance, and wear resistance due to the unique multiprincipal element composition [80]. It is promising for fabricating coating materials on engineering parts for wear and oxidation resistance. As the flexible mixture of powders from at least five principal elements, different atomic ratios can be varied to study the element effect on the as-fabricated HEAs, such as $Al_xCrCuFeNi$ [81], where x varies while the atomic percentage of all five elements are between 5 at.% and 35 at.%.

Coating HEAs on conventional structural alloys are highly attractive due to the potential high hardness of HEAs. Cui et al. [82] applied DED to coat AlCoCrFeNi on AISI SS316 using the elemental powders. An intermediate $CoFe_2Ni$ layer was applied between the AlCoCrFeNi HEA coating and the SS316 substrate to avoid cracking. The intermediate $CoFe_2Ni$ was also synthesized by elemental powder, which has the purpose of providing an average coefficient of thermal expansion (CTE) that does not differ greatly from the SS316 substrate and the HEA coating. Chao et al. [83] applied DED to coat $Al_xCoCrFeNi$ on a 253MA steel substrate, where the value of the Al mole fraction x was taken as 0.3, 0.6, and 0.85. Elemental powders were utilized, and the composition change of Al can be adjusted. The effect of the Al mole fraction on the crystal structure was revealed by material characterization. Chen et al. [84] varied two types of elements (Al and Cu) to study the influence on the structure and properties of $Al_xCoFeNiCu_{1-x}$, where x = 0.25, 0.5, and 0.75. It was found that crystal structure and hardness varied significantly from 0.25 Al/0.75 Cu to 0.75 Al/0.25 Cu. In another work, the hardness of Al alloy was improved by depositing $Al_{0.5}FeCu_{0.7}NiCoCr$ HEA coating [85]. The average value of hardness reached about eight times higher than the Al alloy substrate.

In addition, HEAs that possess high erosion and oxidation resistance can be synthesized by elemental powder mixture. Siddiqui et al. [86] coated $Al_xCu_{0.5}FeNiTi$ HEA on Al alloy AA1050 by elemental powder blend for erosion resistance. It was stated that the erosion rate was decreased mainly due to the improved microhardness of tough grains formed in HEA. The HEA coating using elemental powder is also studied for the potentially high-temperature oxidation resistance. Huang et al. [87] studied that depositing TiVCrAlSi on Ti-6Al-4V can improve the oxidation resistance of Ti-6Al-4V at 800 °C.

As high temperature fields can be created by the high-power laser beam, some works that focus on combining a series of metal elements with high melting points to produce refractory HEAs were also carried out. Dobbelstein et al. [88] produced TiZrNbHfTa from elemental powder blends for the first time. The mixing was homogeneous, and a high hardness was achieved. The effect of the mole fraction of one specific element was also studied. The $W_xNbMoTa$ HEA with the composition change in W was fabricated by Li et al. [89]. The mole fraction x was taken as 0, 0.16, 0.33, and 0.53. It was found the microhardness increases with the increase in W. Based on the flexibility in modifying the mole fraction, different HEAs with different mole fractions can also be fabricated together to make FGMs, which then becomes compositionally graded HEAs. By taking advantage of FGM, Dobbelstein et al. [90] also fabricated compositionally graded TiZrNbTa refractory HEAs using elemental powder blends. Gwalani et al. [91] deposited $AlCrFeMoV_x$ compositionally graded HEA, where the mole fraction of V varies from 0 to 1. The lattice parameter decreased, and an improvement was found on hardness with the increase in V content. DED with elemental powders was regarded as a high-throughput method to study the composition-microstructure-hardness relationship of novel HEAs.

3. Deposition Control

In Section 2, major types of alloys which have been fabricated by elemental powder based DED technique are introduced and discussed. The classification and examples with the reference are tabulated in Table 2. Using elemental powders to fabricate various types of alloys using DED is now very promising. However, as a novel technique, there are still a lot of unsolved issues beyond feasibility studies to popularize this concept. Whether the composition of the deposited part matches up with the originally designed composition is a big issue for this technique. Also, the final deposition is expected to be homogeneous. As the entire feedstock delivery and manufacturing system are highly complex, multiple key factors should be taken into consideration. The stability and repeatability are of great importance to promote this process into a new stage in various industries.

Table 2. Classification of alloys synthesized by elemental powder-based DED.

Alloy Types	Examples	Ref.
Industrial Alloys	Ti-6Al-4V	[33–36]
	Stainless Steel	[37]
Intermetallics	FeAl	[39]
	NiTi	[44–49]
	Ni$_3$Al	[50]
	Hard Coadings: TiCo, Cr$_3$Si, NiSi, etc.	[51–55]
FGMs	Ti-based: Ti-Mo, Ti-V, Ti-Ta, etc.	[68–72]
	Cu-Ni	[73]
	Fe-Cr-Ni	[74,75]
Magnetic Alloys	Fe73.5Si13.5B9Nb3Cu1 etc.	[76,77]
Metallic Glass	78Fe10BC9Si2Al1C etc.	[78,79]
HEAs	Al$_x$CrCuFeNi etc.	[81–91]

3.1. Enthalpy of Mixing

Mixing is important to maintain a good homogeneity during the deposition. To ensure a uniform mixing, the enthalpy of mixing of the system should be negative. Schwendner et al. [92] used Ti-Cr and Ti-Nb binary elemental powder blend to study the effect of mixing enthalpy on the homogeneity of mixing. The results showed that Ti-10%Cr alloy has a negative enthalpy of mixing and has a homogeneous intermixing result within the melt pool. While the Ti-10%Nb system has an inhomogeneous mixing and a slow cooling rate. The mixing of enthalpy can be a useful criterion to guide the design of chemical elements in the alloy. For some of the alloy systems, the constituent elements are in the positive enthalpy of mixing, such as Cu-Ni [56] and Cu-Fe, the adjustment of laser power to generate additional energy for mixture homogenization is needed. For example, Karnati et al. [56] applied pulse width modulated laser power to induce vibrations in the melt pool to avoid segregation during the deposition of the Cu-Ni system.

3.2. Powder Delivery

Although using the elemental powder blend is straightforward to understand, the accuracy of the chemical composition of the as-deposited part is a big challenge. The pre-designed chemical compositions or elemental composition should match well with the deposited part. Element deviations are often observed from most of the previously reported works. Among those works, few of them made an in-depth investigation in this aspect. In [56], blending Ni and Cu resulted in the part with a 4% error with respect to the pre-designed composition. In [82], the atomic percentage of Al in AlCoCrFeNi HEA was between 10 at.% and 15 at.%, while the pre-designed atomic percentage was 20 at.%. In some cases, fabricating the material system with high compositional accuracy is of great importance. A small composition error can cause a significant change in microstructure, phase, and mechanical properties. For example, a slight increase in the Mo composition of Ti-Mo alloy will result in a bimodal distribution of α lath in the β matrix [66]. The control of chemical composition needs to be solved in order to improve the manufacturing quality of the elemental DED process and push this method entirely in the industrial applications.

Powders are the commonly used feedstock material for the DED process. Most of the previous works applied the method of blown powder blend of pre-mixed elemental powders to deliver powders into the melt pool with the carrier gas. Since elemental powders made by different metals are mixed and delivered by carrier gas, the powder properties, and flow behavior are important to control. Collins et al. [93] mentioned that powder size and density could affect the actual composition of the as-deposited part in their work of fabricating Ti-Al-Mo and Ti-Al alloys. Li et al. systematically studied powder separation using both numerical and experimental methods [94–96]. As the powder mixture is delivered by the carrier gas, the velocity of each particle mainly depends on its shape, density, and size. Li performed studies in exploring the powder segregation in the blown powder DED process with pre-mixed spherical elemental powder as feedstocks, as seen in Figure 7. The larger the difference in density, the greater the segregation phenomenon will be. Cu and Al powders were applied for the experimental investigation of the flow behavior. In the investigation, it was found that significant segregation was observed in un-sieved powder mixtures without size adjustment. The segregation issue was resolved after the sieving process under the guidance of the density and size relation. The limitation of this model is it only works for spherical powder particles. For non-spherical powders, the flowability will be changed, and more studies are needed.

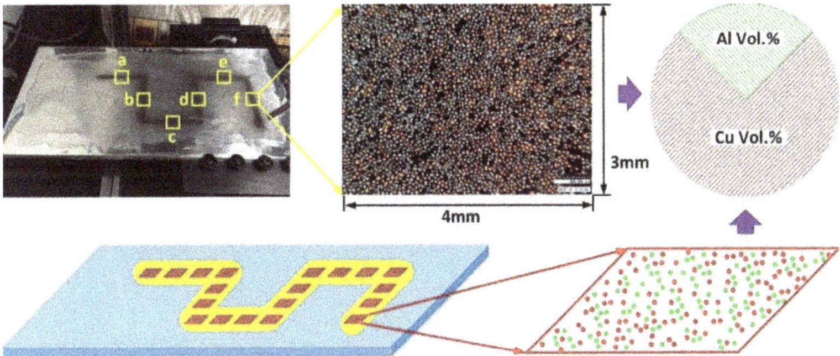

Figure 7. Investigation of powder segregation between Al powder and Cu powder by spraying powder mixture on a glue plate and calculating the volume ratio of Al powder and Cu powder at different locations [96]. (Reproduced with permission from Elsevier).

3.3. Capturing and Melting

When a real deposition is performed, a stream of powder is blown into the melt pool, and only part of the powder can be captured by the melt pool. The capture rate is also a factor to affect the chemical composition, as for different types of powders, the probability of falling into the melt pool also varies. Chen et al. [36] performed an experiment to investigate the powder capture rate of Ti, Al, and V at different size levels. For each metal powder, a constant capture rate can be determined, and the corresponding size level can be picked for the same capture rate. Another perspective for keeping a constant capture rate for different metal powders is to maintain a constant divergence angle to maintain a consistent composition before and after deposition. A mathematical model was worked out by Zhang et al. [97]. The constant divergence is a reflection of constant speed out of the nozzle. The optimized powder particle size of Ti, Al, and V used to maintain a constant divergence angle match nicely with Li's work [94].

The study of the flow behavior by Li et al. [96] revealed the relationship between powder properties and the avoidance of powder separation. However, it was limited in powder spray. The actual laser deposition was not performed to check the exact elemental composition of the as-deposited part. As when an actual deposition is performed, different metals may have different composition loss due to evaporation. The melt pool capture rate of powders can also be different. In the research [74,75] of the Fe-Cr-Ni FGM, Li et al. applied this method and found that the deposition works better when the powder size was sieved and controlled according to the physical properties of Fe, Cr, and Ni, as shown in Figure 8. Fe, Cr, and Ni have relatively similar density and melting point, the effect of evaporation and capture rate will be less obvious. For other combinations with larger differences in properties, such as Cu and Al differ greatly in density and melting temperature, further works should be done for this model extension.

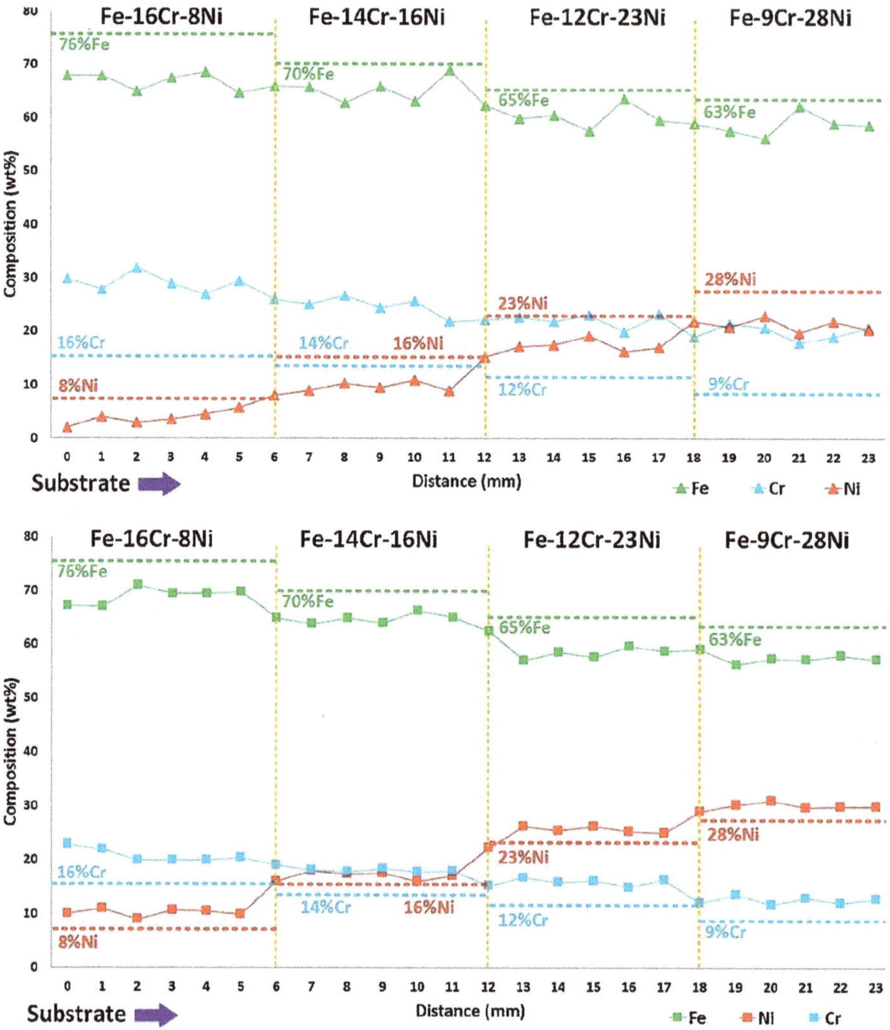

Figure 8. Element composition control of Fe-Cr-Ni FGM with un-sieved (the upper figure) and sieved (the lower figure) elemental powder mixture. The FGM fabricated by sieved powder mixture gives more accurate composition [74]. The percentage indicates wt.% in this figure. (Reproduced with permission from Elsevier).

Therefore, apart from the flow properties of powder particles, the thermal interactions are also complex, which may affect the composition, microstructure, and performance. Metal elements cover a wide range of physical properties such as density, melting point, and laser absorption rate. For instance, among engineering structural alloys, the melting point can span from 600 °C in Al to 3400 °C in W [91]. The energy absorption rate is also worth considering when a highly reflected element, such as Cu, is part of the alloy [98]. In SLM work, a refractory HEA comprises of NbMoTaW was deposited [99]. A 3.5% composition deviation was found. Although the element composition of 5%–35% can all be regarded as HEA, how this deviation can affect the phase evolution and the final properties are still unknown at this moment. In [90], a new processing of DED was applied to deposit a functionally graded refractory HEA, which includes a wide range of melting points. The newly designed process route consists of a low power step to yield low powder evaporation and a high power step to remelt the previous layer and homogenize the elements. In Figure 9, Figure 9a,b are single track deposition and deposition track after the second remelting step, respectively. Figure 9d,e are EDS mapping of elements (Zr, Nb, Ti, Ta, and the substrate Mo) of Figure 9a,b respectively. From the EDS mapping, it can be seen that the remelting step strongly homogenizes the refractory elements within the entire deposition.

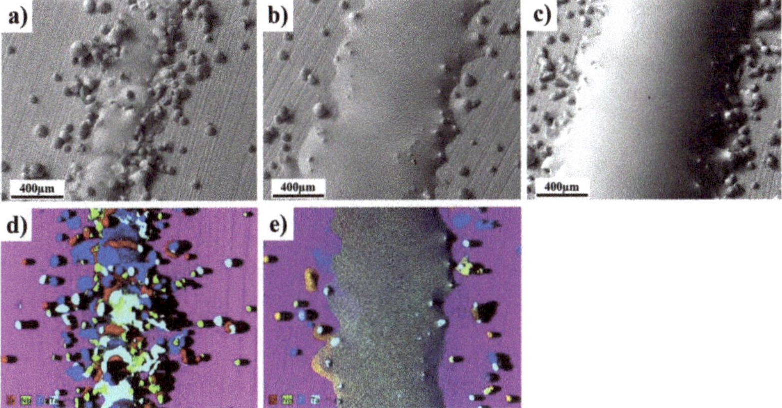

Figure 9. Deposition of TiZrNiTa refractory HEA on Mo substrate: (**a**) Single-track deposition; (**b**) Deposition after the second remelting step; (**c**) Deposition after four deposition and remelting steps; (**d**) The EDS mapping of deposition in (**a**) and (**e**) The EDS mapping of deposition in (**b**) [90]. (Reproduced with permission from Elsevier).

4. Outlook

Based on the major factors discussed in Section 3, it can be concluded that keeping an accurate composition in elemental powder deposition is still challenging. Those factors were partially studies in a couple of previous works. Each factor needs more experimental and theoretical studies. For blown powder-based DED, it was reported that linearly varying flow parameters can still result in nonlinear compositional grading and material behaviors [66]. The model of powder delivery using carrier gas needs to be further improved. Multiple factors, including the elemental powder properties and the processing parameters, are quite relevant, and the interactions among them are not negligible. It was reported that nano-sized powders were applied in high velocity oxygen fuel (HVOF) coating technique [100], which applies powder spray to fabricate thin layers. The delivery of powders with a more tiny size can be further studied that the DED process can be extended to more precise manufacturing applications. For thin layer deposition, preplacing powders on the processing surface before laser melting is another way to obtain a strong metallurgical bonding [50–54]. The dilution analysis can be further investigated to control the resultant phases after the deposition process.

There are very few works that cover the tolerance of chemical compositions of alloys fabricated by DED using elemental powders. Then, the compositional sensitivity study can be an important topic to study for the industry to adopt this processing method. It is worth mentioning that one advantage for elemental powder DED is that the mass loss due to evaporation can be compensated [36]. For pre-alloyed powders, since all the particles are in the same composition, if elements such as Mg, Zn, and Al are included, the favorable evaporation during the laser processing will change the overall composition. So, there is a need to compensate for this loss in pre-alloyed powders. However, for elemental powder mixture, more volatile elements can be prepared in the pre-designed powder mixture by adding more volatile element powders. Mukherjee et al. [101] listed some examples of the most volatile elements in some common alloys. Depositing pre-alloyed Ti-6Al-4V has a large loss in Al. It was reported the percentage of Al loss in pre-alloyed Ti-6Al-4V is higher than Cr loss in Inconel 625. If using elemental powders, various compositions can be pre-mixed to compensate for easy-to-evaporate alloy elements. An early compensation study was performed by Yan et al. [34]. During the first experiment, the pre-designed composition was equal to Ti-6Al-4V, the ideal value. However, the result gives Ti-5Al-2V. Later, a Ti-8Al-8V composition was pre-designed, and the result matched well with Ti-6Al-4V. More details for the compensation study in different alloy systems should be performed in future works, which is a challenging issue. A more mature relationship between the pre-mixed composition and the deposited composition should be studied. Another example is NiTi, which is very sensitive when there is a slight deviation in compositions. There is a need to figure out how to adjust the pre-designed composition in order to get an acceptable chemical composition.

Apart from obtaining an accurate composition, since the process is layer by layer, the heating and cooling history varies at different locations. Also, to build a bulk part, overlaps between tracks cannot be avoided. These are the main features in DED, and the grain structure and the anisotropic mechanical behavior resulted from the layer by layer heating and cooling history have been a hot topic. These features in DED lead to highly dynamic phenomena including dynamic melt pool, particle vaporization, rapid solidification and phase transition. In traditional manufacturing processes, highly dynamic phenomena was found to result in large scattering in mechanical properties [102], and when it comes to DED, the scattering can be more serious. When the feedstock materials become elemental powder mixture, the process will subject to more complex changes and may cause more discrepancies that have not been understood very well. The spatial difference sensitivity is worth learning, and a comprehensive thermophysical model is needed to control the temperature heating and cooling to guide the processing.

As a promising technology which uses elemental powders as feedstock materials, the effect of powder chemistry should be considered in the future study. Powder quality affects the final deposition, and for the DED process, little has known in this area. Powder chemistry, such as the composition of oxygen affects the final part of oxidation sensitive materials such as titanium alloys. Also, the oxidation of powder can induce porosity in the as-built parts. Karnati et al. [56] found the porosity issue in Cu-Ni FGM when using elemental Ni powder. A Ni alloy powder with 96 wt.% Ni and a small amount of Si was used as a substitution of pure Ni powder since Si can consume oxygen and relieve the porosity issue. Powder recycling is attracting research interest in recent years, which is also relevant to this area.

Last but not least, other related AM processing methods based on elemental powders, and the difference between elemental powder-based DED alloys and conventionally manufactured alloys are also worth further study. Nowadays, many important industrial alloys, such as Ti-6Al-4V and NiTi can be manufactured by both conventional methods and AM. The room-temperature mechanical properties of the as-fabricated Ti-6Al-4V using powder mixture were better than wrought counterparts [33]. In AM, selective laser melting (SLM) and electron beam melting (EBM) can also apply elemental powder mixtures. Wang et al. [49] found out the result of using SLM to fabricate NiTi alloy through elemental powders was not similar to DED. It was reported that there was a significant loss in Ti, which resulted in Ni-rich intermetallics as the predominant phase. The fabrication using EBM was not successful, which shows a low printability. Mechanisms and parametric study of SLM and EBM can be very different from DED based elemental powder manufacturing [49]. Studying in this aspect will give more understanding of the differences between additively manufactured alloys and conventionally manufactured alloys. It will also figure out more advantages of using multiple AM techniques to fabricate alloys by elemental powders in the corresponding industrial area.

5. Conclusions

In this paper, the current research status of using the DED process to fabricate metal alloys through elemental powder mixture was summarized based on different types of alloys, including industrial alloy and intermetallics synthesis, alloy modification, FGMs, metallic glass, and HEAs. The main issues and challenges are also summarized.

Various types of alloys, including industrial alloys, FGMs, metallic glass, and HEAs, have been synthesized by DED through elemental powders. Many of those works show good feasibility, and the mechanical properties of the deposited parts are comparable to conventionally manufactured alloys. As a new technique, numerous topics are still unsolved. Main scientific issues like overcoming the entropy of mixing, studying the physical and chemical properties of powders, the flow behavior for different metal powders, and how the laser-material interaction affect the final composition of the as-built part need systematic and in-depth research. Also, a relationship between the initial designed atomic or weight composition and the final composition is needed, and it will be integrated with the knowledge of materials science, dynamics, thermomechanical interaction, and the manufacturing system. Numerical simulation and more experimental results can be done in the future, which will significantly extend this new area to the metal manufacturing industry not only in DED, but also in other AM areas such as SLM and EBM.

Author Contributions: Conceptualization, literature review, and writing original draft, Y.C.; writing—review and editing, Y.C., X.Z., M.M.P., F.L.; project administration, F.L.; funding acquisition, F.L. All authors have read and agreed to the published version of the manuscript.

Funding: This project was supported by National Science Foundation Grants CMMI 1625736, Department of Energy Grant #DE-SC0018879, Intelligent Systems Center, and Material Research Center at Missouri S&T. Their financial support is greatly appreciated.

Conflicts of Interest: The authors declare no conflict of interest.

References

1. Frazier, W.E. Metal Additive Manufacturing: A Review. *J. Mater. Eng. Perform.* **2014**, *23*, 1917–1928. [CrossRef]
2. Huang, Y.; Leu, M.; Mazumder, J.; Donmez, M.A. Additive Manufacturing: Current State, Future Potential, Gaps and Needs, and Recommendations. *J. Manuf. Sci. Eng.* **2015**, *137*, 014001. [CrossRef]
3. Paris, H.; Mokhtarian, H.; Coatanéa, E.; Museau, M.; Ituarte, I.F. Comparative environmental impacts of additive and subtractive manufacturing technologies. *CIRP Ann.* **2016**, *65*, 29–32. [CrossRef]
4. Debroy, T.; Wei, H.; Zuback, J.; Mukherjee, T.; Elmer, J.; Milewski, J.; Beese, A.M.; Wilson-Heid, A.; De, A.; Zhang, W. Additive manufacturing of metallic components—Process, structure and properties. *Prog. Mater. Sci.* **2018**, *92*, 112–224. [CrossRef]
5. Zhang, Y.; Jarosinski, W.; Jung, Y.-G.; Zhang, J. Additive manufacturing processes and equipment. In *Additive Manufacturing*; Elsevier BV: Oxford, UK, 2018; pp. 39–51.
6. F42 Committee. *Standard Guide for Directed Energy Deposition of Metals*; ASTM International: West Conshohocken, PA, USA, 2016.
7. Gibson, I.; Rosen, D.; Stucker, B. Directed Energy Deposition Processes. In *Additive Manufacturing Technologies*; Springer Science and Business Media LLC: New York, NY, USA, 2015; pp. 245–268.
8. Wang, Z.; Palmer, T.A.; Beese, A.M. Effect of processing parameters on microstructure and tensile properties of austenitic stainless steel 304L made by directed energy deposition additive manufacturing. *Acta Mater.* **2016**, *110*, 226–235. [CrossRef]
9. Melia, M.A.; Nguyen, H.-D.A.; Rodelas, J.M.; Schindelholz, E.J. Corrosion properties of 304L stainless steel made by directed energy deposition additive manufacturing. *Corros. Sci.* **2019**, *152*, 20–30. [CrossRef]
10. Sciammarella, F.M.; Najafabadi, B.S. Processing Parameter DOE for 316L Using Directed Energy Deposition. *J. Manuf. Mater. Process.* **2018**, *2*, 61. [CrossRef]
11. Weng, F.; Gao, S.; Jiang, J.; Wang, J.; Guo, P. A novel strategy to fabricate thin 316L stainless steel rods by continuous directed energy deposition in Z direction. *Addit. Manuf.* **2019**, *27*, 474–481. [CrossRef]
12. Ning, F.; Cong, W. Microstructures and mechanical properties of Fe-Cr stainless steel parts fabricated by ultrasonic vibration-assisted laser engineered net shaping process. *Mater. Lett.* **2016**, *179*, 61–64. [CrossRef]
13. Kistler, N.A.; Nassar, A.R.; Reutzel, E.W.; Corbin, D.J.; Beese, A.M. Effect of directed energy deposition processing parameters on laser deposited Inconel®718: Microstructure, fusion zone morphology, and hardness. *J. Laser Appl.* **2017**, *29*, 022005. [CrossRef]

14. Jinoop, A.; Paul, C.; Mishra, S.; Bindra, K. Laser Additive Manufacturing using directed energy deposition of Inconel-718 wall structures with tailored characteristics. *Vacuum* **2019**, *166*, 270–278. [CrossRef]

15. Kakinuma, Y.; Mori, M.; Oda, Y.; Mori, T.; Kashihara, M.; Hansel, A.; Fujishima, M. Influence of metal powder characteristics on product quality with directed energy deposition of Inconel 625. *CIRP Ann.* **2016**, *65*, 209–212. [CrossRef]

16. Fujishima, M.; Oda, Y.; Ashida, R.; Takezawa, K.; Kondo, M. Study on factors for pores and cladding shape in the deposition processes of Inconel 625 by the directed energy deposition (DED) method. *CIRP J. Manuf. Sci. Technol.* **2017**, *19*, 200–204. [CrossRef]

17. Carroll, B.E.; Palmer, T.A.; Beese, A.M. Anisotropic tensile behavior of Ti–6Al–4V components fabricated with directed energy deposition additive manufacturing. *Acta Mater.* **2015**, *87*, 309–320. [CrossRef]

18. Prabhu, A.W.; Vincent, T.; Chaudhary, A.; Zhang, W.; Babu, S.S. Effect of microstructure and defects on fatigue behaviour of directed energy deposited Ti–6Al–4V. *Sci. Technol. Weld. Join.* **2015**, *20*, 659–669. [CrossRef]

19. Beese, A.M.; Carroll, B.E. Review of Mechanical Properties of Ti-6Al-4V Made by Laser-Based Additive Manufacturing Using Powder Feedstock. *JOM* **2015**, *68*, 724–734. [CrossRef]

20. Saboori, A.; Gallo, D.; Biamino, S.; Fino, P.; Lombardi, M. An Overview of Additive Manufacturing of Titanium Components by Directed Energy Deposition: Microstructure and Mechanical Properties. *Appl. Sci.* **2017**, *7*, 883. [CrossRef]

21. Onuike, B.; Bandyopadhyay, A. Additive manufacturing in repair: Influence of processing parameters on properties of Inconel 718. *Mater. Lett.* **2019**, *252*, 256–259. [CrossRef]

22. Wilson, J.M.; Piya, C.; Shin, Y.C.; Zhao, F.; Ramani, K. Remanufacturing of turbine blades by laser direct deposition with its energy and environmental impact analysis. *J. Clean. Prod.* **2014**, *80*, 170–178. [CrossRef]

23. Zhang, X.; Cui, W.; Li, W.; Liou, F. Effects of tool path in remanufacturing cylindrical components by laser metal deposition. *Int. J. Adv. Manuf. Technol.* **2018**, *100*, 1607–1617. [CrossRef]

24. Zhang, X.; Pan, T.; Li, W.; Liou, F. Experimental Characterization of a Direct Metal Deposited Cobalt-Based Alloy on Tool Steel for Component Repair. *JOM* **2018**, *71*, 946–955. [CrossRef]

25. Zhang, X.; Li, W.; Chen, X.; Cui, W.; Liou, F. Evaluation of component repair using direct metal deposition from scanned data. *Int. J. Adv. Manuf. Technol.* **2017**, *95*, 3335–3348. [CrossRef]

26. Yu, J.; Rombouts, M.; Maes, G.; Motmans, F. Material Properties of Ti6Al4V Parts Produced by Laser Metal Deposition. *Phys. Procedia* **2012**, *39*, 416–424. [CrossRef]

27. Jia, M.; Blanchard, C.; Bolzoni, L. Microstructure and Mechanical Properties of Ti-5Al-2.5Fe Alloy Produced by Powder Forging. *Key Eng. Mater.* **2018**, *770*, 39–44. [CrossRef]

28. Liu, Y.; Chen, L.; Tang, H.; Liu, C.-T.; Liu, B.; Huang, B. Design of powder metallurgy titanium alloys and composites. *Mater. Sci. Eng. A* **2006**, *418*, 25–35. [CrossRef]

29. Azarniya, A.; Garmendia, X.; Mirzaali, M.J.; Sovizi, S.; Bartolomeu, F.; Mare, K.S.W.; Wits, W.W.; Yap, C.Y.; Ahn, J.; Miranda, G.; et al. Additive manufacturing of Ti–6Al–4V parts through laser metal deposition (LMD): Process, microstructure, and mechanical properties. *J. Alloys Compd.* **2019**, *804*, 163–191. [CrossRef]

30. Saresh, N.; Pillai, M.G.; Mathew, J. Investigations into the effects of electron beam welding on thick Ti–6Al–4V titanium alloy. *J. Mater. Process. Technol.* **2007**, *192*, 83–88. [CrossRef]

31. Ivasishin, O.; Anokhin, V.; Demidik, A.; Savvakin, D. Cost-Effective Blended Elemental Powder Metallurgy of Titanium Alloys for Transportation Application. *Key Eng. Mater.* **2000**, *188*, 55–62. [CrossRef]

32. Wang, H.; Fang, Z.Z.; Sun, P. A critical review of mechanical properties of powder metallurgy titanium. *Int. J. Powder Metall.* **2010**, *46*, 45–57.

33. Hua, T.; Jing, C.; Fengying, Z.; Xin, L.; Huang, W. Microstructure and Mechanical Properties of Laser Solid Formed Ti-6Al-4V from Blended Elemental Powders. *Rare Met. Mater. Eng.* **2009**, *38*, 574–578. [CrossRef]

34. Yan, L.; Chen, X.; Li, W.; Newkirk, J.; Liou, F. Direct laser deposition of Ti-6Al-4V from elemental powder blends. *Rapid Prototyp. J.* **2016**, *22*, 810–816. [CrossRef]

35. Chen, X.; Yan, L.; Li, W.; Wang, Z.; Liou, F.; Newkirk, J. Effect of Powder Particle Size on the Fabrication of Ti-6Al-4V Using Direct Laser Metal Deposition from Elemental Powder Mixture. *J. Mech. Eng. Autom.* **2016**, *6*, 348. [CrossRef]

36. Chen, X. Fabrication and Characterization of Advanced Materials Using Laser Metal Deposition from Elemental Powder Mixture. Ph.D. Thesis, Missouri University Of Science And Technology, Rolla, MO, USA, 2018.

37. Clayton, R.M. The Use of Elemental Powder Mixes in Laser-Based Additive Manufacturing. Master's Thesis, Missouri University Of Science And Technology, Rolla, MO, USA, 2013.
38. Karczewski, K.; Pęska, M.; Ziętala, M.; Polański, M.; Dąbrowska, M. Fe-Al thin walls manufactured by Laser Engineered Net Shaping. *J. Alloys Compd.* **2017**, *696*, 1105–1112. [CrossRef]
39. Pęska, M.; Karczewski, K.; Rzeszotarska, M.; Polański, M. Direct Synthesis of Fe-Al Alloys from Elemental Powders Using Laser Engineered Net Shaping. *Materials* **2020**, *13*, 531. [CrossRef]
40. Mwangi, J.W.; Nguyen, L.T.; Bui, V.D.; Berger, T.; Zeidler, H.; Schubert, A. Nitinol manufacturing and micromachining: A review of processes and their suitability in processing medical-grade nitinol. *J. Manuf. Process.* **2019**, *38*, 355–369. [CrossRef]
41. Krishna, B.V.; Bose, S.; Bandyopadhyay, A. Laser Processing of Net-Shape NiTi Shape Memory Alloy. *Met. Mater. Trans. A* **2007**, *38*, 1096–1103. [CrossRef]
42. Marattukalam, J.J.; Singh, A.K.; Datta, S.; Das, M.; Balla, V.K.; Bontha, S.; Kalpathy, S.K. Microstructure and corrosion behavior of laser processed NiTi alloy. *Mater. Sci. Eng. C* **2015**, *57*, 309–313. [CrossRef]
43. Baran, A.; Polański, M. Microstructure and properties of LENS (laser engineered net shaping) manufactured Ni-Ti shape memory alloy. *J. Alloys Compd.* **2018**, *750*, 863–870. [CrossRef]
44. Halani, P.R.; Shin, Y.C. In Situ Synthesis and Characterization of Shape Memory Alloy Nitinol by Laser Direct Deposition. *Met. Mater. Trans. A* **2011**, *43*, 650–657. [CrossRef]
45. Shiva, S.; Palani, I.; Mishra, S.; Paul, C.; Kukreja, L.M. Investigations on the influence of composition in the development of Ni–Ti shape memory alloy using laser based additive manufacturing. *Opt. Laser Technol.* **2015**, *69*, 44–51. [CrossRef]
46. Bimber, B.A.; Hamilton, R.F.; Palmer, T.A. Ni-Concentration Dependence of Directed Energy Deposited NiTi Alloy Microstructures. *Shape Mem. Superelasticity* **2019**, *5*, 182–187. [CrossRef]
47. Bimber, B.A.; Hamilton, R.F.; Keist, J.; Palmer, T.A. Anisotropic microstructure and superelasticity of additive manufactured NiTi alloy bulk builds using laser directed energy deposition. *Mater. Sci. Eng. A* **2016**, *674*, 125–134. [CrossRef]
48. Hamilton, R.F.; Bimber, B.A.; Palmer, T.A. Correlating microstructure and superelasticity of directed energy deposition additive manufactured Ni-rich NiTi alloys. *J. Alloys Compd.* **2018**, *739*, 712–722. [CrossRef]
49. Wang, C.; Tan, X.; Du, Z.; Chandra, S.; Sun, Z.; Lim, C.; Tor, S.; Lim, C.; Wong, C.H. Additive manufacturing of NiTi shape memory alloys using pre-mixed powders. *J. Mater. Process. Technol.* **2019**, *271*, 152–161. [CrossRef]
50. Yu, Y.; Zhou, J.; Chen, J.; Zhou, H.; Guo, C.; Guo, B. Preparation, microstructure and tribological properties of Ni3Al intermetallic compound coating by laser cladding. *Intermetallics* **2010**, *18*, 871–876. [CrossRef]
51. Xue, Y.; Wang, H. Microstructure and wear properties of laser clad TiCo/Ti$_2$Co intermetallic coatings on titanium alloy. *Appl. Surf. Sci.* **2005**, *243*, 278–286. [CrossRef]
52. Wang, H.M.; Liu, Y.F. Microstructure and wear resistance of laser clad Ti5Si3/NiTi2 intermetallic composite coating on titanium alloy. *Mat. Sci. Eng. A-Struc.* **2002**, *338*, 126–132. [CrossRef]
53. Wang, H.; Duan, G. Wear and corrosion behavior of laser clad Cr3Si reinforced intermetallic composite coatings. *Intermetallics* **2003**, *11*, 755–762. [CrossRef]
54. Wang, H.; Wang, C.; Cai, L. Wear and corrosion resistance of laser clad Ni$_2$Si/NiSi composite coatings. *Surf. Coat. Technol.* **2003**, *168*, 202–208. [CrossRef]
55. Zhong, M.; Liu, W.; Zhang, Y.; Zhu, X. Formation of WC/Ni hard alloy coating by laser cladding of W/C/Ni pure element powder blend. *Int. J. Refract. Met. Hard Mater.* **2006**, *24*, 453–460. [CrossRef]
56. Karnati, S.; Liou, F.F.; Newkirk, J.W. Characterization of copper–nickel alloys fabricated using laser metal deposition and blended powder feedstocks. *Int. J. Adv. Manuf. Technol.* **2019**, *103*, 239–250. [CrossRef]
57. Li, C.; Ma, S.; Liu, X.; Li, J.; Le, G. Microstructures and properties of 80W-20Fe alloys prepared using laser melting deposition process. *Int. J. Refract. Met. Hard Mater.* **2018**, *77*, 113–119. [CrossRef]
58. Fallah, V.; Corbin, S.F.; Khajepour, A. Process optimization of Ti–Nb alloy coatings on a Ti–6Al–4V plate using a fiber laser and blended elemental powders. *J. Mater. Process. Technol.* **2010**, *210*, 2081–2087. [CrossRef]
59. Zhang, F.; Yang, M.; Clare, A.; Lin, X.; Tan, H.; Chen, Y. Microstructure and mechanical properties of Ti-2Al alloyed with Mo formed in laser additive manufacture. *J. Alloys Compd.* **2017**, *727*, 821–831. [CrossRef]
60. Chen, Y.; Liou, F. Additive Manufacturing of Metal Functionally Graded Materials: A Review. In Proceedings of the 29th Annual International Solid Freeform Fabrication Symposium, Austin, TX, USA, 13–15 August 2018.

61. Carroll, B.E.; Otis, R.; Borgonia, J.P.; Suh, J.-O.; Dillon, R.P.; Shapiro, A.A.; Hofmann, U.C.; Liu, Z.-K.; Beese, A.M. Functionally graded material of 304L stainless steel and inconel 625 fabricated by directed energy deposition: Characterization and thermodynamic modeling. *Acta Mater.* **2016**, *108*, 46–54. [CrossRef]

62. Liu, Y.; Liu, C.; Liu, W.; Ma, Y.; Zhang, C.; Cai, Q.; Liu, B. Microstructure and properties of Ti/Al lightweight graded material by direct laser deposition. *Mater. Sci. Technol.* **2017**, *34*, 945–951. [CrossRef]

63. Schneider-Maunoury, C.; Weiss, L.; Acquier, P.; Boisselier, D.; Laheurte, P. Functionally graded Ti6Al4V-Mo alloy manufactured with DED-CLAD® process. *Addit. Manuf.* **2017**, *17*, 55–66. [CrossRef]

64. Bobbio, L.D.; Otis, R.; Borgonia, J.P.; Dillon, R.P.; Shapiro, A.A.; Liu, Z.-K.; Beese, A.M. Additive manufacturing of a functionally graded material from Ti-6Al-4V to Invar: Experimental characterization and thermodynamic calculations. *Acta Mater.* **2017**, *127*, 133–142. [CrossRef]

65. Banerjee, R.; Collins, P.; Bhattacharyya, D.; Fraser, H. Microstructural evolution in laser deposited compositionally graded α/β titanium-vanadium alloys. *Acta Mater.* **2003**, *51*, 3277–3292. [CrossRef]

66. Collins, P.; Banerjee, R.; Fraser, H. Laser deposition of compositionally graded titanium–vanadium and titanium–molybdenum alloys. *Mater. Sci. Eng. A* **2003**, *352*, 118–128. [CrossRef]

67. Nag, S.; Banerjee, R.; Fraser, H.L. A novel combinatorial approach for understanding microstructural evolution and its relationship to mechanical properties in metallic biomaterials. *Acta Biomater.* **2007**, *3*, 369–376. [CrossRef] [PubMed]

68. Zhang, Y.; Meacock, C.; Vilar, R. Laser powder micro-deposition of compositional gradient Ti–Cr alloy. *Mater. Des.* **2010**, *31*, 3891–3895. [CrossRef]

69. Shishkovsky, I.; Missemer, F.; Smurov, I. Direct Metal Deposition of Functional Graded Structures in Ti- Al System. *Phys. Procedia* **2012**, *39*, 382–391. [CrossRef]

70. Mendoza, M.Y.; Samimi, P.; Brice, D.A.; Martin, B.; Rolchigo, M.R.; Lesar, R.; Collins, P. Microstructures and Grain Refinement of Additive-Manufactured Ti-xW Alloys. *Met. Mater. Trans. A* **2017**, *48*, 3594–3605. [CrossRef]

71. Mendoza, M.Y.; Samimi, P.; Brice, D.A.; Ghamarian, I.; Rolchigo, M.; Lesar, R.; Collins, P. On the role of composition and processing parameters on the microstructure evolution of Ti-xMo alloys. *BMC Chem.* **2019**, *13*, 1–8. [CrossRef]

72. Banerjee, R.; Bhattacharyya, D.; Collins, P.; Viswanathan, G.; Fraser, H. Precipitation of grain boundary α in a laser deposited compositionally graded Ti–8Al–xV alloy—An orientation microscopy study. *Acta Mater.* **2004**, *52*, 377–385. [CrossRef]

73. Karnati, S.; Zhang, Y.; Liou, F.; Newkirk, J. On the Feasibility of Tailoring Copper–Nickel Functionally Graded Materials Fabricated through Laser Metal Deposition. *Metals* **2019**, *9*, 287. [CrossRef]

74. Li, W.; Yan, L.; Chen, X.; Zhang, J.; Zhang, X.; Liou, F. Directed energy depositing a new Fe-Cr-Ni alloy with gradually changing composition with elemental powder mixes and particle size' effect in fabrication process. *J. Mater. Process. Technol.* **2018**, *255*, 96–104. [CrossRef]

75. Li, W.; Chen, X.; Yan, L.; Zhang, J.; Zhang, X.; Liou, F. Additive manufacturing of a new Fe-Cr-Ni alloy with gradually changing compositions with elemental powder mixes and thermodynamic calculation. *Int. J. Adv. Manuf. Technol.* **2017**, *95*, 1013–1023. [CrossRef]

76. Conteri, R.; Borkar, T.; Nag, S.; Jaeger, D.; Chen, X.; Ramanujan, R.; Banerjee, R. Laser additive processing of Fe-Si-B-Cu-Nb magnetic alloys. *J. Manuf. Process.* **2017**, *29*, 175–181. [CrossRef]

77. Borkar, T.; Conteri, R.; Chen, X.; Ramanujan, R.; Banerjee, R. Laser additive processing of functionally-graded Fe–Si–B–Cu–Nb soft magnetic materials. *Mater. Manuf. Process.* **2016**, *32*, 1581–1587. [CrossRef]

78. Manna, I.; Majumdar, J.D.; Chandra, B.R.; Nayak, S.; Dahotre, N.B. Laser surface cladding of Fe–B–C, Fe–B–Si and Fe–BC–Si–Al–C on plain carbon steel. *Surf. Coat. Technol.* **2006**, *201*, 434–440. [CrossRef]

79. Hou, X.; Du, D.; Wang, K.; Hong, Y.; Chang, B. Microstructure and Wear Resistance of Fe-Cr-Mo-Co-C-B Amorphous Composite Coatings Synthesized by Laser Cladding. *Metals* **2018**, *8*, 622. [CrossRef]

80. Tsai, M.-H.; Yeh, J.-W. High-Entropy Alloys: A Critical Review. *Mater. Res. Lett.* **2014**, *2*, 107–123. [CrossRef]

81. Cui, W.; Zhang, X.; Li, L.; Chen, Y.; Pan, T.; Liou, F. Fabrication and Characterization of AlxCrCuFeNi2 High-Entropy Alloys Coatings by Laser Metal Deposition. *Procedia Manuf.* **2019**, *39*, 509–518. [CrossRef]

82. Cui, W.; Karnati, S.; Zhang, X.; Burns, E.; Liou, F. Fabrication of AlCoCrFeNi High-Entropy Alloy Coating on an AISI 304 Substrate via a CoFe2Ni Intermediate Layer. *Entropy* **2018**, *21*, 2. [CrossRef]

83. Chao, Q.; Guo, T.; Jarvis, T.; Wu, X.; Hodgson, P.; Fabijanic, D. Direct laser deposition cladding of AlxCoCrFeNi high entropy alloys on a high-temperature stainless steel. *Surf. Coat. Technol.* **2017**, *332*, 440–451. [CrossRef]

84. Chen, X.; Yan, L.; Karnati, S.; Zhang, Y.; Liou, F. Fabrication and Characterization of Al$_x$CoFeNiCu$_{1-x}$ High Entropy Alloys by Laser Metal Deposition. *Coatings* **2017**, *7*, 47. [CrossRef]

85. Ni, C.; Shi, Y.; Liu, J.; Huang, G. Characterization of Al0.5FeCu0.7NiCoCr high-entropy alloy coating on aluminum alloy by laser cladding. *Opt. Laser Technol.* **2018**, *105*, 257–263. [CrossRef]

86. Siddiqui, A.; Dubey, A.; Paul, C.P. A study of metallurgy and erosion in laser surface alloying of AlxCu0.5FeNiTi high entropy alloy. *Surf. Coatings Technol.* **2019**, *361*, 27–34. [CrossRef]

87. Huang, C.; Zhang, Y.; Shen, J.; Vilar, R. Thermal stability and oxidation resistance of laser clad TiVCrAlSi high entropy alloy coatings on Ti–6Al–4V alloy. *Surf. Coat. Technol.* **2011**, *206*, 1389–1395. [CrossRef]

88. Dobbelstein, H.; Gurevich, E.L.; George, E.P.; Ostendorf, A.; Laplanche, G. Laser metal deposition of a refractory TiZrNbHfTa high-entropy alloy. *Addit. Manuf.* **2018**, *24*, 386–390. [CrossRef]

89. Li, Q.; Zhang, H.; Li, D.; Chen, Z.; Huang, S.; Lu, Z.; Yan, H. WxNbMoTa Refractory High-Entropy Alloys Fabricated by Laser Cladding Deposition. *Materials* **2019**, *12*, 533. [CrossRef] [PubMed]

90. Dobbelstein, H.; Gurevich, E.L.; George, E.P.; Ostendorf, A.; Laplanche, G. Laser metal deposition of compositionally graded TiZrNbTa refractory high-entropy alloys using elemental powder blends. *Addit. Manuf.* **2019**, *25*, 252–262. [CrossRef]

91. Gwalani, B.; Soni, V.; Waseem, O.A.; Mantri, S.A.; Banerjee, R. Laser additive manufacturing of compositionally graded AlCrFeMoVx (x = 0 to 1) high-entropy alloy system. *Opt. Laser Technol.* **2019**, *113*, 330–337. [CrossRef]

92. Schwendner, I.K.; Banerjee, R.; Collins, P.; Brice, C.A.; Fraser, H.L. Direct laser deposition of alloys from elemental powder blends. *Scr. Mater.* **2001**, *45*, 1123–1129. [CrossRef]

93. Collins, P.C. A Combinatorial Approach to the Development of Composition-Microstructure-Property Relationships in Titanium Alloys Using Directed Laser Deposition. Ph.D. Thesis, The Ohio State University, Columbus, OH, USA, December 2004.

94. Wei, L. Modeling and Experimental Investigations on Particle Dynamic Behaviors in Laser 3D Printing with Blown Powder. Ph.D. Thesis, Missouri University of Science and Technology, Rolla, MO, USA, 2018.

95. Wei, L.; Zhang, X.; Liou, F. Modeling analysis of argon gas flow rate's effect on pre-mixed powder separation in laser metal deposition process and experimental validation. *Int. J. Adv. Manuf. Technol.* **2018**, *96*, 4321–4331. [CrossRef]

96. Li, W.; Karnati, S.; Zhang, Y.; Liou, F. Investigating and eliminating powder separation in pre-mixed powder supply for laser metal deposition process. *J. Mater. Process. Technol.* **2018**, *254*, 294–301. [CrossRef]

97. Zhang, F.; Chen, J.; Tan, H.; Lin, X.; Huang, W. Composition control for laser solid forming from blended elemental powders. *Opt. Laser Technol.* **2009**, *41*, 601–607. [CrossRef]

98. Asano, K.; Tsukamoto, M.; Sechi, Y.; Sato, Y.; Masuno, S.-I.; Higashino, R.; Hara, T.; Sengoku, M.; Yoshida, M. Laser metal deposition of pure copper on stainless steel with blue and IR diode lasers. *Opt. Laser Technol.* **2018**, *107*, 291–296. [CrossRef]

99. Zhang, H.; Zhao, Y.; Huang, S.; Zhu, S.; Wang, F.; Li, D. Manufacturing and Analysis of High-Performance Refractory High-Entropy Alloy via Selective Laser Melting (SLM). *Materials* **2019**, *12*, 720. [CrossRef] [PubMed]

100. Guzanová, A.; Brezinová, J.; Draganovská, D.; Maruschak, P. Properties of coatings created by HVOF technology using micro-and nano-sized powder. *Koroze Ochr. Mater.* **2019**, *63*, 86–93. [CrossRef]

101. Mukherjee, T.; Zuback, J.S.; De, A.; Debroy, T. Printability of alloys for additive manufacturing. *Sci. Rep.* **2016**, *6*, 19717. [CrossRef] [PubMed]

102. Maruschak, P.; Konovalenko, I.; Chausov, M.G.; Pylypenko, A.P.; Panin, S.V.; Vlasov, I.V.; Prentkovskis, O. Impact of Dynamic Non-Equilibrium Processes on Fracture Mechanisms of High-Strength Titanium Alloy VT23. *Metals* **2018**, *8*, 983. [CrossRef]

Article

Investigation of Microstructure and Mechanical Properties for Ti-6Al-4V Alloy Parts Produced Using Non-Spherical Precursor Powder by Laser Powder Bed Fusion

Jaime Varela [1,2], Edel Arrieta [1,2], Muktesh Paliwal [3], Mike Marucci [3], Jose H. Sandoval [4], Jose A. Gonzalez [4], Brandon McWilliams [5], Lawrence E. Murr [2,6], Ryan B. Wicker [1,2] and Francisco Medina [1,2,*]

[1] Department of Mechanical Engineering, The University of Texas at El Paso, El Paso, TX 79938, USA; jvarela11@miners.utep.edu (J.V.); egarrieta@utep.edu (E.A.); rwicker@utep.edu (R.B.W.)

[2] W.M. Keck Center for 3D Innovation, University of Texas at El Paso, El Paso, TX 79968, USA; lemurr@utep.edu

[3] Kymera International—Reading Alloys, Robesonia, PA 19551, USA; Muktesh.paliwal@kymerainternational.com (M.P.); mike.marucci@kymerainternational.com (M.M.)

[4] Lockheed Martin Missiles and Fire Control, Dallas, TX 75051, USA; jose.h.sandoval@lmco.com (J.H.S.); jose.a1.gonzalez@lmco.com (J.A.G.)

[5] CCDC Army Research Laboratory, Aberdeen Proving Ground, MD 21005, USA; brandon.a.mcwilliams.civ@mail.mil

[6] Department of Metallurgical, Materials and Biomedical Engineering, The University of Texas at El Paso, El Paso, TX 79968, USA

* Correspondence: frmedina@utep.edu

check for **updates**

Citation: Varela, J.; Arrieta, E.; Paliwal, M.; Marucci, M.; Sandoval, J.H.; Gonzalez, J.A.; McWilliams, B.; Murr, L.E.; Wicker, R.B.; Medina, F. Investigation of Microstructure and Mechanical Properties for Ti-6Al-4V Alloy Parts Produced Using Non-Spherical Precursor Powder by Laser Powder Bed Fusion. *Materials* 2021, 14, 3028. https://doi.org/10.3390/ma14113028

Academic Editor: Tuhin Mukherjee

Received: 9 May 2021
Accepted: 29 May 2021
Published: 2 June 2021

Publisher's Note: MDPI stays neutral with regard to jurisdictional claims in published maps and institutional affiliations.

Abstract: An unmodified, non-spherical, hydride-dehydride (HDH) Ti-6Al-4V powder having a substantial economic advantage over spherical, atomized Ti-6Al-4V alloy powder was used to fabricate a range of test components and aerospace-related products utilizing laser beam powder-bed fusion processing. The as-built products, utilizing optimized processing parameters, had a Rockwell-C scale (HRC) hardness of 44.6. Following heat treatments which included annealing at 704 °C, HIP at ~926 °C (average), and HIP + anneal, the HRC hardnesses were observed to be 43.9, 40.7, and 40.4, respectively. The corresponding tensile yield stress, UTS, and elongation for these heat treatments averaged 1.19 GPa, 1.22 GPa, 8.7%; 1.03 GPa, 1.08 GPa, 16.7%; 1.04 GPa, 1.09 GPa, 16.1%, respectively. The HIP yield strength and elongation of 1.03 GPa and 16.7% are comparable to the best commercial, wrought Ti-6Al-4V products. The corresponding HIP component microstructures consisted of elongated small grains (~125 microns diameter) containing fine, alpha/beta lamellae.

Keywords: non-spherical; hydride-dehydride (HDH) Ti-6Al-4V powder; laser powder bed fusion; post-process heat treatment; microstructure; mechanical properties

1. Introduction

Additive manufacturing/3D printing has become a key enabling manufacturing process which has been characterized as underpinning the so-called Fourth Industrial Revolution. Powder-bed fusion processes such as selective laser melting (SLM) and electron beam melting (EBM) in particular provide a wide range of cost savings, high precision and speed of production for complex product shapes applied to aerospace automotive, biomedical and related applications, including maintenance, repair and sustainment. Qualification and certification of optimized parts, especially aerospace and aircraft components are also important issues [1–6].

Due to its low density, high mechanical strength, excellent corrosion resistance, and related properties, Ti-6Al-4V alloy has become one of the most widely used titanium alloys for additive manufacturing of a wide range of components [7,8], especially for the laser-based or laser beam powder-bed fusion (LBPBF) process. Since the laser-based (SLM) process utilizes an inert gas atmosphere (argon or nitrogen), this shield gas flow reduces oxidation of the alloy powder and the melted layers. However, the morphology, size and size distribution of the precursor powder are also important factors because they affect powder flowability, laser beam energy absorption, and conductivity of the powder bed which change as the bed consolidates and melts. In addition, the laser process parameters have a controlling effect on layer building as well as the microstructure and properties of the as-built product. These generally include laser beam energy and energy density which is related to the absorbed energy in the powder layer, the scan speed, and the beam size or scan spacing [8,9].

It is apparent, as noted above, that the initial powder bed particle packing or packing density has an effect on the laser beam energy absorption and melt efficiency of the layers, and this will in fact change as the powder bed layer melts. While spherical powders having a wider distribution of particle sizes can optimize bed packing and densification by more effectively filling void spaces with smaller particles requisite flowability, these powders generally have a high cost since their production involves gas or plasma atomization [9]. In contrast, Ti-6Al-4V powders having non-spherical shapes are easily produced by forming stable, brittle hydrides that can be crushed, milled, and screened to produce fine powders which are dehydrided to form non-spherical alloy powder. This hydride-dehydride (HDH) process is a long-established process for Ti and Ti alloy powder production [10,11], and represents a significant economic advantage over spherical, atomized powder production [12–14].

The challenge for powder bed fusion fabrication utilizing non-spherical precursor powder is the achievement of requisite flowability, packing, and melt efficiency in order to achieve optimized part production characterized by short production times to achieve requisite mechanical properties and associated microstructures. Jaber et al. [12] have recently demonstrated that for L-PBF of a hybrid-50% spherical and 50% non-spherical (HDH)-Ti-6Al-4V powder, the flowability was the crucial parameter governing the residual tensile properties of fabricated components. Hou et al. [15] also recently modified HDH Ti-6Al-4V powder by ball milling and were able to produce 99% dense products for this modified powder using laser beam powder-bed fusion processing. Narra et al. [16] have also compared melt pool porosity for electron beam powder-bed fusion processing of spherical and non-spherical Ti-6Al-4V alloy powders. Microstructures observed for HDH Ti-6Al-4V alloy builds were observed to be similar to those for parts fabricated using spherical, atomized powders.

In this investigation, laser beam powder-bed fusion process parameters were systematically varied in order to find optimized conditions to fabricate an assortment of complex Ti-6Al-4V alloy products and test components; utilizing 100% non-spherical HDH Ti-6Al-4V alloy powder. While the as-fabricated components were near full density, conventional post-process annealing relieved process-induced internal stress, while HIP reduced remaining porosity and produced components with microstructures and mechanical properties compatible with those characteristic of Ti-6Al-4V alloy products fabricated using spherical, atomized precursor powder and heat treated. In fact, as-built and post process HIP components fabricated from non-spherical, HDH Ti-6Al-4V alloy powder using laser beam powder-bed fusion exhibited tensile properties as good as the best, commercial Ti-6Al-4V wrought products.

2. Materials and Methods

The recent work of Jaber et al. [12] as noted above compared laser beam powder-bed fusion components of Ti-6Al-4V utilizing 100% commercial, spherical powder and a 1-to-1 mixture of spherical and non-spherical (HDH) Ti-6Al-4V alloy powder. The powder mixture exhibited a 30% reduction in flowability while the corresponding, as-built components exhibited porosity and large, unmelted fusion zones ~100 microns in size. The tensile strength and elongation for the mixed powder components were observed to be ~17% and 31% lower, respectively, than for the 100% spherical powder-fabricated components. It is notable that the build process parameters were constant at 125 W beam energy and 10^3 mm/s laser beam scan speed for a layer thickness of ~20 μm.

2.1. Powder and Feedstock

In the present work, the laser powder-bed fusion process parameters were systematically varied in fabricating simple test blocks or cubes to achieve a wide range of laser beam scan speeds and volumetric energy densities in non-spherical, HDH Ti-6Al-4V alloy powder layers ~30 microns thick. Optimum process parameters were as selected on the basis of minimum, as-built test cube porosities and corresponding maximum densities. Additionally, post-process heat treatments were also utilized to create essentially full density, low porosity products. Unlike the work of Hou et al. [15] utilizing modified (ball-milled) HDH powder Ti-6Al-4V powder in electron beam powder-bed fusion processing, the current study utilized as produced Ti-6Al-4V (Ti64) HDH powder as shown in the SEM images in Figure 1 and provided by Reading Alloys-Kymera International (Raleigh, NC, USA). Particle size and morphology analysis was performed through the Retsch Camsizer X2 (Haan, Germany) Dynamic Image Analysis system. Analysis revealed a particle size distribution of D10: 54.9 μm, D50: 72.7 μm, and D90: 88.9 μm as shown in Figure 2. The scanning electron microscope (SEM) images of the powder shown in Figure 1 were taken with a JEOL JSM-IT500 SEM (Tokyo, Japan).

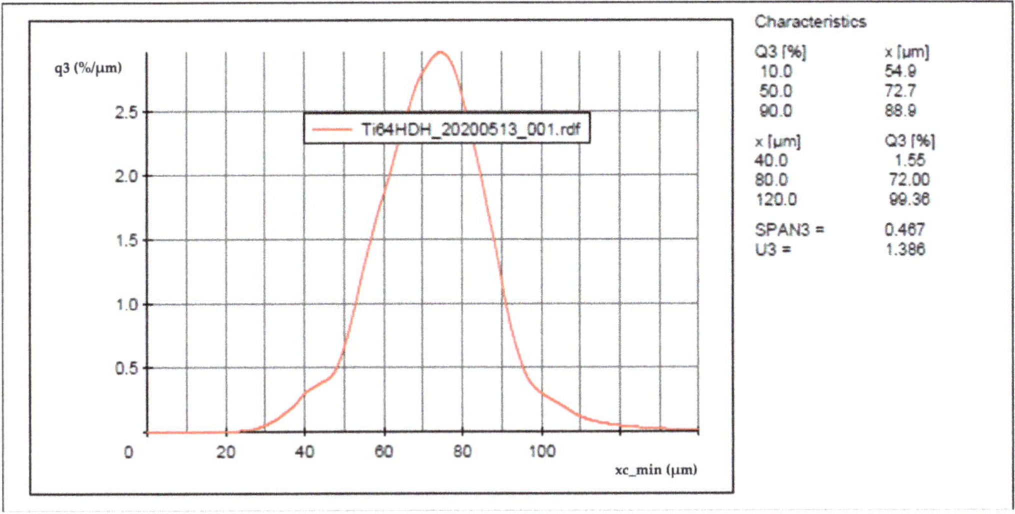

Figure 1. Particle size distribution.

Figure 2. Ti64 HDH powder SEM images. (**a**) Low magnification. (**b**) High magnification.

2.2. Process Parameters

Ti-6Al-4V HDH samples were fabricated on a preheated bedplate at 175 °C. A design of experiments was performed to develop the optimal printing parameters for the HDH powder. Fourteen 15 mm × 15 mm × 15 mm test cubes were printed with various parameter changes. A layer thickness of 30 μm and a stripe width of 10 mm were kept constant. Scan speed, laser power, hatch distance all changed to yield different volumetric energy densities as shown below in Table 1.

Table 1. Process parameters design of experiments.

Cube -	Scan Speed (mm/s)	Laser Power (W)	Hatch (mm)	Layer Thickness (µm)	Stripe Width (mm)	e (J/mm^3)
1	1200	280	0.14	30	5	55.6
2	1200	310	0.14	30	5	61.5
3	1200	340	0.14	30	5	67.5
4	1100	280	0.14	30	5	60.6
5	1000	280	0.14	30	5	66.7
6	1000	280	0.12	30	5	77.8
7	1000	280	0.1	30	5	93.3
8	400	110	0.12	30	5	76.4
9	400	120	0.12	30	5	83.3
10	400	130	0.12	30	5	90.3
11	360	110	0.12	30	5	84.9
12	320	110	0.12	30	5	95.5
13	320	110	0.1	30	5	114.6
14	320	110	0.08	30	5	143.2

Cube one yielded the least noticeable porosity from optical micrographs as well as the highest density from pycnometry. These were the chosen printing parameters moving forward.

2.3. Laser Powder Bed Fusion System, Setup, and Fabrication

The fabrication of the Ti64 HDH pieces was completed with an EOS M290 (Krailling, Germany). The M290 is an industrial production LPBF system equipped with 400 W Ytterbium fiber laser and a 250 mm × 250 mm × 325 mm build volume. Sixty Ø 14 mm × 80 mm vertical cylinders and eleven 77 mm × 14 mm × 60 mm bars (A–J in Figure 3) were fabricated as shown in the layout in Figure 3. Of the sixty successfully printed cylinders eleven were allocated for the annealing heat treatment described in Section 2.4, six for HIP, and eleven for the combination of HIP and annealing.

In addition to the various test geometries fabricated as shown in Figure 3, a series of complex, aerospace components were fabricated from the non-spherical HDH Ti64 powder utilizing the optimized process parameters described above. These are illustrated in the examples shown in Figures 4 and 5. Figure 5b,c depicts the internal porosity and microstructure representations, respectively, of the electronics box enclosure section at the indicated red circle (Figure 5a).

The parts fabricated were an electronics box enclosure and a topologically optimized bracket. The bracket and electronics box enclosure were designed by incorporating elements from a Design for Additive Manufacturing (DfAM) perspective and took into account L-PBF process capabilities. Both parts are representative of use cases for Department of Defense and Aerospace industry applications. The parts produced under this program using Ti64 HDH metal powder feedstock was selected as a case study and allow for a one-to-one comparison against the same geometries that were built using Gas Atomized or Plasma Atomized metal powders in L-PBF.

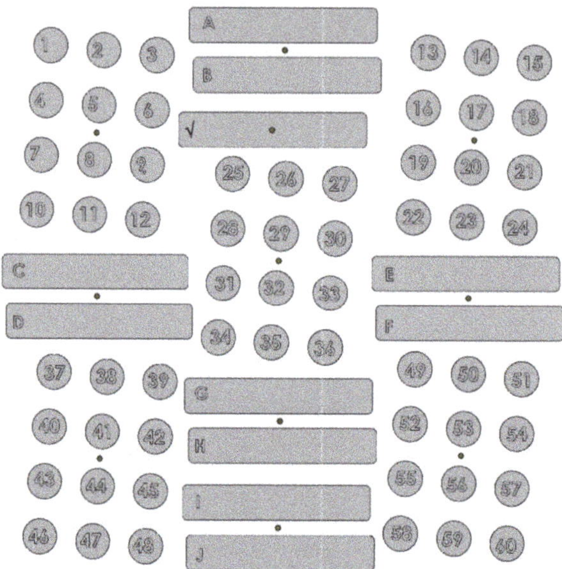

Figure 3. Build plate print layout.

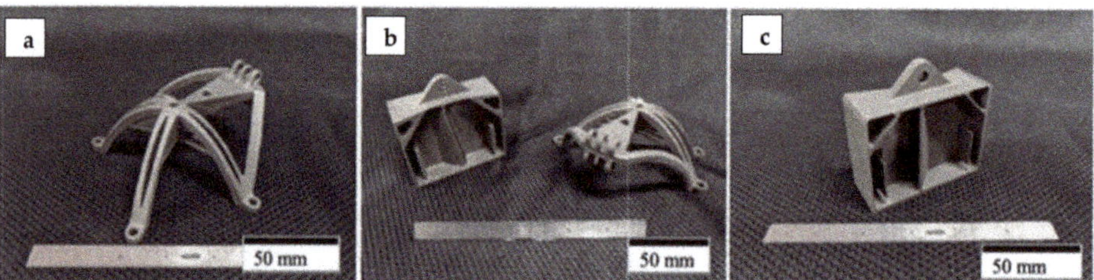

Figure 4. Annealed parts successfully printed from Ti64 HDH. (**a**) Topologically optimized bracket. (**b**) Side to side printed geometries. (**c**) Electronics box enclosure.

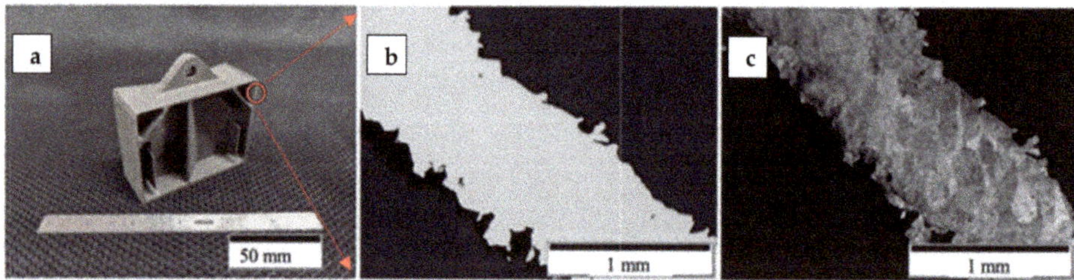

Figure 5. Sectioned printed part. (**a**) Sectioned electronic box enclosure at the red circle. (**b**) Polished micrograph. (**c**) Etched micrograph.

2.4. Heat Treatment Parameters

All tensile samples were heat treated prior to machining. The annealing was performed in accordance with the SAE Aerospace AMS 2801B parameters. HIP parameters followed the ASTM F2924-14 Standard. Detailed heat treatment parameters for three variants are shown in Table 2 below.

Table 2. Hot isostatic pressing and annealing parameters.

Variant	Process	Pressure (MPa)	Temperature (°C)	Hold Time (min)	Cooling
1	Anneal	None	704 ± 14	120 ± 15	Air or furnace cooled
2	HIP	100	896–955 (±15 of selected temp)	180 ± 60	Under inert atmosphere below 425 °C
3	HIP	100	896–955 (±15 of selected temp)	180 ± 60	Under inert atmosphere below 425 °C
	Anneal	None	704 ± 14	120 ± 15	Air or furnace cooled

2.5. Tensile Testing

Following the heat treatment process all parts were machined in accordance with ASTM E8 standard. An MTS Landmark (Eden Prairie, MN, USA) servo-hydraulic tensile test system was utilized for monotonic uniaxial tensile strength tests. The testing was performed with threaded grips and an MTS 30 mm axial clip extensometer for axial strain measurements. Samples were strained at a rate of 0.47625 mm/min. Results were averaged from 11 specimens in the Anneal (Variant 1) and the HIP + Anneal (Variant 3); and averaged from six specimens in the HIP variant.

2.6. Microstructure Characterization

Each tensile specimen had two threaded sections of an approximate of 12 mm in diameter and 15 mm of length. These unstrained threaded sections were sectioned both at the top and the bottom for metallographic analysis as shown in Figure 6. This method was performed in order to study the sections of the tested samples in a location that was not affected by the tensile test therefore not disrupting the microstructure representation. Each sample was sectioned as to reveal the X, Y, and Z planes with the X and Y planes being in accordance with the printing orientation.

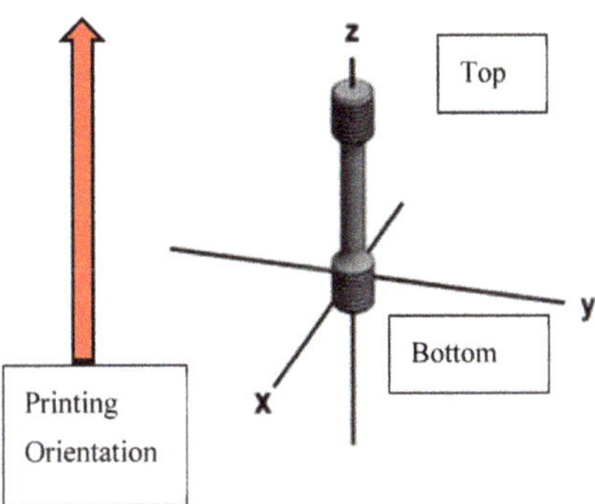

Figure 6. Machined tensile specimen sectioned at the top and bottom threaded sections for metallography.

All metallographic samples were created with the ATM OPAL 460 (Haan, Germany) hot mounting press and black epoxy. Metallographic samples were then ground and polished using the ATM SAPHIR 530 (Haan, Germany) semiautomatic system. The grinding process began with Silicon Carbide abrasive paper with 320 grit at 300 rpm with a force of 35 N until plane. Samples were subsequently more finely grinded with a 9 µm diamond suspension on a fine grinding disc at 150 rpms with a force of 25 N for 5 min. The final polishing step was performed on a neoprene polishing pad with a 0.2 µm fumed silica suspension at 150 rpm with a force of 25 N for 5 min.

Microstructure was revealed using Kroll's Reagent consisting of 92 mL of distilled water, 6 mL of nitric acid and 2 mL of hydrofluoric acid. Samples were submerged in the reagent for 10–15 s depending on the variant. Optical microscopy was performed in an Olympus™ GX53 (Olympus Inc., Tokyo, Japan).

2.7. Density Measurements

Gas displacement pycnometry was performed on an AccuPyc II 1340 (Norcross, GA, USA) to obtain volume measurement. The AccuPyc performed five measurements of every sample giving an average. Mass measurements were obtained from a Sartorius CP124S weight balance (Sartorius AG, Göttingen, Germany). Density was then calculated from the resulting volume and mass measurements. In addition, each variant was sectioned and polished to obtain visual representation of internal porosity as demonstrated in Figure 7.

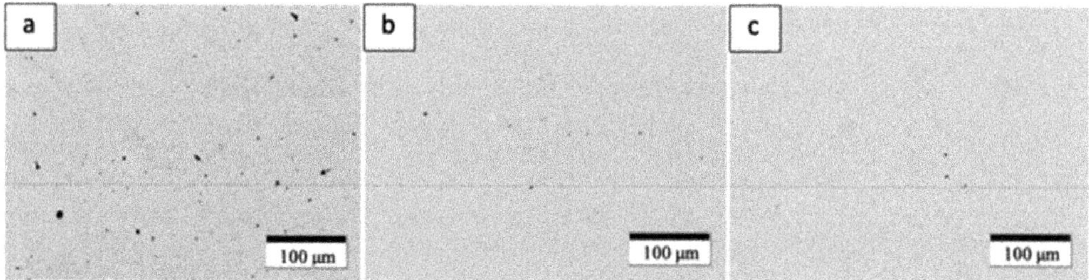

Figure 7. Polished optical micrographs demonstrating porosity across variants. (**a**) Annealed, (**b**) HIP, (**c**) HIP + Annealed.

2.8. Hardness Testing

Hardness measurements were performed on the Wilson® Rockwell® 2000 (Canton, MA, USA) hardness tester. Samples were indented with a Brale indenter on the Rockwell C scale (HRC). This was done with a pre-load of 10 Kgf and a main-load of 150 Kgf. Measurements were performed on two samples from each variant at the top and bottom portions and the X, Y, and Z surfaces. Three indentations were made on every surface separated by at least one millimeter. An average from all these measurements were then reported.

2.9. Fracture Surface Analysis

Following the tensile tests, the fractures of each variant were analyzed using the JEOL JSM-IT500 SEM (JEOL, Tokyo, Japan) scanning electron microscope. One end of the fracture surface from each variant was mounted for observation and comparison.

2.10. Chemical Analysis

Chemical analysis was performed by Reading Alloys-Kymera International. O, N, H analysis was performed on an 836 Series Elemental Analyzer from the LECO corporation (St. Joseph, MI, USA). C and S measurements were done on an CS744 Series Carbon/Sulfur Analyzer from LECO (St. Joseph, MI, USA). Al, V, Fe, Si, and Ca results were obtained from the Ciros Vision Inductively Coupled Plasma Spectrometer from SPECTRO Analytical Instruments (Kleve, Germany). Results met grade 5 chemistry for Ti64 as shown in Table 3.

Table 3. Chemical composition of as-built parts.

Ti64 HDH	Al	V	Fe	Si	Ca	O	N	C	S	H
	5.92	4.06	0.20	0.014	0.006	0.17	0.042	0.033	0.0009	0.009

3. Results and Discussion

3.1. Microstructure Analysis and Discussion

Figure 8 shows the typical microstructure for the optimized, as-built Ti-6Al-4V alloy product utilizing the non-spherical, HDH powder in Figure 1. The low magnification image in Figure 8a shows varying sizes of grains elongated in the build direction along with layer-related melt bands. Build direction for all micrograph images is oriented from bottom to top. The higher magnification views in Figure 8b,c show a preponderance of alpha-prime martensite represented by the black lamellae which are variously etched in optimally oriented grains. These martensite lamellae, having widths of ~2 microns, result from the rapid cooling associated with the laser beam processing. The corresponding HRC hardness average characteristic of the test components represented in Figure 8 was 44.6, which along with the martensitic microstructure in Figure 8b,c is typical for LPBF as-built Ti-6Al-4V alloy components utilizing spherical, atomized precursor powder [2,7–9,17]. The alpha-prime (martensite) microstructure shown in Figure 8b,c results from the diffusion-less, composition invariant beta → alpha-prime martensite transformation, and was also observed by Jaber et al. [12] using an HDH-non-spherical/spherical Ti-6Al-4V powder mixture for laser beam powder-bed fusion processing, as well as the more recent work of Narra et al. [16] using a modified, non-spherical HDH Ti-6Al-4V alloy precursor powder.

By comparison with the as-built microstructure represented typically in Figure 8, the non-spherical, HDH Ti64 as-built and annealed Ti-6Al-4V product microstructure typical for the complex component fabrications shown in Figures 4 and 5 is illustrated in Figures 9 and 10a,b at the bottom and top for test rods (Figure 6), respectively. There is little difference for both the low and high magnification images in Figures 9 and 10a,b in contrast to the corresponding as-built microstructures shown in Figure 8a,c. It is also noted in Figure 10b that there is some residual porosity at the top of the fabricated component where the temperature is highest. The optical micrographs in Figures 9b and 10b show the black, lamellar contrast characteristic of the alpha-prime martensite shown in Figure 8c. The Rockwell C-scale hardness corresponding to Figures 9b and 10b averaged HRC 43.9; this reduction from the as-built component hardness of HRC 44.6 noted above attests to the stress-relief provided by the anneal, which ideally involves the annihilation of process-induced dislocations.

Following HIP of the LPBF as-built Ti-6Al-4V components using the non-spherical HDH precursor powder (Figure 1), the variously elongated grain structure, with average grain sizes of ~125 microns, was essentially unchanged (Figures 9c and 10c) while the black, lamellar (and acicular) alpha-prime martensite shown in Figures 9b and 10b was replaced by varying sizes and distributions of lamellar alpha (alpha/beta), along with non-lamellar and globular alpha which is rendered white in the etching to produce the optical micrographs shown in Figures 9d and 10d. Some very small residual alpha-prime is also observed as tiny black dots. It is also notable that the widths of the lamellar alpha segments are ~2 microns. The corresponding hardness characteristic of the HIP microstructures shown in Figures 9 and 10c,d averaged HRC 40.7; representing a reduction of ~7% from the annealed components noted above. These results are similar to components fabricated from commercial, spherical Ti64 powder by LPBF where the HIPed (at 920 °C) microhardness dropped by ~13% in contrast to as-built and stress-relief annealed (at 704 °C) components [18].

As shown in Figure 9e,f and Figure 10e,f HIPed (at ~926 °C) and annealed (at 704 °C) components exhibited little change in the residual microstructure from the HIPed components (Figure 9c,d and Figure 10c,d). This feature was also attested to by the characteristic, average hardness of HRC 40.4; essentially unchanged from HRC 40.7 for HIPed components.

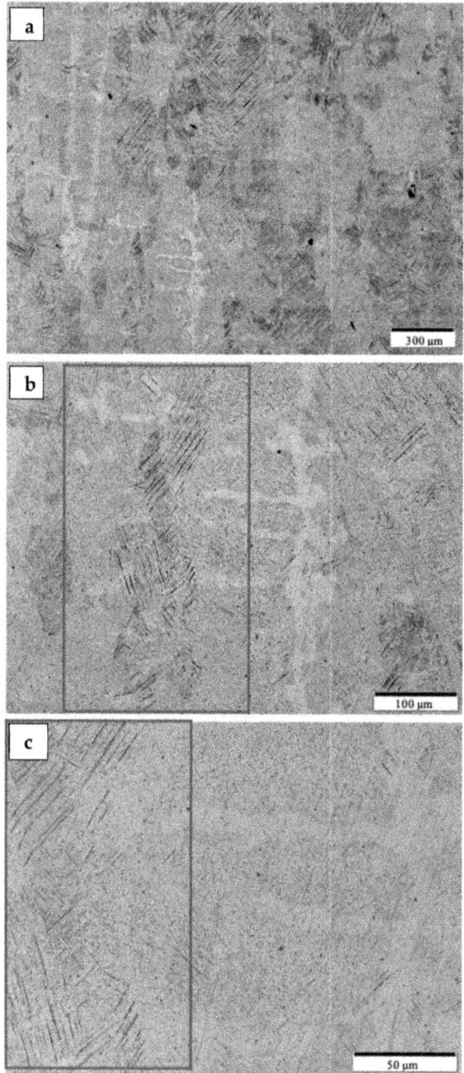

Figure 8. Optical micrographs of as-built DOE cubes from the build middle. (**a**) Low magnification. (**b**) Medium magnification. Alpha-prime enclosed in the blue rectangle. (**c**) High magnification. Alpha-prime enclosed in the blue rectangle.

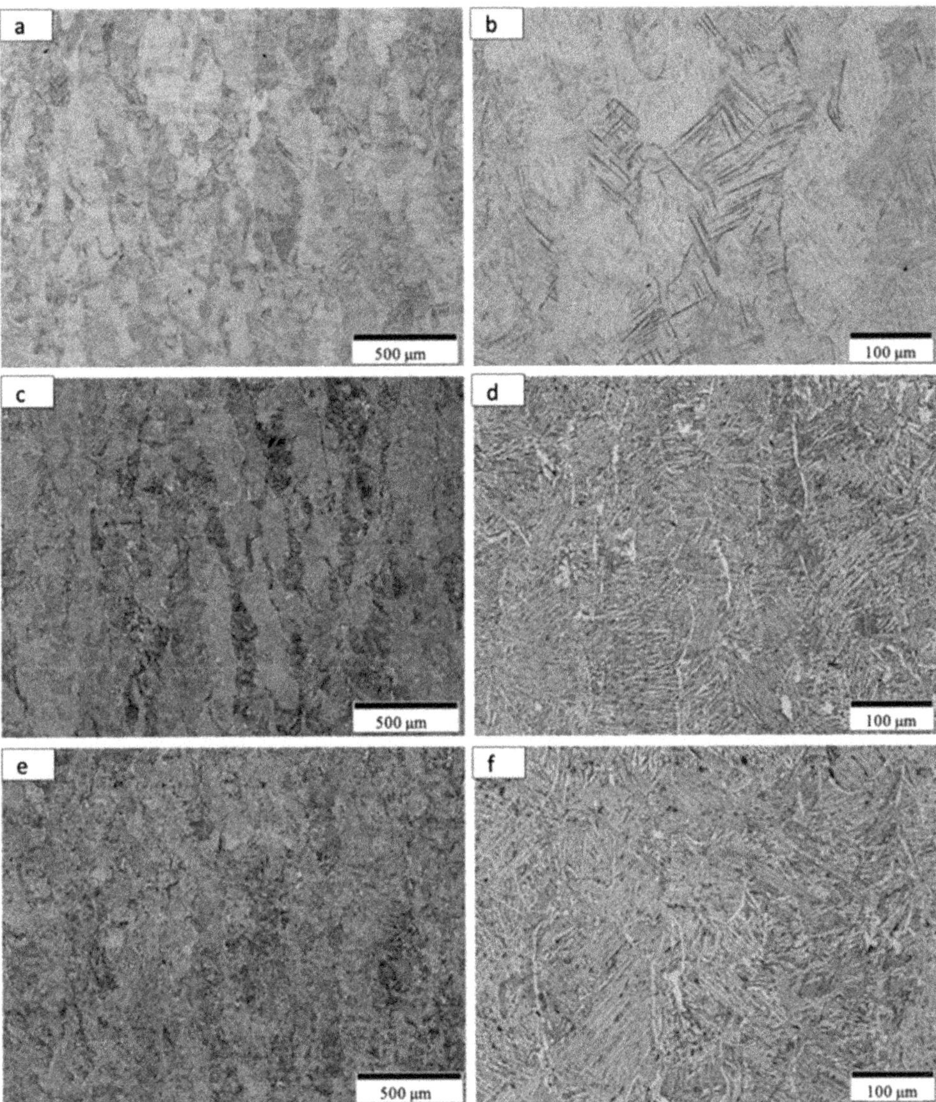

Figure 9. Optical micrographs for Anneal, HIP and HIP + Anneal samples from the build bottom. (**a**,**b**) show low and high magnification images for anneal samples. (**c**,**d**) show low and high magnification images for HIP samples. (**e**,**f**) show low and high magnification image for HIP + Anneal samples.

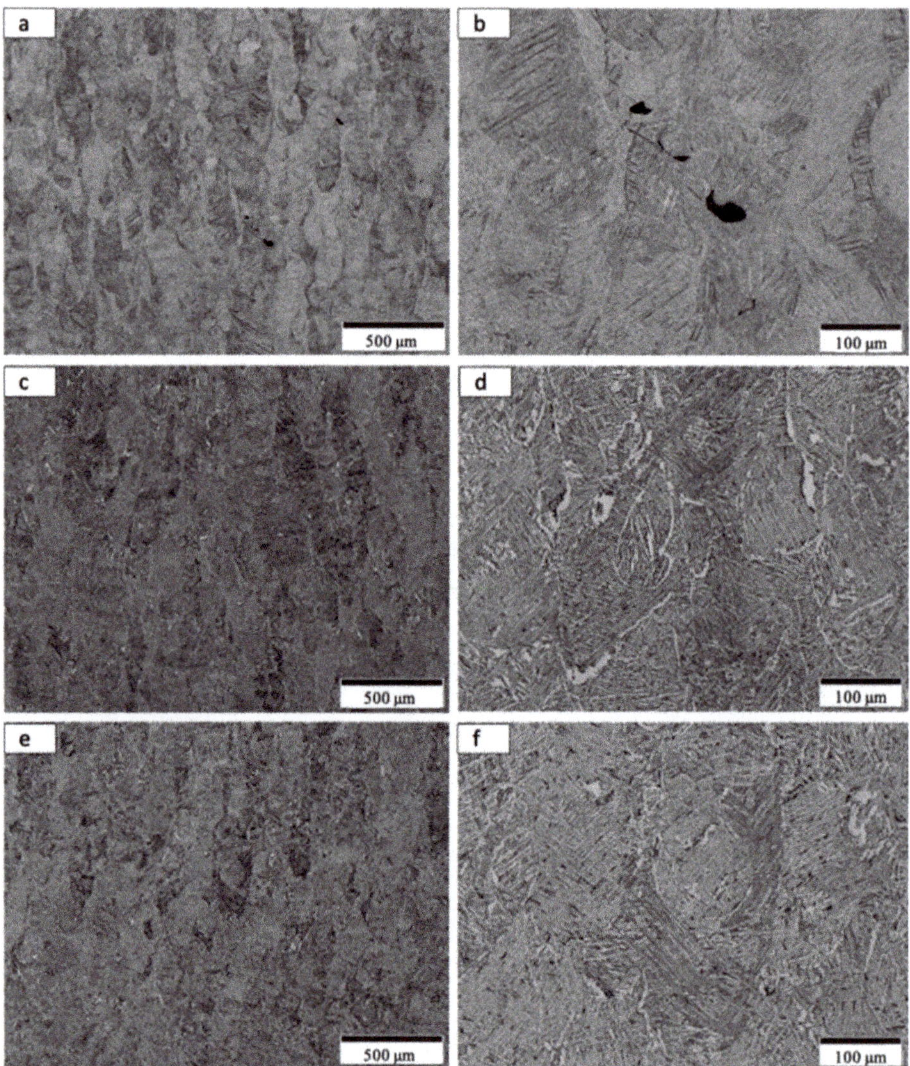

Figure 10. Optical micrographs for Anneal, HIP and HIP + Anneal samples from the build top. (**a,b**) show low and high magnification images for anneal samples. (**c,d**) show low and high magnification images for HIP samples. (**e,f**) show low and high magnification image for HIP + Anneal samples.

It can be observed on comparing Figures 9 and 10b,d,f that the lamellar martensite in Figures 9b and 10b, having a thickness of ~2 microns, is essentially unchanged for the transformed, lamellar alpha (alpha/beta) microstructure in Figures 9 and 10d,f. There is also some non-lamellar alpha and globular alpha. This microstructure variation as a consequence of HIP treatment accounts for the hardness variation.

3.2. Tensile Testing and Mechanical Property Comparisons and Discussion

The results of tensile tests for the heat-treated, non-spherical HDH Ti-6Al-4V powder-fabricated components, and corresponding to microstructures presented in Figures 9 and 10, are summarized in Table 4. The average yield stress and UTS of ~1.2 GPa and elongation

(ϵmax) of ~8.7% for the annealed components is characteristic of mill-annealed and solution-treated and aged bars and billets of Ti-6Al-4V; as well as laser beam powder-bed fusion fabricated Ti-6Al-4V components using spherical, atomized precursor powder [7–9,15,19–22]. Figure 11 compares tensile stress–strain diagrams for test specimens exhibiting properties closest to the averages shown in Table 4. Table 4 also compares the average component densities and the Rockwell C-scale (HRC) hardnesses. Figure 12 compares the fracture surface features corresponding to the stress–strain diagrams for heat-treated specimens shown in Figure 11. Ductile dimples 1–2 microns in diameter dominate the annealed specimens shown in Figure 12b, while somewhat larger dimple sizes averaging 2–3 microns characterize the HIP and HIP + anneal specimens shown in Figure 12d,f, respectively. These variations in ductile dimple sizes roughly corroborate the ductility values shown in Table 4 and Figure 11. There is also notable porosity associated with the annealed component shown in the image in Figure 10b, and this may contribute to the reduced ductility in contrast to the HIP and HIP + anneal components.

Table 4. Mechanical properties of tensile samples using HDH Ti-6Al-4V.

Variant	Yield Stress (MPa)		UTS (MPa)		ϵmax (%)		Density (g/cm^3)		Relative Density *	Hardness (HRC)	
	Mean	SD **	Mean	SD **	Mean	SD **	Mean	SD **	Mean	Mean	SD **
Anneal	1185	15	1217	13	8.7	1.3	4.39	0.004	99.2%	43.9	1.36
HIP	1025	17	1083	13	16.7	0.8	4.41	0.006	99.5%	40.7	0.93
HIP + Anneal	1039	9	1089	9	16.1	1.1	4.40	0.005	99.4%	40.4	0.97
As-built	–	–	–	–	–	–	4.38	0.002	98.9%	44.6	0.71

* 4.43 g/cm^3 as full density. ** Standard deviation.

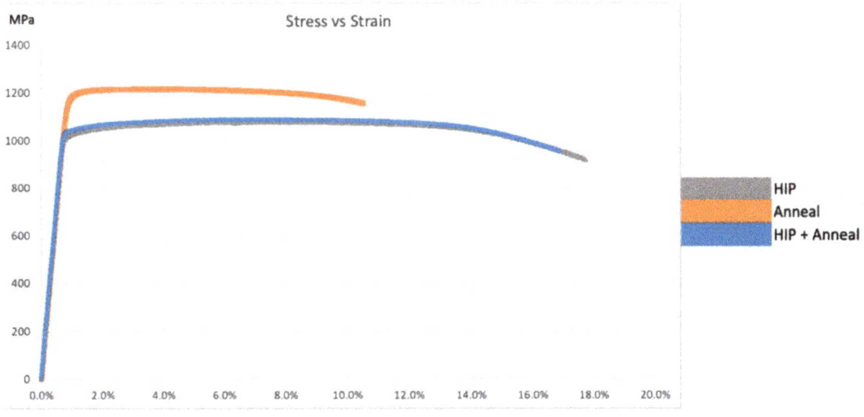

Figure 11. Stress vs. strain curves of individual samples closest to the average.

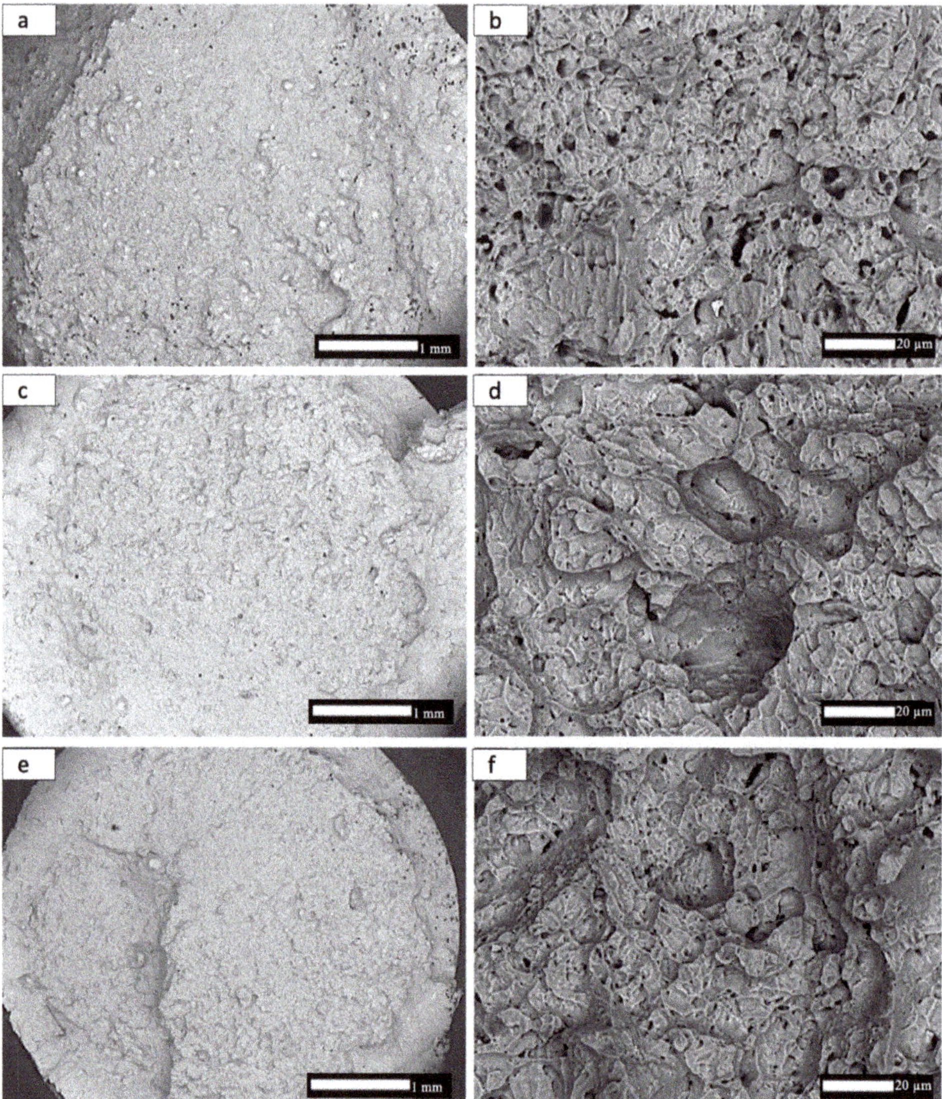

Figure 12. SEM images of fracture surfaces. (**a**,**b**) show low and high magnification of the Anneal fracture. (**c**,**d**) show low and high magnification of the HIP fracture. (**e**,**f**) show low and high magnification of the HIP + Anneal fracture.

The yield stress, UTS, and elongation associated with the HIPed components (Variant 2 in Table 4) are exceptional and exceed laser powder bed fusion Ti64 components fabricated from traditional, spherical, atomized alloy powders [21,22]. This is due in part to the closing of small pores as shown in Figure 7a,b, and increased density implicit in Table 4. The fine, lamellar alpha/beta microstructure shown in Figures 9d and 10d also facilitates the resulting high yield stress and ductility; which are equivalent to the best commercial, wrought Ti-6Al-4V products.

4. Conclusions

Heat treatment of Ti-6Al-4V components, especially HIP of as-built components fabricated from 100% non-spherical, HDH precursor powder by laser beam powder-bed fusion, has produced nearly fully dense, high strength (yield stress > 1 GPa) and high ductility (~17% elongation) products. The resulting mechanical properties rival the best commercial, wrought Ti-6Al-4V products, and exceed Ti-6Al-4V products fabricated by laser powder bed fusion utilizing more conventional, spherical, atomized precursor powders following heat treatment, including HIP. It is especially notable that superior Ti-6Al-4V alloy products have been fabricated using 100% non-spherical, HDH which represents a considerable economic advantage over conventional spherical, atomized precursor powder. This represents a milestone development in Ti-6Al-4V additive manufacturing.

Author Contributions: Conceptualization, J.V., F.M.; experiments and data, J.V., E.A., L.E.M.; research administration, E.A., F.M., R.B.W.; writing and editing, L.E.M., J.V., E.A., J.H.S., J.A.G., F.M., R.B.W. funding acquisition and resources B.M., M.M., M.P., J.H.S., J.A.G. All authors have read and agreed to the published version of the manuscript.

Funding: This research was sponsored by the DEVCOM Army Research Laboratory and was accomplished under cooperative agreement no. W911NF-20-2-0194.

Data Availability Statement: Not applicable.

Acknowledgments: The research presented here was conducted at The University of Texas at El Paso within the W.M. Keck Center for 3D Innovation (Keck Center), a 13,000-sq. ft. state-of-the-art additive manufacturing facility. Additionally, equipment housed within the Department of Metallurgical, Materials and Biomedical Engineering was utilized. The authors are grateful to Joe Capone for his participation and efforts to advance Ti-6Al-4V HDH feedstock toward Powder Bed Fusion applications. Thanks is extended to Adam Roberts for his contributions.

Conflicts of Interest: The authors declare no conflict of interest.

References

1. Frazer, W.E. Metal additive manufacturing. *J. Mater. Engr. Perform.* **2014**, *23*, 1917–1928. [CrossRef]
2. Nie, Z.; Jung, S.; Kara, L.B.; Whitefoot, K.S. Optimization of Part Consolidation for Minimum Production Costs and Time Using Additive Manufacturing. *J. Mech. Des.* **2019**, *142*, 1–16. [CrossRef]
3. Popov, V.; Grilli, M.; Koptyug, A.; Jaworska, L.; Katz-Demyanetz, A.; Klobčar, D.; Balos, S.; Postolnyi, B.O.; Goel, S. Powder Bed Fusion Additive Manufacturing Using Critical Raw Materials: A Review. *Materials* **2021**, *14*, 909. [CrossRef] [PubMed]
4. Albu, M.; Mituche, S.; Nachtnabel, M.; Krisper, R.; Dienstleder, M.; Schrottner, H.; Kothleitner, G. Microstrucrure investigations of powder and additive manufactured parts. *Berg. Huttenmaenn Monatsh* **2020**, *165*, 169–178. [CrossRef]
5. Gaytan, S.M.; Murr, L.E.; Medina, F.; Martinez, E.; Lopez, M.I.; Wicker, R.B. Advanced metal powder based manufacturing of complex components by electron beam melting. *Mater. Technol.* **2009**, *24*, 180–190. [CrossRef]
6. Medina, F. Reducing Metal Alloy Powder Costs for Use in Powder Bed Fusion Additive Manufacturing: Improving the Economics for Production. Ph.D. Thesis, University of Texas at El Paso, El Paso, TX, USA, 2013.
7. Liu, S.; Shin, Y.C. Additive manufacturing of Ti6Al4V alloy: A review. *Mater. Des.* **2019**, *164*, 107552. [CrossRef]
8. Debroy, T.; Wei, H.L.; Zuback, J.S.; Mukherjee, T.; Elmer, J.W.; Milewski, J.O.; Beese, A.M.; Wilson-Heid, A.; De, A.; Zhang, W. Additive manufacturing of metallic components—Process, structure and properties. *Prog. Mater. Sci.* **2018**, *92*, 112–224. [CrossRef]
9. Sun, S.; Brandt, M.; Easton, M. Powder bed fusion processes: An overview. In *Laser Additive Manufacturing: Materials Design, Technology and Applications*, 1st ed.; Brandt, M., Ed.; Elsevier: Amsterdam, The Netherlands, 2017; pp. 55–79.
10. Williams, J.C. (Ed.) *Titanium and Titanium Alloys*; Scientific and Technology Aspects; Springer: New York, NY, USA, 1982; Volume 3.
11. Qian, M.; Froes, F.H. (Eds.) *Titanium Powder Metallurgy, Science, Technology and Applications*, 1st ed.; Elsevier: Amsterdam, The Netherlands, 2015.
12. Jaber, H.; Kovacs, T.; János, K. Investigating the impact of a selective laser melting process on Ti6Al4V alloy hybrid powders with spherical and irregular shapes. *Adv. Mater. Process. Technol.* **2020**, 1–17. [CrossRef]
13. Yan, M.; Yu, P. *An Overview of Densification, Microstructure and Mechanical Property of Additively Manufactured Ti-6Al-4V—Comparison among Selective Laser Malting, Electron Beam Melting, Laser Metal Deposition and Selective Laser Sintering, and with Conventional Powder*; ItechOpen: London, UK, 2015.
14. Dong, Y.; Li, Y.; Zhou, S.; Zhou, Y.; Dargusch, M.; Peng, H.; Yan, M. Cost-affordable Ti-6Al-4V for additive manufacturing: Powder modification, compositional modulation and laser in-situ alloying. *Addit. Manuf.* **2021**, *37*, 101699. [CrossRef]

15. Hou, Y.; Liu, B.; Liu, Y.; Zhou, Y.; Song, T.; Zhou, Q.; Sha, G.; Yan, M. Ultra-low cost Ti powder for selective laser melting additive manufacturing and superior mechanical properties associated. *Opto-Electron. Adv.* **2019**, *2*, 18002801–18002808. [CrossRef]
16. Narra, S.P.; Wu, Z.; Patel, R.; Capone, J.; Paliwal, M.; Beuth, J.; Rollett, A. Use of Non-Spherical Hydride-Dehydride (HDH) Powder in Powder Bed Fusion Additive Manufacturing. *Addit. Manuf.* **2020**, *34*, 101188. [CrossRef]
17. Karimi, J.; Suryanarayana, C.; Okulov, I.; Prashanth, K. Selective laser melting of Ti6Al4V: Effect of laser re-melting. *Mater. Sci. Eng. A* **2021**, *805*, 140558. [CrossRef]
18. Eshawish, N.; Malinov, S.; Sha, W.; Walls, P. Microstructure and Mechanical Properties of Ti-6Al-4V Manufactured by Selective Laser Melting after Stress Relieving, Hot Isostatic Pressing Treatment, and Post-Heat Treatment. *J. Mater. Eng. Perform.* **2021**, 1–7. [CrossRef]
19. Xu, W.; Liu, E.W.; Patres, A.; Qian, M.; Brandt, M. Insitu tailoring of microstructure in additively manufactured Ti-6Al-4V for superior mechanical performance. *Acta Mater.* **2017**, *125*, 390–400. [CrossRef]
20. Ter Haar, G.M.; Becker, T.H. Selective Laser Melting Produced Ti-6Al-4V: Post-Process Heat Treatments to Achieve Superior Tensile Properties. *Materials* **2018**, *11*, 146. [CrossRef] [PubMed]
21. Zafari, A.; Xia, K. High Ductility in a fully martensitic microstructure: A paradox in a Ti alloy produced by selective laser melting. *Mater. Res. Lett.* **2018**, *6*, 627–633. [CrossRef]
22. Qian, M.; Xu, W.; Brandt, M.; Tang, H.P. Additive manufacturing and post processing of Ti-6Al-4V for superior mechanical properties. *MRS Bull.* **2020**, *61*, 775–783.

Article

Influence of Powder Bed Temperature on the Microstructure and Mechanical Properties of Ti-6Al-4V Alloys Fabricated via Laser Powder Bed Fusion

Lei-Lei Xing [1], Wen-Jing Zhang [1,*], Cong-Cong Zhao [2], Wen-Qiang Gao [1], Zhi-Jian Shen [1,3] and Wei Liu [1,*]

1 School of Materials Science and Engineering, Tsinghua University, Beijing 100084, China; xingll16@mails.tsinghua.edu.cn (L.-L.X.); 13269387926@163.com (W.-Q.G.); shenzhijian@tsinghua.edu.cn (Z.-J.S.)
2 Jihua Laboratory, Foshan 528000, China; zhaocc@jihualab.com
3 Arrhenius Laboratory, Department of Materials and Environmental Chemistry, Stockholm University, S-106 91 Stockholm, Sweden
* Correspondence: cxzhangwj@163.com (W.-J.Z.); liuw@mail.tsinghua.edu.cn (W.L.); Tel.: +86-159-0106-5952 (W.-J.Z.); +86-010-62772852 (W.L.)

Abstract: Laser powder bed fusion (LPBF) is being increasingly used in the fabrication of complex-shaped structure parts with high precision. It is easy to form martensitic microstructure in Ti-6Al-4V alloy during manufacturing. Pre-heating the powder bed can enhance the thermal field produced by cyclic laser heating during LPBF, which can tailor the microstructure and further improve the mechanical properties. In the present study, all the Ti-6Al-4V alloy samples manufactured by LPBF at different powder bed temperatures exhibit a near-full densification state, with the densification ratio of above 99.4%. When the powder bed temperature is lower than 400 °C, the specimens are composed of a single α' martensite. As the temperature elevates to higher than 400 °C, the α and β phase precipitate at the α' martensite boundaries by the diffusion and redistribution of V element. In addition, the α/α' lath coarsening is presented with the increasing powder bed temperature. The specimens manufactured at the temperature lower than 400 °C exhibit high strength but bad ductility. Moreover, the ultimate tensile strength and yield strength reduce slightly, whereas the ductility is improved dramatically with the increasing temperature, when it is higher than 400 °C.

Keywords: Ti-6Al-4V alloy; laser powder bed fusion; powder bed temperature; microstructure evolution; mechanical properties

Citation: Xing, L.-L.; Zhang, W.-J.; Zhao, C.-C.; Gao, W.-Q.; Shen, Z.-J.; Liu, W. Influence of Powder Bed Temperature on the Microstructure and Mechanical Properties of Ti-6Al-4V Alloys Fabricated via Laser Powder Bed Fusion. *Materials* **2021**, *14*, 2278. https://doi.org/10.3390/ma14092278

Academic Editor: Tuhin Mukherjee

Received: 4 April 2021
Accepted: 26 April 2021
Published: 28 April 2021

Publisher's Note: MDPI stays neutral with regard to jurisdictional claims in published maps and institutional affiliations.

1. Introduction

Ti-6Al-4V alloy is regarded as an important advanced material and has been widely used in the aerospace industry and biomedical applications because of its low density, high strength, and good corrosion resistance [1–5]. However, their excessive cost and difficulty in processing them severely narrow the expanding application of this alloy. Alternatively, as a near-net forming technique, the laser powder bed fusion (LPBF) can shorten the machining process and improve the mechanical performance; thus, it has attracted extensive attentions from the researchers around the world willing to apply this method to manufacture the Ti-6Al-4V alloy structured parts [6–8].

Low-temperature α phase (Hexagonal Close Packed) and high-temperature β phase (Body-Centered Cubic) are two dominating phases in the Ti-6Al-4V alloy. The differences in morphology, size, and volume fraction of the α and β phase always strongly influence the properties of the alloys [9,10]. The metastable acicular martensitic α' phase with an HCP crystal structure prefers to form during LPBF. This is because the tiny molten pools, formed by the laser melting metal powder during LPBF, undergo rapid solidification $(10^3–10^6 \text{ K/s})$ [11,12]. The martensitic α' phase is bad for ductility, whereas the ductility of the lath-shaped α phase and the intergranular β phase, decomposed from the martensitic

α' phase, is improved by heat treatment [13,14]. To improve the ductility of the Ti-6Al-4V products manufactured by LPBF, the post-LPBF heat treatment is often necessary by the martensitic α' phase decomposing into the $\alpha + \beta$ phase [1,7,15,16]. However, the post-LPBF heat treatment method has the disadvantage of decreasing the production efficiency.

During printing, the previously deposited layers are always influenced by the thermal effect from the heating, melting, and solidification (latent heat) of the successive layers [17–19]. The cumulative heat could lead to the rise in the specimen temperature; however, this increased temperature is still low and hardly causes the α' martensite to decompose into the α and β phase. In this case, some researchers try to heat the power bed to improve the mechanical properties of the LPBFed structure parts. However, most of the conventional available LPBF systems can only heat the powder bed to less than 200 °C, at which the residual stress generated during LPBF can be reduced, whereas this low temperature hardly induces the α' martensite to decompose into the α and β phase.

F. Bruckner et al. [20] proposed a model to describe the influence of pre-heating powder bed temperature on the thermometallurgical phenomena in laser cladding initially. B. Vrancken et al. [21] experimentally studied the effect of powder bed temperature on the microstructure evolution of the Ti-6Al-4V alloy, and found that there was little deposition of the α' martensite when the bed temperature increased to 400 °C. This is the first time that a study reports the α' martensite decomposition of the Ti-6Al-4V alloy during LPBF. In 2017, Haider Ai et al. [19] investigated the martensitic decomposition and mechanical properties of the Ti-6Al-4V alloy using the powder bed with the temperature of up to 770 °C during LPBF. They found that the α' martensite could decompose into the $\alpha + \beta$ phase when the bed temperature increased to 570 °C. However, there are some cavities in the LPBFed Ti-6Al-4V alloy sample, which might damage the mechanical properties. In addition, there is a lack of a systematic study on the microstructural evolution (β phase characteristic) with the increase in in the powder bed temperature during LPBF. Moreover, the relationship between the microstructure and mechanical properties are yet to be further revealed.

Accordingly, in the present study, the Ti-6Al-4V alloy samples were fabricated by LPBF at different powder bed temperatures; subsequently, the microstructure and mechanical properties were detailed, characterized, and measured. All of these were carried out in order to systematically reveal the microstructural evolution with the increasing powder bed temperature during LPBF, in conjunction with the correlation between the microstructure and mechanical properties.

2. Experimental Material and Procedures

The Ti-6Al-4V alloy powder produced by plasma rotation electrode process (Grade 1, BDN Technology, China, 15–53 μm) was used. The specific powder composition (in wt.%), measured by an X-ray fluorescence spectrometer (XRF-1500), is shown in Table 1. Nearly all the particles were spherical, and most of the particles had the diameter of no larger than 53 μm, as shown in Figure 1.

Table 1. Powder elemental composition (weight %).

Elements.	Al	V	Fe	Na	Ti
Contents/wt.%	6.14	3.82	0.183	0.135	balanced

All of the samples were fabricated by the SLM 280 HL facility (SLM Solutions, Lubeck, Germany) with a pre-heating platform. The powder bed can be heated up to 600 °C. The machine is equipped with a fiber laser with a maximum power of 700 W and an operating wavelength of 1075 nm (IPG Photonics Corp, Oxford, MA, USA). The laser beam has a Gaussian profile and a spot diameter of 80 μm. The LPBF processing parameters used in the present study are the laser power (P) of 200 W and the scanning speed (v) of 758 mm/s. The powder layer thickness (t) of 30 μm and a stripe filling strategy with the hatch distance (h) of 110 μm were applied. The laser energy density (E) was 80 J/mm^3, and this value was obtained using the following equation: $E = \frac{P}{vth}$. The samples were fabricated at unheated, 200 °C, 300 °C, 400 °C, 500 °C, and 600 °C. Multiple-cube specimens with a dimension of 10 mm × 10 mm × 10 mm were obtained. The specific experimental parameters are shown in Table 2. Argon was used to ensure that the oxygen content was below 100 ppm during printing. Three samples were fabricated for each parameter in the present study to reduce the experimental error. And the powder around the fabricated sample was vacuumed between the two processes to reduce the effect of oxidized powder and impurities on the subsequent manufacturing experiments.

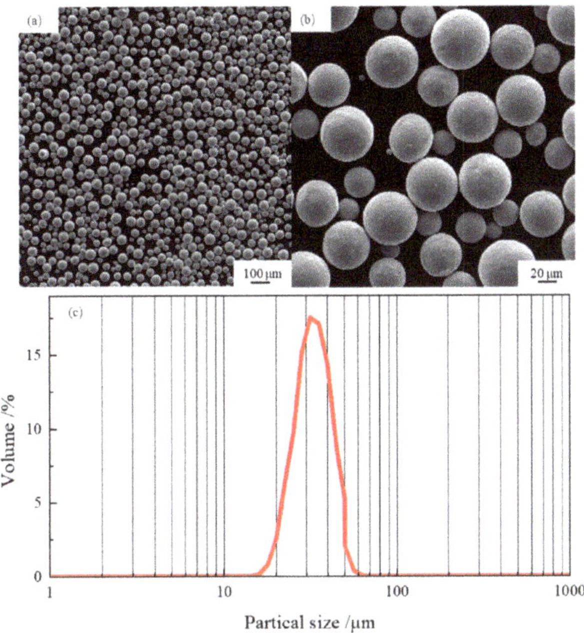

Figure 1. (**a,b**) morphology; (**c**) partial size distribution of the Ti-6Al-4V alloy powder.

Table 2. Laser powder bed melting fabricated parameters

Bed Temperature	Laser Power/W	Scanning Speed/mm/s	Powder Layer Thickness/μm	Hatch Distance/μm	Energy Density/J/mm³
Unheated					
200 °C					
300 °C					
400 °C	200	758	30	110	80
500 °C					
600 °C					

The relative density of the LPBFed samples was calculated using the Archimedes method, according to which a body that is immersed in a liquid exhibits an apparent loss in weight equal to the weight of the liquid it displaces. The weight of the samples in ethanol and in the air were tested, respectively. Based on the Archimedes principle, it was as follows:

$$m_{air}g - m_{ethanol}g = \rho_{ethanol}gV \tag{1}$$

where m_{air} is the weight of the sample tested in the air, m_{ethanol} is the weight tested in ethanol, g is the gravitational constant, and V is the volume of the sample, which can be gained by Equation (2), as follows:

$$V = \frac{m_{\mathrm{air}}}{\rho_{\mathrm{sample}}} \tag{2}$$

where ρ_{sample} is the density of the samples, which can be calculated based on Equations (1) and (2).

$$\rho_{\mathrm{sample}} = \frac{m_{\mathrm{air}}\rho_{\mathrm{ethanol}}}{m_{\mathrm{air}} - m_{\mathrm{ethanol}}} \tag{3}$$

And the relative density can be calculated by the dividing of theoretical densities (4.43 g/mm^3) and measured densities of the Ti-6Al-4V sample.

The morphology and size of the powders and the microstructures of the LPBFed samples were characterized by the scanning electron microscope (SEM, TESCAN) equipped with an electron backscatter diffraction detector (EBSD, Oxford Instruments, Oxford, UK). The samples for SEM and BSE (Backscattered Electron) analysis were prepared by standard mechanical polishing with 400, 800, 1500, and 2000 grit SiC paper, and then polished with alumina solution, with an average diameter of alumina particles of 50 nm. The samples for EBSD analysis were electropolished in a solution of 97% ethanol and 3% perchloric acid. Phase identification was conducted using X-ray diffraction (XRD, D/Max 2500 V) with a Cu target in a conventional X-ray tube, operating at an accelerating voltage of 40 kV and an electron beam current of 30 mA. Further microstructural observation was conducted using transmission electron microscope (TEM, JEOL-2100). The TEM samples were preliminarily prepared by grinding a 3 mm-diameter disk to approximately 30 μm in thickness and then finally thinned by a Gatan-made Precision Ion Polishing System (PIPS) at 5 keV, with a gun angle of ±8°.

Dog-bone shaped specimens with a gauge-section size of 5 mm in length, 2 mm in width, and 1 mm in thickness were cut from the LPBFed samples. Subsequently, all the specimens were mechanically polished to obtain mirror-like surfaces. The cutting position of the tensile sample along with its shape and size are shown in Figure 2. The tensile tests were uniaxially performed at room temperature with an initial strain rate of 10^{-3} s^{-1} using the INSTRON-5969 machine. Each tensile test was repeated for three times to minimize the effect of experimental error on the tensile results.

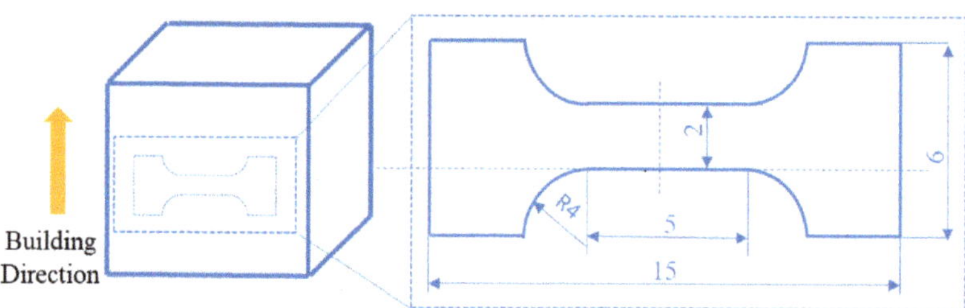

Figure 2. The cutting position with the shape and size of the tensile specimen for mechanical tests (dimensions of the sample given in mm).

3. Results

3.1. Densification Ratio

It is well accepted that the densification ratio plays an important role in determining the ultimate mechanical performance. To eliminate or minimize the effect of fabricated defect on the mechanical properties and to optimize the LPBF parameters, the densification ratios of the LPBFed specimens at different bed temperatures were calculated using the Archimedes method, and the calculated results show that all the LPBFed specimens obtained in the present printing parameters are higher than 99.4%, as shown in Figure 3. Generally, when the densification ratio is higher than 99%, it is recognized as a near-full densification state [22]; thus, in the present study, the effect of the fabricated defect on the mechanical properties can be ignored to some extent.

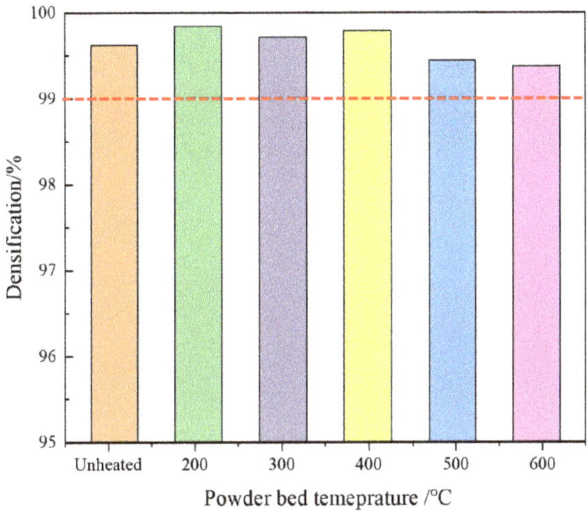

Figure 3. The densification ratio of the Ti-6Al-4V alloy samples manufactured at different powder bed temperatures.

3.2. Phase Characterization

To investigate the phase constitution of the LPBFed Ti-6Al-4V alloy at different powder bed temperatures, the XRD results are shown in Figure 4. The diffraction peaks of the α/α' phase with HCP structure are found in the sample in the unheated powder bed. It should be noted that it is difficult to differentiate the α and α' phase by the diffraction peaks, due to their similar crystallographic structure [23]. It is well known that the α' martensite prefers to form in the Ti-6Al-4V alloy when the high-temperature β phase is rapidly cooled because the atomic diffusion would be inhibited by high cooling rate (over 10^4 K/s) [11]. As a consequence, the phase transition of $\beta \rightarrow \alpha'$ occurred and the α' phase formed in the unheated bed [24,25]. The XRD spectrums located in the left were obtained between 32 and 65°, with a step size of 0.02° and a counting time of 1 s/step, which is a coarse scanning strategy and displays a severe signal-to-noise ratio. The β phase is difficult to be detected in this speed. However, the XRD spectrums located in the right were obtained between 55° and 60°, with a step size of 0.01° and a counting time of 5 s/step, which is a fine scanning strategy and displays a slight signal-to-noise ratio. Moreover, the β phase could be well detected under this condition. When the powder bed temperature increases to 200 °C and 300 °C, there is still only α/α' phase and no β phase detected. However, as the powder bed temperature increases to 400 °C, the diffraction peaks of the β phase appear, indicating that the α' martensite decomposed into the α and β phase. More β phase is found at the powder bed temperatures of 500 °C and 600 °C. It should be noted that the diffraction peaks cannot be found when the phase content is less than 5%, and thus, it is not sufficient to reveal the phase decomposition only by XRD; more characterization work should be performed.

3.3. Microstructure Characterization

A typical microstructure of the Ti-6Al-4V alloy sample manufactured using an unheated powder bed is given in Figure 5a, exhibiting martensitic α' laths, which is consistent with the XRD results (Figure 4), which does not exist in the β phase. The martensitic α' appears not to be homogeneous, and the thickness of the α' lath exhibits a large difference.

The microstructure at the powder bed temperature of 200 °C (Figure 5b) is almost similar to the unheated bed sample. However, as the powder bed temperature increases to 300 °C, compared with the microstructure at 200 °C, the large difference in the morphology takes place, even though there is still no β phase observed, as shown in Figure 5c. When the powder bed temperature increases to 400 °C, the β phase (the white area) appears, as highlighted by the red arrow in Figure 5d. Figure 5e shows the microstructure at the powder bed temperature of 500 °C, and more β phase is observed along the boundaries of the α phase laths, which is highlighted by the blue arrow. The increased powder bed temperature can enhance the decomposing martensitic α′ laths and the coarsening α lath. Thus, much thicker α phase laths and more β phase (highlighted by the purple arrows) are found at the powder bed temperature of 600 °C, as shown in Figure 5f.

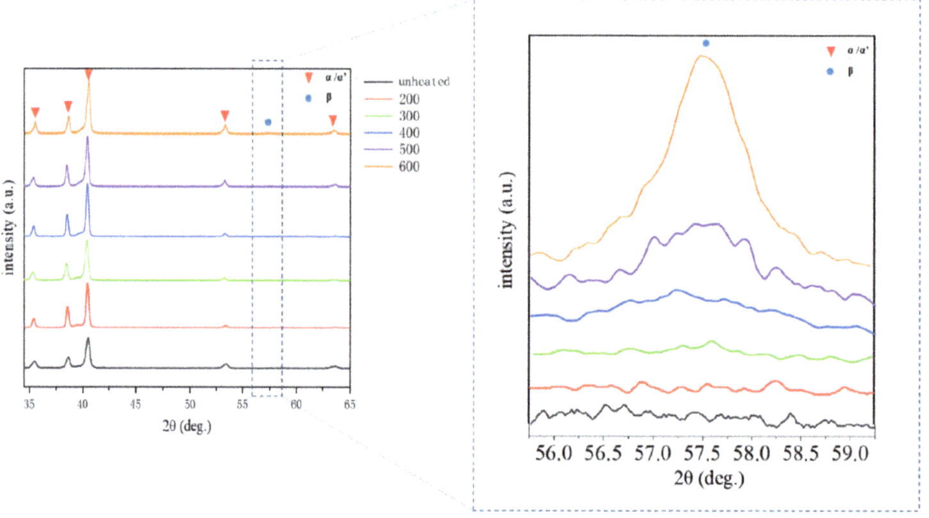

Figure 4. The XRD patterns of the Ti-6Al-4V alloy samples manufactured at different powder bed temperatures.

For further understanding the phase transformation during LPBF, TEM was performed to observe the microstructures at different bed temperatures from a more microscopic perspective, as shown in Figure 6. It can be seen from Figure 6a that the microstructure exhibits lath-shaped morphology with HCP crystal structure, which can be attested by the corresponding selected area electron diffraction patterns (SADP) displayed at the bottom right corner of Figure 6a. Moreover, the high-density dislocations existing inside the laths reveal that a large amount of strain and lattice defects were generated during LPBF. These laths are recognized as α′ martensite, which is consistent with the previous results reported by other researchers [1–3]. During LPBF, the rapid cooling leads to the formation of metastable martensitic with an HCP crystal structure, which is bad for ductility [26]. To solve this problem, the preheated powder bed works as an option [19]. When the powder bed temperatures of 200 °C and 300 °C were utilized, the microstructure after manufacturing still appeared as martensitic α′ laths, but the amount of the dislocations inside α′ laths decreased, as shown in Figure 6b,c. As the powder bed temperature increased to 400 °C, the boundary became clear, and a thin layer between the α laths formed; this thin layer can be proved as a β phase by the SADP displayed at the bottom left corner in Figure 6d. This is consistent with the SEM result shown in Figure 5. The width of the α laths increases with the powder bed temperature increasing to 500 °C and 600 °C, as shown in Figure 6e,f. The thickness of the β phase layer is approximately 5–10 nm, as highlighted by the blue arrows in Figure 6d–f. Scan transmission electron microscope

(STEM) was performed to reveal the element content in the β phase layer distributing along the boundaries at the powder bed temperatures of 400 °C, 500 °C, and 600 °C, and the results are shown in Table 3. It can be found that the V elements enrich in the β phase layer, and this phenomenon is more obvious with the increasing powder bed temperature. By the same token, it is reasonable to conclude by the TEM results that the martensitic α' laths decompose into α laths and the intragranular β phase by the V element partitioning, when the powder bed temperature is equal to and above 400 °C; meanwhile, the α laths coarsen with the increasing powder bed temperature.

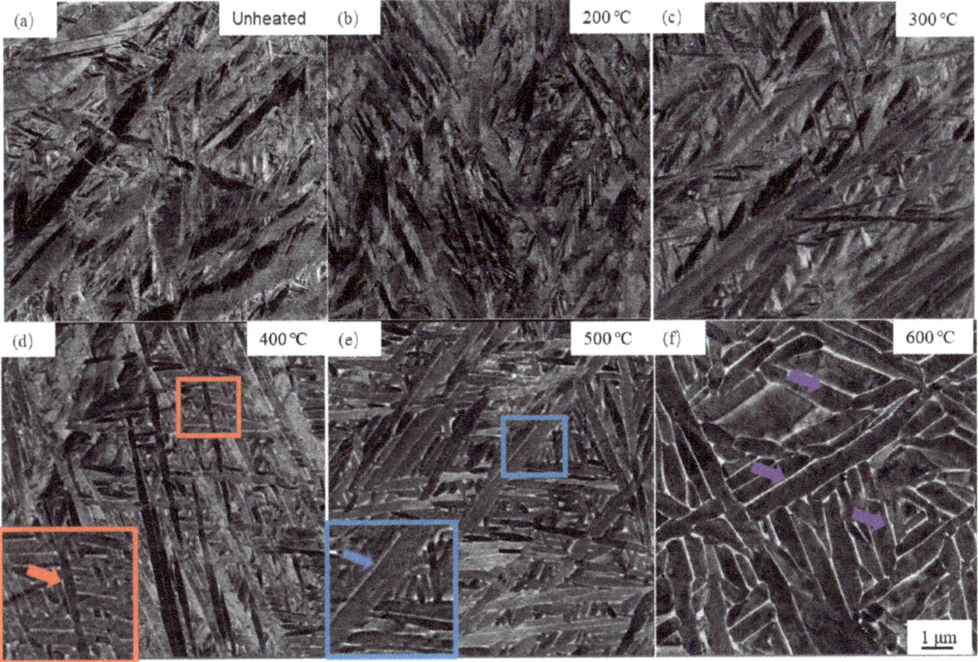

Figure 5. BSE micrographs of the Ti-6Al-4V alloy produced by LPBF in the case of (**a**) unheated, (**b**) 200 °C, (**c**) 300 °C, (**d**) 400 °C, (**e**) 500 °C, and (**f**) 600 °C.

Table 3. Element content in the intragranular β phase of Figure 5d–f.

Element Content, wt.%	400 °C	500 °C	600 °C
Ti	87.65	83.80	82.32
Al	3.43	3.28	4.65
V	8.92	12.91	13.03

To reveal the orientation relationship of the α and β phase, the high resolution TEM (HRTEM) image was obtained, as shown in Figure 7. Figure 7b presents the Inverse Fast Fourier Transition (IFFT) image of the boundaries between the α and β phase. The (1(—)100) atomic row spacing in the α grain is 0.255 nm, and the (011(—)) atomic row spacing in the β grain is 0.299 nm. Figure 7c reveals the FFT image corresponding to Figure 7b. It can be concluded that the β grain displays a conventional orientation relationship (OR) of $[011]_\beta // [0001]_\alpha$ in the α grain.

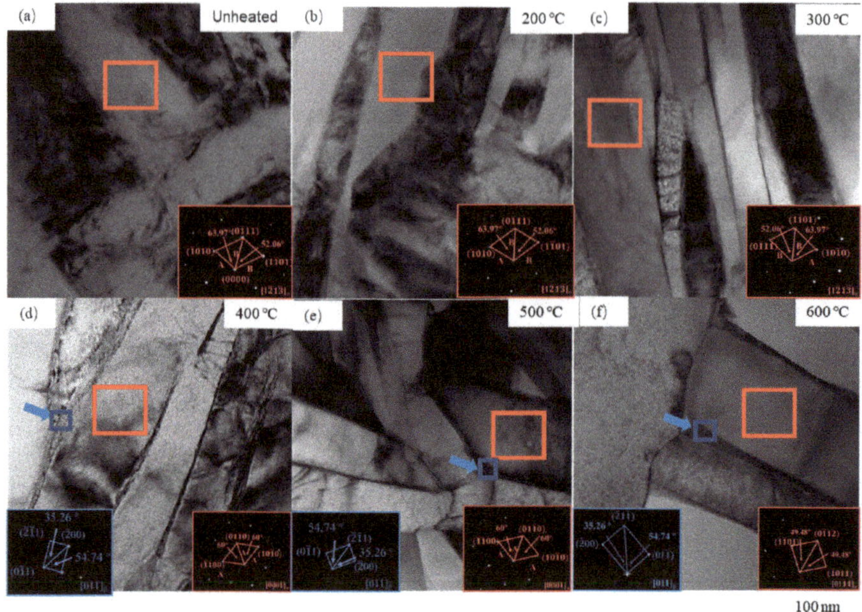

Figure 6. Bright field images and the corresponding selected area electron diffraction of the Ti-6Al-4V alloy after LPBF in the case of (**a**) unheated, (**b**) 200 °C, (**c**) 300 °C, (**d**) 400 °C, (**e**) 500 °C, and (**f**) 600 °C.

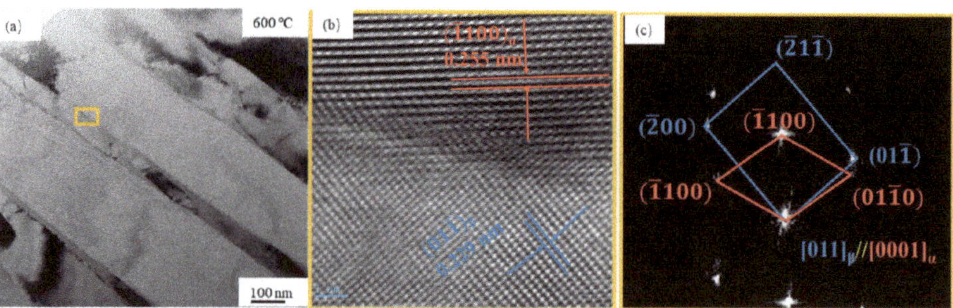

Figure 7. (**a**) Bright field image, (**b**) HRTEM IFFT image, and (**c**) FFT image of the Ti-6Al-4V alloy after LPBF at the powder bed temperature of 600 °C.

To further identify the characteristics of α/α' laths in statistics, the EBSD maps of Ti-6Al-4V alloy after LPBF at different powder bed temperatures are shown in Figure 8. It is obvious that these microstructures consist of a large amount of long and straight laths. As described above, these laths have been confirmed as α/α' phase with an HCP structure. It should be noted that the intragranular β phase with the BCC structure is difficult to identify due to its thin thickness (Figure 6). Here, to quantitatively analyze the lath-shaped microstructure evolution with the increasing powder bed temperature, the aspect ratio (it is defined as the ratio of length to width of the lath) was calculated by analyzing the EBSD data. It is found that the average aspect ratios of laths at different powder bed temperatures are 5.06, 4.77, 4.46, 4.05, 3.67, and 2.75, respectively, and they decrease with the increasing powder bed temperature, as shown in Table 4.

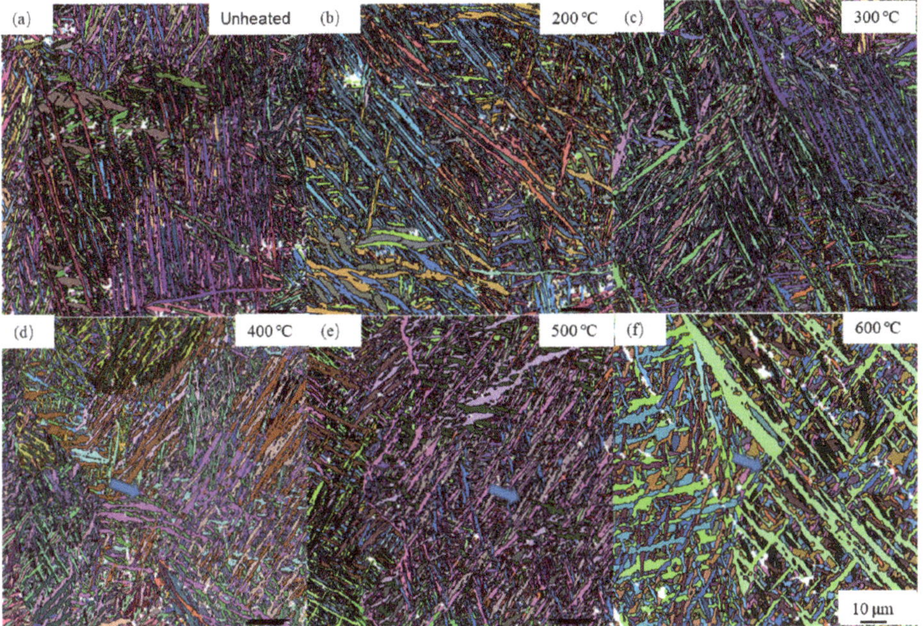

Figure 8. EBSD maps of the Ti-6Al-4V alloy after LPBF in the case of (**a**) unheated, (**b**) 200 °C, (**c**) 300 °C, (**d**) 400 °C, (**e**) 500 °C, and (**f**) 600 °C.

Table 4. The aspect ratios at different powder bed temperatures.

Grain Statistics	Unheated	200 °C	300 °C	400 °C	500 °C	600 °C
Aspect ratio	5.06	4.77	4.46	4.05	3.67	2.75

3.4. Mechanical Properties

The room-temperature tensile results of the Ti-6Al-4V alloy after LPBF at different powder bed temperatures are shown in Figure 9. The specimens fabricated in unheated powder bed show the highest strength and the worst ductility. As the powder bed temperature increases, the tensile performance exhibits a trade-off phenomenon in strength and ductility; specifically, the ultimate tensile and yield strengths decrease, whereas ductility improves. The ultimate tensile and yield strengths decreased by 6.7%and 1.5%, respectively, and the elongation increased by 66.0%, when the powder bed was heated to 600 °C. It should be noted that, although the strengths (especially the ultimate tensile strength) decrease, they do not decrease noticeably and still maintain at a high level with the increasing powder bed temperature. By contrast, ductility apparently improves as the powder bed temperature increases, especially when the temperature is above 400 °C, as shown in Figure 9c. In addition, in all cases, the elastic modulus remains unchanged, with a value of around 1168 GPa. Based on what is described above, the negative mechanical properties can be eliminated by heating the powder bed, and it is reasonable to conclude that, in the present study, the excellent comprehensive mechanical properties are obtained at the powder bed temperature of 600 °C.

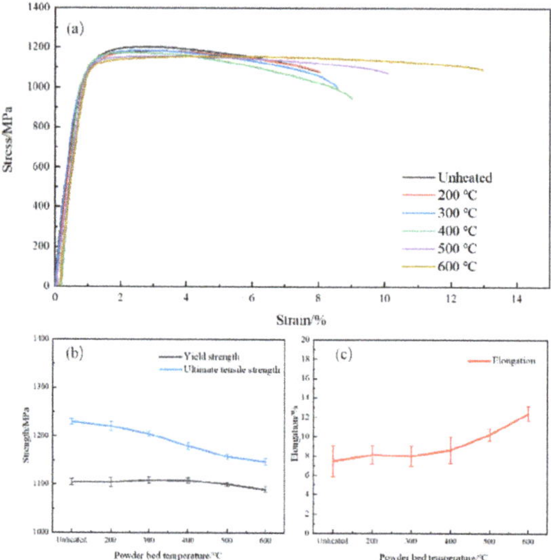

Figure 9. (**a**) Engineering stress-strain curves, (**b**) strengths, and (**c**) elongation of the Ti-6Al-4V alloy after LPBF at different powder bed temperatures.

4. Discussion

4.1. Microstructure Evolution with the Increasing Powder Bed Temperature during LPBF

In the present study, it is clear that the powder bed temperature plays an important role in determining the microstructure after LPBF. The ultimate microstructure is dominated by the α' martensite, with a large aspect ratio of 5.06, in the un-preheated powder bed. It is also well known that the structure parts are built layer by layer during LPBF, and it is reported that the cooling rate during LPBF is approximately 10^3–10^8 K/s [11,12], which is much higher than the 1500 K/s detected during water quenching [27]. This high cooling rate during LPBF suppresses the atomic diffusion and the phase transformation of $\beta \rightarrow \alpha$ (occurring in the equilibrium state) is inhibited; alternatively, the $\beta \rightarrow \alpha'$ occurs during LPBE.

When the powder temperature is elevated to 200 °C, the sample temperature during manufacturing is still low, which means that the V element is hardly diffused [18,28–30], which is regarded as the indicator of the occurrence of martensite decomposition. However, the increasing powder bed temperature widens the laths slightly, which leads to a decrease in the aspect ratio (4.77), lower than 5.06 obtained in the un-preheated condition [17,31,32].

As the powder bed temperature increases up to 300 °C, the boundaries between the martensitic α' laths become clear. However, the specimen temperature during printing is still lower than the martensitic decomposition temperature of ~550 °C for the Ti-6Al-4V alloy [33]. In addition, the higher temperature decreases the aspect ratio due to the widening of the α' lath.

During LPBF, the previous deposited layers are affected by the thermal-cycle effect from the laser heating as well as the melting and solidification latent heat of the subsequent successive layers. As the powder bed temperature increases to 400 °C, though it is lower than the martensitic decomposition temperature of ~550 °C for the Ti-6Al-4V alloy, the cumulative heat generated due to the thermal-cycle history during LPBF could lead to the specimen temperature rising above 550 °C [17–19,34]. It is difficult to directly detect the temperature evolution process of the specimens during LPBF. Xu et al. [34] drew the temperature evolution schematic diagrams of the Ti-6Al-4V alloy during LPBF and thought the temperature of the sample gradually increases at the initial stage, and then decreases to the powder bed temperature. Based on this observation, it can be concluded that the sample during LPBF can be heated, while the peak temperature must be higher than the pre-heated temperature of powder bed. Consequently, the peak temperature of the sample during LPBF could be above 550 °C, though the powder bed temperature is only 400 °C. In this case, the diffusion of the V element in α' laths was motivated and the α phase initiated to nucleate along the α' boundaries, during which and the subsequent process, the V atoms was expulsed from the newly formed α phase and preferred to diffuse towards the lath boundaries [2,18]. The enrichment of the V element caused the formation of the β phase. Ultimately, the martensitic α' phase decomposed into the $\alpha + \beta$ phase. However, the decomposition of the α' phase is not fully completed and the amount of the β phase is small at the powder bed temperature of 400 °C, which is attributed to the short-residence time in the decomposition temperature range [34]. In addition, the aspect ratio of laths decreases due to the widening of the α' lath at the powder bed temperature of 400 °C.

When the powder bed temperature is elevated to 500 °C and 600 °C, more energy is supplied, which tells us that the peak temperature of the sample during LPBF is much higher than the martensitic decomposition temperature of ~550 °C. It also shows that the cumulative residence time of the previous deposited layer in the α' martensitic decomposition temperature range is much longer, resulting in more β phase forming and the widening of the laths [17,31].

4.2. Influence of Microstructure on Mechanical Properties

It is well known that the microstructure determines the mechanical properties. In the present study, it is evident that the mechanical properties are strictly related to the microstructure of the LPBFed Ti-6Al-4V alloy, at different powder bed temperatures.

The thickness of the martensitic α' laths after LPBF in the unheated powder bed is shown to be relatively fine and un-homogeneous, which makes it easy to prevent the dislocation slip, leading to the crack initiation and propagation at the weak location in the microstructure during deformation [19,35]. In addition, there exist high-density dislocations induced by the high residual stress in the α' phase, as shown in Figure 5a, which could generate stress accumulation. As a result, the microstructure obtained in the unheated powder bed shows the highest strength but the worst ductility.

When the powder bed temperature is below 400 °C, the widening and decomposition of the α' martensite laths are limited, which should be responsible for the slight decrease in strength and the improvement of ductility.

As the powder bed temperature increases to 500 °C and 600 °C, the microstructure becomes coarsening and the β phase increases. The widening laths would increase the dislocation slip length and decrease the stress accumulation during deformation [36]. Moreover, the slip resistance between the α and β phase is lower than that between the α' laths [9,10,35,37]. As a result, ductility is improved, but the strength decreases with the increasing powder bed temperature [37,38]. It should be noted that the ductility of the samples manufactured at the powder bed temperatures of 500 °C and 600 °C is superior to that reported by other researchers using the same method of heating the powder bed [19]. This may be due to the fact that the densification of the present manufactured samples is better than that in ref. [19]. In a word, increasing the powder bed temperature enhances the comprehensive mechanical properties due to the β phase's precipitating and the α lath's widening, which is better suited for the various applications of the LPBFed Ti-6Al-4V alloy.

5. Conclusions

In the present study, Ti-6Al-4V alloy samples were fabricated by laser powder bed fusion, at different powder bed temperatures. The microstructure and the corresponding mechanical properties of the LPBFed Ti-6Al-4V alloy were investigated. The main conclusions are as follows:

All of the Ti-6Al-4V alloy samples fabricated by laser powder bed fusion at different powder bed temperatures exhibit a near-full densification state with the densification ratio of above 99.4%.

When the powder bed is unheated or the heating temperature is lower than 400 °C, the microstructure is dominated by the α' martensite with a large aspect ratio. When the increasing temperature is equal to or above 400 °C, the diffusion and redistribution of the β-phase stabilizer of the V element in the α' laths matrix occurs, resulting in the enrichment of V atoms along the grain boundaries. Thus, ultimately, the α' lath decomposes into a lath-shaped α and an intergranular β phase. In addition, with the increase in the powder bed temperature, the α/α' laths widen, and the aspect ratio decreases.

When the powder bed is unheated or the heating temperature is lower than 400 °C, the specimens exhibit high strength but bad ductility because of the existence of the α' martensite. As the temperature elevates to a value higher than 400 °C, the ultimate tensile strength and yield strength reduce slightly, whereas the ductility is dramatically improved with the increasing powder bed temperature, which is attributed to the α' martensite decomposing into the α and β phase and the widening of α laths.

The powder bed temperature has a significant influence on microstructure evolution and mechanical properties of the Ti-6Al-4V alloy during LPBF. The excellent combination of mechanical properties, including strength and ductility, appears at the powder bed temperature of 600 °C.

Author Contributions: Conceptualization, L.-L.X.; Data curation, W.-J.Z.; Formal analysis, L.-L.X., C.-C.Z., Z.-J.S. and W.L.; Funding acquisition, W.L.; Investigation, L.-L.X. and W.-J.Z.; Methodology, L.-L.X.; Supervision, W.L.; Writing—original draft, L.-L.X.; Writing—review & editing, W.-J.Z., W.-Q.G. and W.L. All authors have read and agreed to the published version of the manuscript.

Funding: This work was financially supported by National Key Program of China (No. 2018YFB1106900 and 2019YFE03130003).

Institutional Review Board Statement: Not applicable.

Informed Consent Statement: Not applicable.

Data Availability Statement: Not applicable.

Conflicts of Interest: The authors declare no conflict of interest.

References

1. Lu, Y.; Aristizabal, M.; Wang, X.; Pang, B.; Chiu, Y.L.; Kloenne, Z.T.; Fraser, H.L.; Loretto, M.H. The influence of heat treatment on the microstructure and properties of HIPped Ti-6Al-4V. *Acta Mater.* **2020**, *165*, 520–527. [CrossRef]
2. Yan, X.; Yin, S.; Chen, C.; Huang, C.; Bolot, R.; Lupoi, R.; Kuang, M.; Ma, W.; Coddet, C.; Liao, H.; et al. Effect of heat treatment on the phase transformation and mechanical properties of Ti6Al4V fabricated by selective laser melting. *J. Alloys Compd.* **2018**, *764*, 1056–1071. [CrossRef]
3. Lan, L.; Xin, R.; Jin, X.; Gao, S.; He, B.; Rong, Y.; Min, N. Effects of Laser Shock Peening on Microstructure and Properties of Ti-6Al-4V Titanium Alloys Fabricated via Selective Laser Melting. *Materials* **2020**, *13*, 3261. [CrossRef] [PubMed]
4. Brandl, E.; Palm, F.; Michailov, V.; Viehweger, B.; Leyens, C. Mechanical properties of additive manufactured titanium (Ti–6Al–4V) blocks deposited by a solid-state laser and wire. *Mater. Des.* **2011**, *32*, 4665–4675. [CrossRef]
5. Erdakov, I.; Glebov, L.; Pashkeev, K.; Bykov, V.; Bryk, A.; Lezin, V.; Radionova, L. Effect of the Ti6Al4V Alloys Track Trajectories on Mechanical Properties in Direct Metal Deposition. *Machines* **2020**, *8*, 79. [CrossRef]
6. Huang, Q.; Hu, N.; Yang, X.; Zhang, R.; Feng, Q. Microstructure and inclusion of Ti–6Al–4V fabricated by selective laser melting. *Front. Mater. Sci.* **2016**, *10*, 428–431. [CrossRef]
7. Ter Haar, G.M.; Becker, T.H. Selective Laser Melting Produced Ti-6Al-4V: Post-Process Heat Treatments to Achieve Superior Tensile Properties. *Materials* **2018**, *11*, 146.
8. Murr, L.E.; Quinones, S.A.; Gaytan, S.M.; Lopez, M.I.; Rodela, A.; Martinez, E.Y.; Hernandez, D.H.; Martinez, E.; Medina, F.; Wicker, R.B. Microstructure and mechanical behavior of Ti-6Al-4V produced by rapid-layer manufacturing, for biomedical applications. *J. Mech. Behav. Biomed. Mater.* **2009**, *2*, 20–32. [CrossRef]
9. Zhang, W.; Ding, H.; Cai, M.; Yang, W.; Ji, J. Ultra-grain refinement and enhanced low-temperature superplasticity in a friction stir-processed Ti-6Al-4V alloy. *Mater. Sci. Eng.* **2018**, *727*, 90–96. [CrossRef]
10. Zhang, W.; Liu, H.; Ding, H.; Fujij, H. Superplastic deformation mechanism of the friction stir processed fully lamellar Ti-6Al-4V alloy. *Mater. Sci. Eng.* **2020**, *785*, 139390. [CrossRef]

11. Barriobero-Vila, P.; Gussone, J.; Stark, A.; Schell, N.; Haubrich, J.; Requena, G. Peritectic titanium alloys for 3D printing. *Nat. Commun.* **2018**, *9*, 3426. [CrossRef] [PubMed]
12. Huang, Q.L.; Yang, X.; Zhang, R.R.; Shen, Z.J.; Feng, Q.L. Specific heat treatment of selective laser melted Ti–6Al–4V for biomedical applications. *Front. Mater. Sci.* **2015**, *9*, 4. [CrossRef]
13. Wu, S.Q.; Lu, Y.J.; Gan, Y.L.; Huang, T.T.; Zhao, C.Q.; Lin, J.J.; Guo, S.; Lin, J.X. Microstructural evolution and microhardness of a selective-laser-melted Ti–6Al–4V alloy after post heat treatments. *J. Alloys Compd.* **2016**, *672*, 643–652. [CrossRef]
14. Kumar, P.; Ramamurty, U. Microstructural optimization through heat treatment for enhancing the fracture toughness and fatigue crack growth resistance of selective laser melted Ti 6Al 4V alloy. *Acta Mater.* **2019**, *169*, 45–59. [CrossRef]
15. Cao, S.; Chu, R.; Zhou, X.; Yang, K.; Jia, Q.; Lim, C.V.S.; Huang, A.; Wu, X. Role of martensite decomposition in tensile properties of selective laser melted Ti-6Al-4V. *J. Alloys Compd.* **2018**, *744*, 357–363. [CrossRef]
16. Ganor, Y.I.; Tiferet, E.; Vogel, S.C.; Brown, D.W.; Chonin, M.; Pesach, A.; Hajaj, A.; Garkun, A.; Samuha, S.; Shneck, R.Z.; et al. Tailoring Microstructure and Mechanical Properties of Additively-Manufactured Ti6Al4V Using Post Processing. *Materials* **2021**, *14*, 658. [CrossRef]
17. Brückner, F.; Lepski, D.; Beyer, E. Modeling the Influence of Process Parameters and Additional Heat Sources on Residual Stresses in Laser Cladding. *J. Therm. Spray Technol.* **2007**, *16*, 355–373. [CrossRef]
18. Vrancken, B. Influence of preheating and oxygen content on Selective Laser Melting of Ti6Al4V. In Proceedings of the 16th RAPDASA Conference, RAPDASA, Annual International Conference on Rapid Product Development Association of South Africa, Pretoria, South Africa, 4–6 November 2015.
19. Xu, W.; Brandt, M.; Sun, S.; Elambasseril, J.; Liu, Q.; Latham, K.; Xia, K.; Qian, M. Additive manufacturing of strong and ductile Ti–6Al–4V by selective laser melting via in situ martensite decomposition. *Acta Mater.* **2015**, *85*, 74–84. [CrossRef]
20. Xu, W.; Lui, E.W.; Pateras, A.; Qian, M.; Brandt, M. In situ tailoring microstructure in additively manufactured Ti-6Al-4V for superior mechanical performance. *Acta Mater.* **2017**, *125*, 390–400. [CrossRef]
21. Ali, H.; Ma, L.; Ghadbeigi, H.; Mumtaz, K. In-situ residual stress reduction, martensitic decomposition and mechanical properties enhancement through high temperature powder bed pre-heating of Selective Laser Melted Ti6Al4V. *Mater. Sci. Eng. A* **2017**, *695*, 211–220. [CrossRef]
22. Cain, V.; Thijs, L.; Van Humbeeck, J.; Van Hooreweder, B.; Knutsen, R. Crack propagation and fracture toughness of Ti6Al4V alloy produced by selective laser melting. *Addit. Manuf.* **2015**, *5*, 68–76. [CrossRef]
23. Thijs, L.; Verhaeghe, F.; Craeghs, T.; Humbeeck, J.V.; Kruth, J.-P. A study of the microstructural evolution during selective laser melting of Ti–6Al–4V. *Acta Mater.* **2010**, *58*, 3303–3312. [CrossRef]
24. Kazantseva, N.; Krakhmalev, P.; Thuvander, M.; Yadroitsev, I.; Vinogradova, N.; Ezhov, I. Martensitic transformations in Ti-6Al-4V (ELI) alloy manufactured by 3D Printing. *Mater. Charact.* **2018**, *146*, 101–112. [CrossRef]
25. Yang, J.; Yu, H.; Yin, J.; Gao, M.; Wang, Z.; Zeng, X. Formation and control of martensite in Ti-6Al-4V alloy produced by selective laser melting. *Mater. Des.* **2016**, *108*, 308–318. [CrossRef]
26. Liang, Z.; Sun, Z.; Zhang, W.; Wu, S.; Chang, H. The effect of heat treatment on microstructure evolution and tensile properties of selective laser melted Ti6Al4V alloy. *J. Alloys Compd.* **2019**, *782*, 1041–1048. [CrossRef]
27. Vilaro, T.; Colin, C.; Bartout, J.D. As-Fabricated and Heat-Treated Microstructures of the Ti-6Al-4V Alloys Processed by Selective Laser Melting. *Met. Mater. Trans. A* **2011**, *42*, 3190–3199. [CrossRef]
28. Gil Mur, F.X.; Rodríguez, D.; Planell, J.A. Influence of tempering temperature and time on the α'-Ti-6Al-4V martensite. *J. Alloys Compd.* **1996**, *234*, 287–289. [CrossRef]
29. Guo, B.; Fall, A.; Jahazi, M.; Jonas, J.J. Kinetics of Post-dynamic Coarsening and Reverse Transformation in Ti-6Al-4V. *Met. Mater. Trans. A* **2018**, *49*, 5956–5961. [CrossRef]
30. Koike, J.; Shimoyama, Y.; Ohnuma, I.; Okamura, T.; Kainuma, R.; Ishida, K.; Maruyama, K. Stress-induced phase transformation during superplastic deformation in two-phase Ti–Al–Fe alloy. *Acta Mater.* **2000**, *48*, 1. [CrossRef]
31. Sabban, R.; Bahl, S.; Chatterjee, K.; Suwas, S. Globularization using heat treatment in additively manufactured Ti-6Al-4V for high strength and toughness. *Acta Mater.* **2019**, *162*, 239–254. [CrossRef]
32. de Formanoir, C.; Martin, G.; Prima, F.; Allain, S.Y.P.; Dessolier, T.; Sun, F.; Vivès, S.; Hary, B.; Bréchet, Y.; Godet, S. Micromechanical behavior and thermal stability of a dual-phase $\alpha+\alpha'$ titanium alloy produced by additive manufacturing. *Acta Mater.* **2019**, *162*, 149–162. [CrossRef]
33. Kaschel, F.R.; Vijayaraghavan, R.K.; Shmeliov, A.; McCarthy, E.K.; Canavan, M.; McNally, P.J.; Dowling, D.P.; Nicolosi, V.; Celikin, M. Mechanism of stress relaxation and phase transformation in additively manufactured Ti-6Al-4V via in situ high temperature XRD and TEM analyses. *Acta Mater.* **2020**, *188*, 720–732. [CrossRef]
34. Xu, Y.; Zhang, D.; Guo, Y.; Hu, S.; Wu, X.; Jiang, Y. Microstructural tailoring of As-Selective Laser Melted Ti6Al4V alloy for high mechanical properties. *J. Alloys Compd.* **2020**, *816*, 152536. [CrossRef]
35. Matsumoto, H.; Nishihara, T.; Iwagaki, Y.; Shiraishi, T.; Ono, Y.; Chiba, A. Microstructural evolution and deformation mode under high-temperature-tensile-deformation of the Ti-6Al-4V alloy with the metastable α' martensite starting microstructure. *Mater. Sci. Eng. A* **2016**, *661*, 68–78. [CrossRef]

36. Matsumoto, H.; Yoshida, K.; Lee, S.H.; Ono, Y.; Chiba, A. Ti–6Al–4V alloy with an ultrafine-grained microstructure exhibiting low-temperature–high-strain-rate superplasticity. *Mater. Lett.* **2013**, *98*, 209–212. [CrossRef]
37. Park, C.H.; Ko, Y.G.; Park, J.-W.; Lee, C.S. Enhanced superplasticity utilizing dynamic globularization of Ti–6Al–4V alloy. *Mater. Sci. Eng. A* **2008**, *496*, 150–158. [CrossRef]
38. Zhang, W.; Liu, H.; Ding, H.; Fujii, H. The optimal temperature for enhanced low-temperature superplasticity in fine-grained Ti–15V–3Cr–3Sn–3Al alloy fabricated by friction stir processing. *J. Alloys Compd.* **2020**, *832*, 154917. [CrossRef]

 *materials*

Article

Porosity Analysis of Additive Manufactured Parts Using CAQ Technology

Peter Pokorný [1,*], Štefan Václav [1], Jana Petru [2] and Michaela Kritikos [1]

1 Faculty of Materials Science and Technology in Trnava, Institute of Production Technologies, Slovak University of Technology in Bratislava, 917 24 Trnava, Slovakia; stefan.vaclav@stuba.sk (Š.V.); michaela.kritikos@stuba.sk (M.K.)
2 Faculty of Mechanical Engineering, VŠB—Technical University of Ostrava, 70800 Ostrava, Czech Republic; jana.petru@vsb.cz
* Correspondence: peter.pokorny@stuba.sk

Abstract: Components produced by additive technology are implemented in various spheres of industry, such as automotive or aerospace. This manufacturing process can lead to making highly optimized parts. There is not enough information about the quality of the parts produced by additive technologies, especially those made from metal powder. The research in this article deals with the porosity of components produced by additive technologies. The components used for the research were manufactured by the selective laser melting (SLM) method. The shape of these components is the same as the shape used for the tensile test. The investigated parts were printed with orientation in two directions, Z and XZ with respect to the machine platform. The printing strategy was "stripe". The material used for printing of the parts was SS 316L-0407. The printing parameters were laser power of 200 W, scanning speed of 650 mm/s, and the thickness of the layer was 50 µm. A non-destructive method was used for the components' porosity evaluation. The scanning was performed by CT machine METROTOM 1500. The radiation parameters used for getting 3D scans were voltage 180 kV, current 900 µA, detector resolution 1024 × 1024 px, voxel size 119.43 µm, number of projections 1050, and integration time 2000 ms. This entire measurement process responds to the computer aided quality (CAQ) technology. VG studio MAX 3.0 software was used to evaluate the obtained data. The porosity of the parts with Z and XZ orientation was also evaluated for parts' thicknesses of 1, 2, and 3 mm, respectively. It has been proven by this experimental investigation that the printing direction of the part in the additive manufacturing process under question affects its porosity.

Keywords: porosity; additive technology; SLM; computer tomography

 check for **updates**

Citation: Pokorný, P.; Václav, Š.; Petru, J.; Kritikos, M. Porosity Analysis of Additive Manufactured Parts Using CAQ Technology. *Materials* **2021**, *14*, 1142. https://doi.org/10.3390/ma14051142

Academic Editor: Tuhin Mukherjee

Received: 29 January 2021
Accepted: 24 February 2021
Published: 28 February 2021

Publisher's Note: MDPI stays neutral with regard to jurisdictional claims in published maps and institutional affiliations.

1. Introduction

Extensive funds are invested in additive technologies. Large corporations are building laboratories focused on research and the application of additive technologies. For example, it is possible to cite the Centre for Additive Technology [1]. Components made with additive technologies are also implemented in the automotive and aerospace industries; for example, the use of additive manufacturing to manufacture fuel nozzles [2]. Additive technologies for the production of metal components use various metal powders. Bajaj et al. dealt with steels used in the additive manufacturing process in their experimental investigation. The article compares some mechanical properties of components produced by additive technology and conventional technology. It was proven that some properties (hardness, corrosion resistance) are better for components made with additive technologies. Some properties, such as ductility or fatigue strength, are worse for components made by additive technologies in comparison to conventionally produced steel parts [3]. One of the kinds of steel powders used for additive production is 316L. Similarly, other authors have researched and described the change in the mechanical properties and change of the structure of parts produced by the selective laser melting (SLM) method [4]. Through the analysis of mechanical properties, these authors were able to determine the direction in which these properties were changing. Additionally, anisotropic properties of steel were studied in different papers [5]. These papers specifically examined stainless steel. The authors investigated the properties of components produced by additive technology that were oriented at different angles during production. They found out that components made at an angle of 45° had the highest tensile strength. Research has shown that the angle of orientation of a part during its production and the direction of its fibres are influenced by the tensile characteristics. In a similar way, the heat treatment, mechanical, and microstructural properties of stainless steel were discussed in investigations by the authors in [6–8]. On the other hand, the parameters of the production process in the production of components by additive technology are important. These include, for example, distance of points and time of exposure. The combination of these parameters has also an influenced the porosity, component surface, microstructure of the material, density, and hardness. The parameters of the sintering process of the powders were investigated [9]. These authors found that, with the increase of the laser energy, the temperature at the sintering site increases and thus the melt fills the voids, leading to a porosity reduction in the component. Another important parameter that affects the properties of the component is the scanning strategy used in the producing process, which in this case, is basically a computer aided manufacturing (CAM) strategy. CAM strategies are also used in conventional machining methods; for example, in the milling process. One article [10] shows the influence of milling strategies on surface accuracy. In this work, a part with simple shapes was modelled. Three finishing strategies (optimized constant Z, spiral finishing, and offset finishing) were applied to these shapes. Afterwards, the measuring of machined cylindrical surfaces was performed by the optical 3D scanner and then by Contura G2. It is clear from the results that the optimized constant Z milling strategy is the most suitable for a simple cylindrical surface. The smallest deviations were recorded in both types of measurements for this strategy. It is actually a movement along the shape of the part, which removes the material during machining and adds material during scanning in additive production. Similarly, authors [11,12] investigated the influence of scanning strategy parameters on residual stress by SLM technology and the impact of process on the final component properties. Three scanning strategies were used (chessboard, stripes, and the meander strategy). A high density of sintered material of up to 99.695% was found. The effect of the laser power on the residual stress was also proven. In terms of the grain structure, this was observed and detailed in the article [13], where the influence of two scanning strategies on the grain structure in the material was studied. The authors used two scanning strategies (island and back and forth). The study showed that a more homogeneous structure can only be achieved by changing the scanning strategy. Here, scanning strategies were applied to a simple sample shape (block). The structure may develop differently on a sample with more complex shape. Mechanical properties of AISI 316 stainless steel engaging different orientation of parts were presented in [14]. In this case, the samples were made with different orientations with respect to the machine platform. The sintering parameters were constant for all samples, while the orientation was the only parameter changing. Measurements have shown that some mechanical properties (such as strength) of steel produced by additive technology are better than those of steel produced by rolling. Similarly, these samples showed anisotropy of properties with respect to their orientation during production. Also, an important property of components is their porosity. The article [15] investigates porosity and microhardness of 316L. When examining the porosity, the areas on the samples with defects were evaluated. One sample was examined by the non-destructive X-ray computer tomography (XCT) method. The other samples were subjected to metallographic examination. A high sample density of more than 99% and a low porosity of about 0.82% were investigated. The pores were not evenly distributed. The samples had higher microhardness than the parts made by molding. Still, in this regard, the authors in [16] investigated density and porosity of sintered steel. The article analyzed particle size, particle shape, temperature of sintering, and time of sintering. These parameters affect the properties of the parts. They determined the relationships between the parameters of the sintering process and the properties of the parts. The authors [17] described the possibility of using computer tomography (CT) to evaluate the properties of parts produced by additive technology. They analyzed the use of CT from several perspectives (defects, dimensions, density, and roughness). Based on an extensive analysis of the use of CT to measure the properties of parts produced by additive technology, they made suggestions for evaluating the relationship between the production of parts and their mechanical behaviour. In addition, they concluded that international standards for the use of CT for printing technology needed to be set; that it was necessary to specify a reference element for CT calibration; and that due to the high cost of CT machines, it was necessary to consider using other cheaper measurement methods as well to establish standards for examining the surface of parts via CT; and, thus, developing a software tool to simulate the CT process would be of great aid. In the article [18], the authors used CT to investigate the properties of parts. Samples were made by PµLSE stereolithography. They used CT to examine surface defects and lattice defects. Thus, they predicted the behaviour of mechanical properties and defects in the lattice. They performed shearing experiments. They summarized the findings of the experimental investigation in several points, concluding that geometry defects were related to the direction of the part building, and that geometry defects affected the development of defects in the material grid itself. On the other hand, the results of their finite element simulations showed the influence of geometry on the formation of defects in the material lattice.

In this paper, we focused on the research of porosity in components made by additive technology, specifically, SLM technology. The motivation for this research was the study of materials and scientific publications on the porosity of materials produced by SLM. These studies have shown that porosity needs to be deeper examined given that it can significantly affect the mechanical properties of parts. The study of porosity and other properties of parts made by 3D printing must lead to quality products that will be produced in a shorter time.

In our study, the designed parts were manufactured at constant 3D printing parameters, i.e., the sintering conditions were not changing. The influence of the parameters of the sintering process on the porosity was not observed. The shape of these components was the same as the shape used for the tensile test. Samples were manufactured with three different thicknesses (1, 2, 3 mm). The samples were printed in different directions with respect to the machine platform, specifically, in the XZ and Z directions. The CT measurement method was used to measure the porosity of components manufactured by the SLM additive technology. Porosity data were evaluated for individual part thicknesses and for individual part orientations in the machine. The main goal of the study was to determine how the orientation of the part during its production affects the porosity.

2. Materials and Methods

The 316L-0407 powder from Renishaw was used in this study. This is an austenitic stainless steel. The applications of this steel are in the plastic industry and die casting molds, dies for extrusion, instruments for surgery, and parts for the navy. The material composition is in Table 1 [19].

Table 1. Chemical components of the 316L-0407 powder used for experiments.

Element	Fe	Cr	Ni	Mo	Mn	Si	P	C	S
Mass (%)	Balance	16–18	10–14	2–3	≤2	≤1	≤0.045	≤0.03	≤0.03

All parts were produced by using the Renishaw AM400 (Wotton-under-Edge, UK) machine. The parameters of the process used in this article were power of laser 200 W, speed of scanning 650 mm/s, and thickness of layer 50 µm. The investigated parts were printed with orientation in two directions—Z and XZ. Material thicknesses of printed parts were 1, 2, 3 mm. (Figure 1).

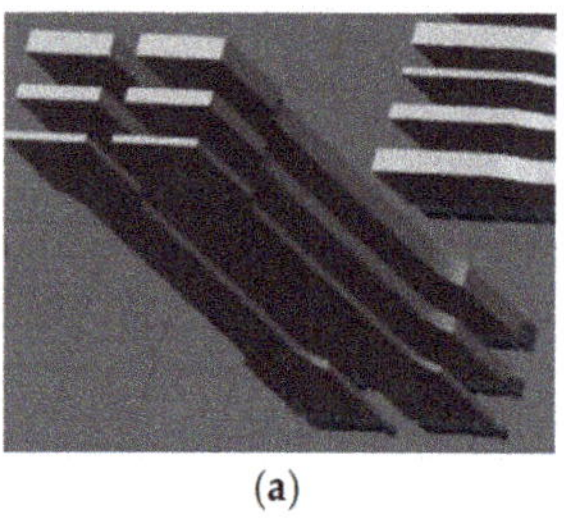 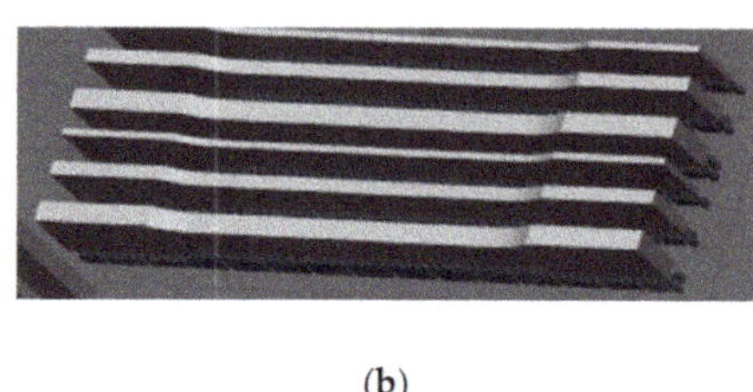

(**a**) (**b**)

Figure 1. Investigated parts printed with different orientation with respect to the machine platform (**a**) Part orientation in Z direction; (**b**) Part orientation in XZ direction.

Different scanning strategies were used to produce components by the SLM method. By applying these strategies, we obtained a finer grain structure [20,21]. In this way, it is possible to produce components with thin walls [22]. Parts for our experiment were made using a "strip" strategy of scanning (Figure 2).

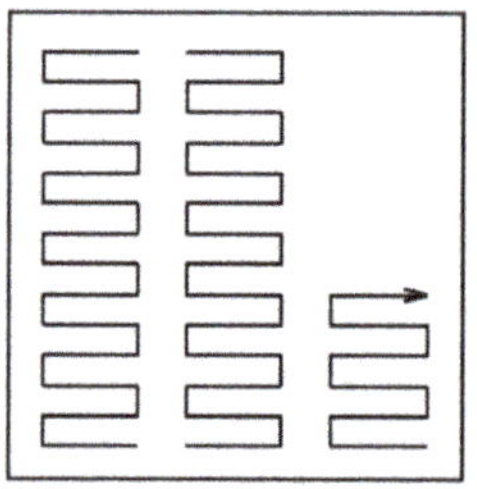

Figure 2. Strip scanning strategy used to build our components by the selective laser melting (SLM) method.

A non-destructive method for evaluation of the components' porosity was used in this study. This method is suitable for measuring dimensions, but also for measuring the structures of materials. Goméz et al. compared measurements by CT techniques and the coordinate measuring machine (CMM). They also discussed standards for estimating measurement uncertainty [23]. The case studies [24,25] compared the CT method and measurement techniques with classical metrology implemented on CMM. They pointed out the advantages of using CT measurement methods. Measurement strategies using CT were investigated and evaluated. The result was evidence of the suitability of CT measurement for production processes, dimensional measurement, and structural control (external, internal).

Experimental investigation was performed on the device METROTOM 1500 from Zeiss (Oberkochen, Germany), using computed tomography. This device consists of these main parts: X-ray tube, rotational table, and detector, which is used for capturing two dimensional images. Software used for scanning and getting data was METROTOM OS 2.8.

The X-ray set ups were made according to producer recommendations and skills of the operator:

- Voltage: 180 kV
- Current: 900 μA
- Resolution: 1024 × 1024 px
- Voxel size: 119.43 μm
- Nr of projections: 1050
- Integration time: 2000 ms

A cupper filter with thickness of 3 mm was used. The distance between the X-ray source and the scanning part was 450 mm.

Three dimensional models were evaluated in the VGStudio MAX 3.0 software (Volume Graphics, Heidelberg, Germany) after reconstruction. The first step was surface determination for recognition of the shape of the part. After that, 2 × compatibility porosity analysis was used. Its application shows pores scanned in each part. The principle of non-destructive measurement by CT is shown in (Figure 3). The position of the parts located in the computed tomography device is shown in (Figure 4).

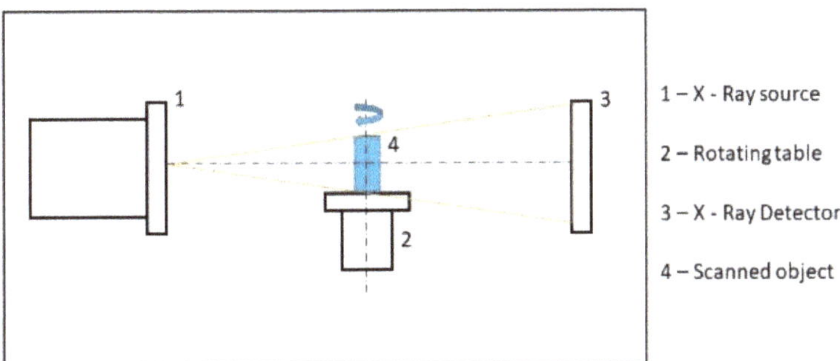

Figure 3. Principle of measurement by industrial computer tomography (CT).

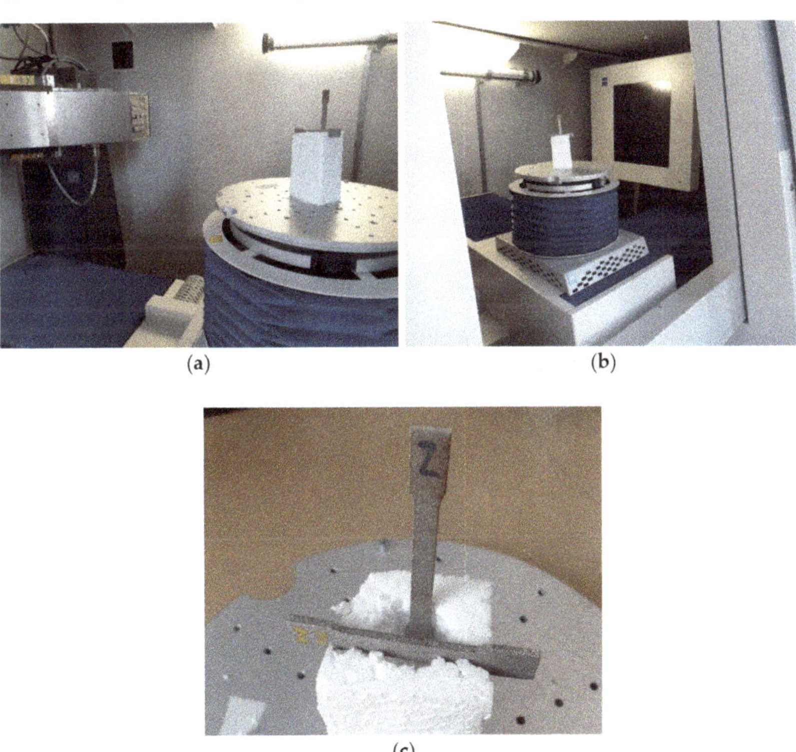

(a)

(b)

(c)

Figure 4. The position of the parts in the computer tomography device (**a**) View of the source of the X-rays, (**b**) View of the X-ray detector, (**c**) View of the experimental parts.

3. Results

The first step was surface determination for recognition of the shape of the part. After that, 2 × compatibility porosity analysis was used. Its application shows pores scanned in each part. Figure 5 shows the pores in the part, which had a thickness of 1 mm and an orientation of layers in the Z direction.

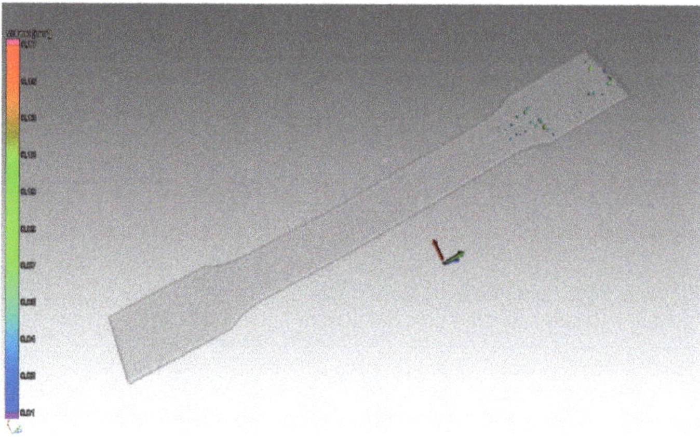

Figure 5. Pores detected in a part, which had a thickness of 1 mm oriented in Z.

Identification of defects at selected locations in the sample, which had a thickness of 1 mm and was oriented in the Z direction, is shown in Figure 6.

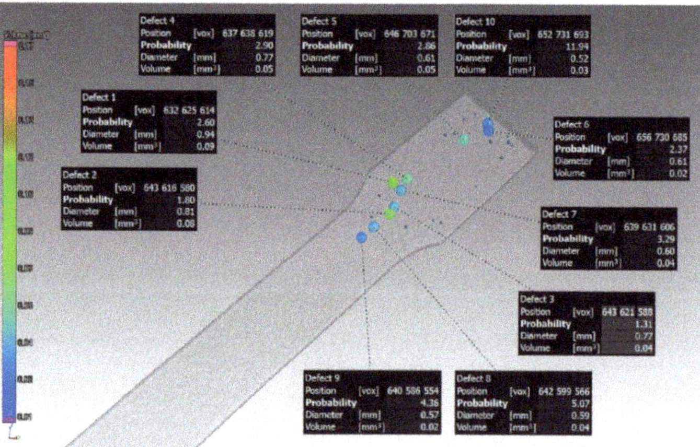

Figure 6. The biggest defects in the sample, which had a thickness of 1 mm and was oriented in Z.

Figure 7 shows the statistics of evaluated values of material volume, defect volume, and defect volume ratio for a sample, which had a thickness of 1 mm and was Z oriented.

Figure 7. Evaluated values for a sample with a thickness of 1 mm, which is oriented in Z.

Figure 8 shows the pores in the 1 mm thick sample with an orientation of layers in the XZ direction.

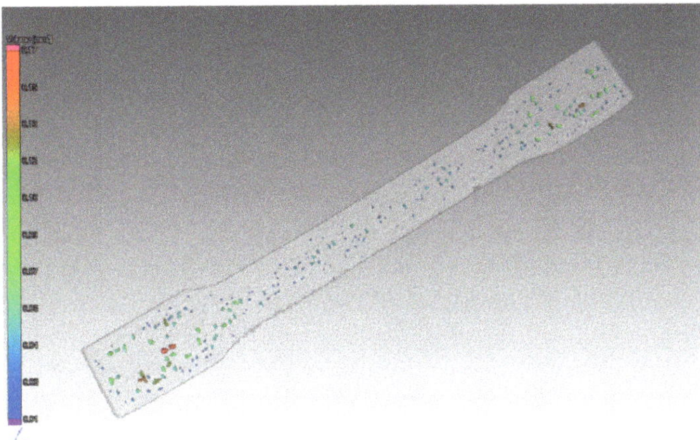

Figure 8. Defects detected in the sample with thickness of 1 mm printed in XZ directions.

The identification of the biggest defects at specific locations of the sample with 1 mm thickness, which was oriented in the XZ direction, is shown in Figure 9.

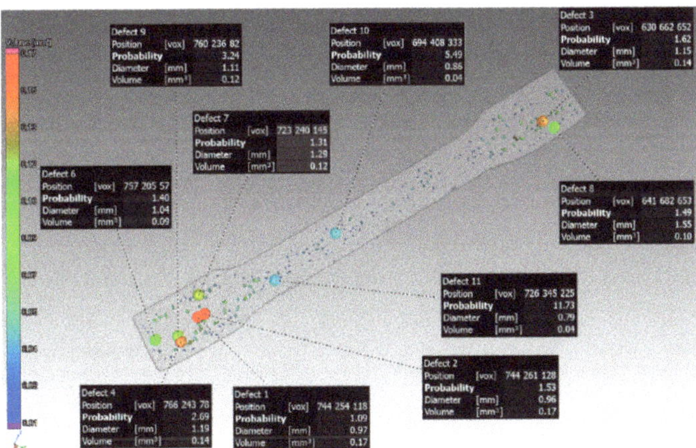

Figure 9. Specific defects evaluated in the sample, which had 1 mm thickness and was oriented XZ.

Figure 10 shows the measured values of material volume, defect volume, and defect volume ratio for a sample, which had 1 mm thickness and was oriented XZ.

Material	
Material volume [mm³]	769.70
Defect volume [mm³]	7.69
Defect volume ratio [%]	0.99

Figure 10. Measured values for a sample with a thickness of 1 mm and oriented XZ.

Parts with 2 and 3 mm thickness with orientation in the Z and X directions were evaluated in an identical way. The identification of defects at specific locations of the sample with 2 mm thickness, which was oriented in the Z direction, is shown in Figure 11. Figure 12 shows samples printed in the XZ direction.

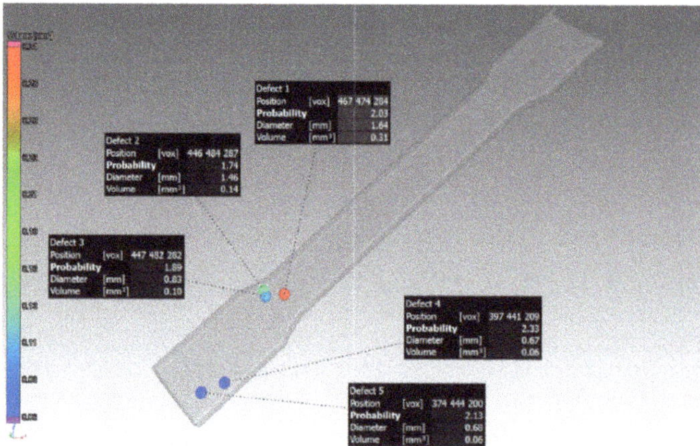

Figure 11. Specific defects on a sample with 2 mm thickness, which was oriented in the Z direction.

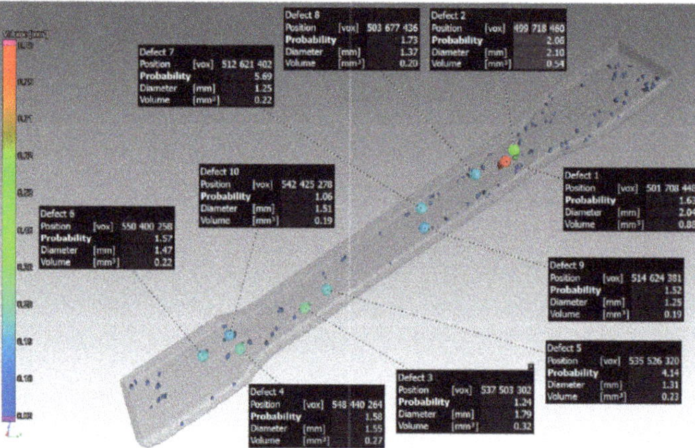

Figure 12. Specific defects on a sample with 2 mm thickness, which was oriented XZ.

Figure 13 shows the measured values of material volume, defect volume, and defect volume ratio for the sample with thickness of 2 mm, which was oriented in the Z direction. The measured values for the sample oriented XZ are shown in Figure 14.

Material	
Material volume [mm³]	1422.96
Defect volume [mm³]	0.67
Defect volume ratio [%]	0.05

Figure 13. Measured values for a sample with 2 mm thickness, which is oriented in Z.

Figure 14. Measured values for a sample with 2 mm thickness, which is oriented in XZ.

The identification of defects at specific locations of the sample with 3 mm thickness, which was oriented in the Z direction, is shown in Figure 15.

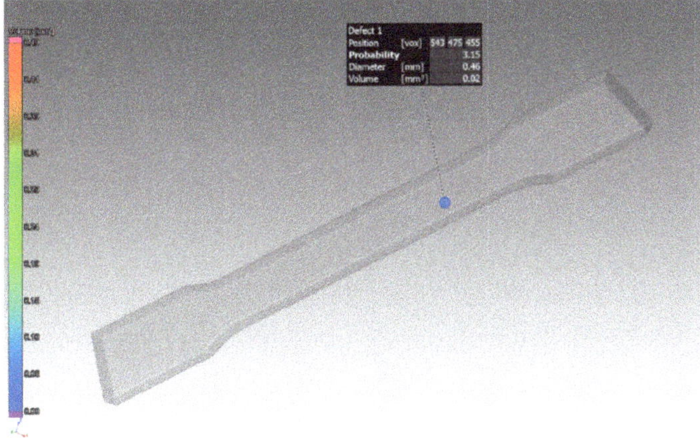

Figure 15. Specific defects in the sample with 3 mm thickness, which was oriented in Z.

The identification of defects at specific locations of the sample with thickness of 3 mm, which was oriented in the XZ direction, is shown in Figure 16.

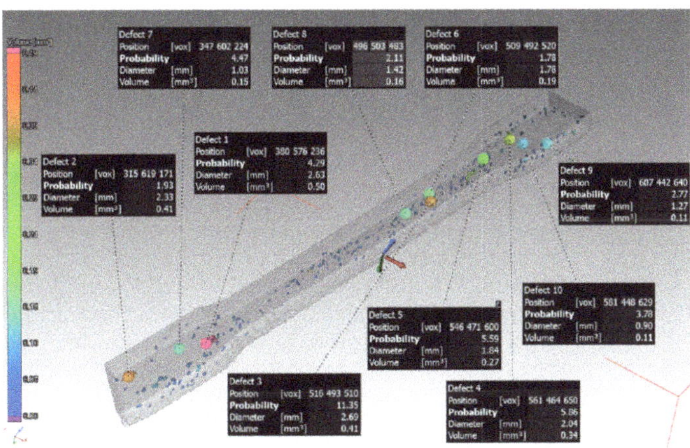

Figure 16. Specific defects in the sample with thickness of 3 mm, which was oriented in XZ.

Figure 17 shows the measured values of material volume, defect volume, and defect volume ratio for a sample with thickness of 3 mm, which was oriented in Z. Figure 18 shows the measured values of material volume, defect volume, and defect volume ratio for the sample with thickness of 3 mm, which was oriented in XZ.

Material	
Material volume [mm³]	2087.67
Defect volume [mm³]	0.02
Defect volume ratio [%]	0.00

Figure 17. Measured values for the sample with thickness of 3 mm, which was oriented in Z.

Material	
Material volume [mm³]	2099.21
Defect volume [mm³]	11.31
Defect volume ratio [%]	0.54

Figure 18. Measured values for the sample with thickness of 3 mm, which was oriented in XZ.

4. Discussion

Porosity is the most common defect observed in components made by SLM technology. The degree of the porosity can be influenced by the parameters of the laser process. Li et al. [26] described the influence of laser processing parameters on the porosity of additive manufactured parts; the material used was powder of 316L steel. The relationship between laser energy density and porosity was investigated. It was found that the porosity distribution is uniform in the layers. Simchi [27] investigated the effect of laser sintering parameters on the properties of different powders (Fe, Fe-C, Fe-Cu), and additionally, for 316L steel and M2 steel. As a result, it was determined that there is a relationship between density and laser power. The research by Kruth et al. [28] into stainless steel components (316L) demonstrated a relationship between scanning speed and grain size. The lower scanning speed causes the formation of larger grains and, thus, large defects. The research by Cherry et al. [9] found that total porosity is strongly influenced by the density of the laser energy. The porosity is high when the laser energy is low. As the energy density increases, the number of pores decreases. The smallest pore volume was observed for an energy of 104.52 J/mm³. Yosuf et al. [15] performed measurements where they applied the Archimedes method and the CT scanning method, which uses X-rays (XCT). The porosity of parts and their microhardness was investigated in this work. The parts were made of 316L steel by the SLM method. High density values (>99%) were found in the samples. A low porosity value was found (~0.82%). It was stated that this low pore volume did not affect the mechanical properties of the parts produced by the additive technology from 316L material. Al Faifi [29] identified the most statistically significant parameters of laser processing. The article identified the relationship between the number of pores and the parameters of the sintering process. It was found that the distance of the point, the time of exposure, and the thickness of the layer affect the parts density. Tolosa et al. [14] examined parts intended for mechanical testing. The parts were made by the additive SLM method. Different production methods were used with different orientations of the parts in space. It was found that the tensile properties of the parts and the strength of the steel parts produced by SLM are better than the properties of the steels produced by rolling. Using additive SLM technology, it is possible to produce complex shapes. Time savings and weight savings can be achieved. One article [13] dealt with the cracking of parts. This effect can be corrected by different part production strategies. This is a parameter of the laser scanning process. The production strategy affects the homogeneity of the structure. There is no need to change other process parameters. Tammas et al. [30] examined the distribution of defects in the volume of parts. They found that defect distribution is related to process parameters, contouring strategies, and section hatching. These parameters can affect the life of the parts. Hajnyš et al. [12] declared the most important parameters determining the strength of the material are scanning speed, then scanning strategy and ultimately laser power. The results of porosity have shown that the most important influences are scanning speed, laser power, and ultimately scanning strategy. In the article [31], the authors used CT method to examine porosity, pore sizes, and orientation. They used confocal microscopy to verify the CT results. The parts were made of 316L steel using SLM technology. Three samples were designed with different production orientations with respect to the machine platform. The smallest pores were detected in the sample oriented at an angle of 45° (0.15%). The highest porosity was detected in the sample oriented in YZ direction (2.97%). The porosity value for the sample oriented in the Z direction was 1.61%. These results are similar to the results from the experimental investigations presented in this paper. Additionally, higher porosity was achieved for the XZ-oriented component than for the Z-oriented component.

All the studies mentioned in the discussion point to how porosity is formed and how the porosity of components manufactured by additive technologies can be affected. There were also studies that report how porosity affects the material and its mechanical properties. Finally, the individual parameters of the laser process were identified here, such as scanning speed, power of laser, and strategy of scanning, which affect the porosity of the part and thus the properties of the part.

In this study, we pointed out the fact that, in addition to the parameters of the laser process, it is necessary to consider the layer orientation of the components in their production. Here, we examined the porosity of the samples and its relationship with respect to the layer orientation of the components.

In further research, we will submit the samples to other tests. We will investigate their properties by analyzing the microstructure and other material properties.

5. Conclusions

In this article, the porosity of components that were made using SLM technology was examined. The samples were produced with layer orientation in the Z and XZ directions with respect to the SLM machine platform. The experimental samples had thicknesses of 1, 2, and 3 mm.

The scan of the sample with thickness of 1 mm, which was printed in the Z direction shows that the pores were detected only on one side of the part. Pores of different sizes were identified, as shown in (Figure 6). The aim of this study was to compare the defect volume ratio for individual components oriented in different directions with respect to the machine platform. In this case, a defect volume ratio of 0.09% was found.

Other findings were shown in the 1 mm thick sample with orientation in the XZ direction. The pores were distributed throughout the entire volume of the part. The pore sizes were different, as shown in (Figure 9). The defect volume ratio for this sample was 0.99%.

Further measurements were performed on the sample with thickness of 2 mm oriented in the Z and XZ directions. Only a small number of pores were detected in the part that was oriented in the Z direction. These pores were located only on one side of the part, as it was in the case of the sample with thickness of 1 mm. Those pores are shown in (Figure 11). The defect volume ratio was 0.05%. Measurements for the part oriented in the XZ direction showed that the pores are distributed throughout the whole volume of the component (Figure 12). The defect volume ratio evaluated for this part was 0.62%.

The last experiments were performed on the samples with a thickness of 3 mm. The orientation of the parts in the machine during their production was in the Z direction and in the XZ direction. In the part oriented in the Z direction, essentially only one pore with a diameter of 0.46 mm was identified (Figure 15). The defect volume ratio was 0.00%. As it was with the previous samples oriented in the XZ direction, the pores in this part were distributed throughout the entire volume of the part. Again, pores of different sizes identified in this sample are shown in (Figure 16). The evaluated defect volume ratio was 0.54%.

Several factors contribute to the formation of porosity in the material. In addition to the parameters of the sintering process, in our case, using stripes as a scanning strategy also contributed. Its characterization is that the laser paths along the surface are divided into a cross-section. The scanning area is thus divided into small strips. These strips should overlap by default. This overlap may not be sufficient. This can cause the formation of pores. The area passed by the laser through the stripe strategy in the printing process of the part oriented in the Z direction is smaller than the area passed by the laser in the printing process of the part oriented in the XZ direction. The distribution of the area into strips is thus greater at XZ than at Z. This could be the cause of the lower porosity of the components produced in Z direction. Another possible cause of the porosity may be the scattering of the energy density. On a larger area, the scattering may not be sufficient, and the surface of the previous layer may melt. This does not create a coherent bond between adjacent surfaces, which can result in the formation of pores. Gases may be also the final factor for higher porosity detection. These gases could be either from the protective atmosphere or a product formed during the evaporation of the material. The assumption is that there is a greater chance of trapping gaseous particles within a larger scanned area. It can cause higher porosity. These stated assumptions need to be confirmed by further experiments.

Another fact we discovered that affects the porosity of the parts (it was not in the original plan and target) during this experimental investigation is the thickness of the sample. It is obvious from the measurements that with increased thickness of the sample, i.e., also with increased part volume, the number of pores in the part decreases. Probably, it is related to the sintering parameters. So far, it can only be expressed that, in the printing process of creating thicker samples, the laser beam acts longer on the surface. It can lead to increasing the melt temperature and improving the flow. The blanks are filled this way. Therefore, the proportion of the pores in the volume of material is reduced. This secondary finding can lead to enhanced experimental investigations about the relationship between the thickness of the part and the number of pores in the part.

It was found that the position of the part during its production affects its porosity. From the measured defect volume ratio values for individual sample thicknesses, it is possible to deduce that:

1. Orientation of the sample during its production in the **Z** direction: as the sample thickness increased, the defect volume ratio in the sample volume decreased. sample thickness 1 mm (Defect volume ratio 0.09%), sample thickness 2 mm (Defect volume ratio 0.05%), sample thickness 3 mm (Defect volume ratio 0.00%).
2. Orientation of the sample during its production in the **XZ**: as the sample thickness increased, the defect volume ratio in the sample volume decreased. sample thickness 1 mm (Defect volume ratio 0.99%), sample thickness 2 mm (Defect volume ratio 0.62%), sample thickness 3 mm (Defect volume ratio 0.54%).

Due to the defect volume ratio, the orientation of the sample in the Z direction during its production was more suitable than the orientation of the sample in the XZ direction during its production. This statement is valid in the relation to porosity evaluation. However, it may not be applied to other material properties.

Author Contributions: Idea: P.P., Š.V., M.K., and J.P., writing (preparation of draft): P.P.; writing (review): P.P., M.K., Š.V., and J.P.; analysis and measurement: M.K.; supervision and funding acquisition: J.P. All authors have read and agreed to the published version of the manuscript.

Funding: This paper has been completed in association with project Innovative and additive manufacturing technology—new technological solutions for 3D printing of metals and composite materials, reg. no. CZ.02.1.01/0.0/0.0/17_049/0008407 financed by Structural Funds of the European Union and project.

Institutional Review Board Statement: Not applicable.

Informed Consent Statement: Not applicable.

Data Availability Statement: Not applicable.

Conflicts of Interest: There is no conflict of interest.

References

1. GE Celebrates Grand Opening of First Additive Manufacturing Center in Pittsburgh. Available online: https://www.ge.com/news/press-releases/ge-celebrates-grand-opening-first-additive-manufacturing-center-pittsburgh (accessed on 7 December 2020).
2. The LEAP Engine. Performance. Execution. Technology. Available online: https://www.cfmaeroengines.com/engines/leap/ (accessed on 7 December 2020).
3. Bajaj, P.; Hariharan, A.; Kini, A.; Kürnsteiner, P.; Raabe, D.; Jagle, E.A. Steels in additive manufacturing: A review of their microstructure and properties. *Mater. Sci. Eng. A* **2020**, *772*, 138633. [CrossRef]
4. Deev, A.A.; Kuznetcov, P.A.; Petrov, S.N. Anisotropy of Mechanical Properties and its Correlation with the Structure of the Stainless Steel 316 L Produced by the SLM Method. *Phys. Procedia* **2016**, *83*, 789–796. [CrossRef]
5. Hitzler, L.; Hirsch, J.; Heine, B.; Merkel, M.; Hall, W.; Ochsner, A. On the anisotropic mechanical properties of selective laser-melted stainless steel. *Materials* **2017**, *10*, 1136. [CrossRef]
6. Chadha, K.; Tian, Y.; Spray, J.G.; Aranas, C. Effect of annealing heat treatment on the microstructural evolution and mechanical properties of hot isostatic pressed 316 L stainless steel fabricated by laser powder bed fusion. *Metals* **2020**, *10*, 753. [CrossRef]
7. Yadollahi, A.; Shamsaei, N.; Thompson, S.M.; Seely, D.W. Effect of process time interval and heat treatment on the mechanical and microstructural properties of direct laser deposited 316 L stainless steel. *Mater. Sci. Eng. A* **2015**, *644*, 171–183. [CrossRef]
8. Casati, R.; Lemke, J.; Vedani, M. Microstructure and fracture behavior of 316 L austenitic stainless steel produced by selective laser melting. *J. Mater. Sci. Technol.* **2016**, *32*, 738. [CrossRef]
9. Cherry, J.A.; Davies, H.M.; Mehmood, S.; Lavery, N.P.; Brown, S.G.R.; Sienz, J. Investigation into the effect of process parameters on microstructural and physical properties of 316 L stainless steel parts by selective laser melting. *Int. J. Adv. Manuf. Technol.* **2015**, *76*, 869–879. [CrossRef]
10. Peterka, J.; Pokorny, P.; Vaclav, S. CAM strategies and surface accuracy. In *Annals of DAAAM & Proceedings*; DAAAM International Vienna: Wolkersdorf, Austria, 2008; pp. 1061–1062.
11. Hajnys, J.; Pagáč, M.; Měsíček, J.; Petru, J.; Król, M. Influence of Scanning Strategy Parameters on Residual Stress in the SLM Process According to the Bridge Curvature Method for AISI 316 L Stainless Steel. *Materials* **2020**, *13*, 1659. [CrossRef]
12. Hajnyš, J.; Pagáč, M.; Kotera, O.; Petrů, J.; Scholz, S. Influence of Basic Process Parameters on Mechanical and Internal Properties of 316L Steel in SLM Process for Renishaw AM400. *MM Sci. J.* **2019**, 2790–2794. [CrossRef]
13. Carter, L.N.; Martin, C.; Withers, P.J.; Attallah, M.M. The influence of the laser scan strategy on grain structure and cracking behaviour in SLM powder-bed fabricated nickel superalloy. *J. Alloy. Compd.* **2014**, *615*, 338–347. [CrossRef]
14. Tolosa, I.; Garciandía, F.; Zubiri, F.; Zapirain, F.; Esnaola, A. Study of mechanical properties of AISI 316 stainless steel processed by "selective laser melting", following different manufacturing strategies. *Int. J. Adv. Manuf. Technol.* **2010**, *51*, 639–647. [CrossRef]
15. Yusuf, S.M.; Chen, Y.; Boardman, R.; Yang, S.; Gao, N. Investigation on porosity and microhardness of 316 L stainless steel fabricated by selective laser melting. *Metals* **2017**, *7*, 64. [CrossRef]
16. Verlee, B.; Dormal, T.; Lecomte-Beckers, J. Density and porosity control of sintered 316L stainless steel parts produced by additive manufacturing. *Powder Metall.* **2012**, *55*, 260–267. [CrossRef]
17. Khosravani, M.R.; Reinicke, T. On the Use of X-ray Computed Tomography in Assessment of 3D-Printed Components. *J. Nondestruct. Eval.* **2020**, *39*, 75. [CrossRef]
18. Qi, D.; Hu, W.; Xin, K.; Zeng, Q.; Xi, L.; Tao, R.; Liao, H.; Deng, Y.; Liao, B.; Wu, W. In-situ synchrotron X-ray tomography investigation of micro lattice manufactured with the projection micro-stereolithography (PµSL) 3D printing technique: Defects characterization and in-situ shear test. *Compos. Struct.* **2020**, *252*, 112710. [CrossRef]
19. Data Sheets—Additive Manufacturing, SS 316L-0407 Powder for Additive Manufacturing. Available online: https://resources.renishaw.com/en/details/data-sheet-ss-316l-0407-powder-for-additive-manufacturing--90802 (accessed on 10 December 2020).
20. Arısoy, Y.M.; Criales, L.E.; Özel, T.; Lane, B.; Moylan, S.; Donmez, A. Influence of scan strategy and process parameters on microstructure and its optimization in additively manufactured nickel alloy 625 via laser powder bed fusion. *Int. J. Adv. Manuf. Technol.* **2017**, *90*, 1393–1417. [CrossRef]
21. Keshavarzkermani, A.; Esmaeilizadeh, R.; Ali, U.; Enrique, P.D.; Mahmoodkhani, Y.; Zhou, N.Y.; Bonakdar, A.; Toyserkani, E. Controlling mechanical properties of additively manufactured hastelloy X by altering solidification pattern during laser powder-bed fusion. *Mater. Sci. Eng. A* **2019**, *762*, 138081. [CrossRef]
22. Leicht, A.; Klement, U.; Hryha, H. Effect of build geometry on the microstructural development of 316L parts produced by additive manufacturing. *Mater. Charact.* **2018**, *143*, 137–143. [CrossRef]
23. Gómez, H.V.; Lee, C.H.; Smith, S.T. Dimensional metrology with X-ray CT: A comparison with CMM measurements on internal features and compliant structures. *Precis. Eng.* **2018**, *51*, 291–307. [CrossRef]
24. Gómez, H.V.; Morse, E.P.; Hocken, R.J.; Smith, S.T. Dimensional metrology of internal features with X-ray computed tomography. In Proceedings of the 29th ASPE Annual Meeting, Boston, MA, USA, 9–14 November 2014; pp. 684–689.
25. Gómez, H.V.; Thousand, J.D.; Morse, E.P.; Smith, S.T. CT measurements and their estimated uncertainty: The significance of temperature and bias determination. In Proceedings of the 15th International Conference on Metrology and Properties of Engineering Surfaces, Charlotte, NC, USA, 2–5 March 2015; pp. 1–8.
26. Li, R.; Shi, Y.; Wang, Z.; Wang, L.; Liu, J.; Jiang, W. Densification behavior of gas and water atomized 316L stainless steel powder during selective laser melting. *Appl. Surf. Sci.* **2010**, *256*, 4350–4356. [CrossRef]

27. Simchi, A. Direct laser sintering of metal powders mechanism, kinetics and microstructural features. *Mater. Sci. Eng. A* **2006**, *428*, 148–158. [CrossRef]
28. Kruth, J.P.; Badrossamay, M.; Yasa, E.; Deckers, J.; Thijs, L.; Humbeeck, J.V. Part and material properties in selective laser melting of metals. In *Proceedings of the 16th International Symposium on Electromachining (ISEM XVI), Shanghai, China, 19–23 April 2010*; SHANGHAI JIAO TONG UNIV PRESS: Shanghai, China, 2010; pp. 3–14.
29. AlFaify, A.; Hughes, J.; Ridgway, K. Controlling the porosity of 316L stainless steel parts manufactured via the powder bed fusion process. *Rapid Prototyp. J.* **2019**, *25*, 162–175. [CrossRef]
30. Tammas, W.S.; Zhao, H.; Léonard, F.; Derguti, F.; Todd, I.; Prangnell, P.B. XCT analysis of the influence of melt strategies on defect population in Ti-6Al-4V components manufactured by Selective Electron Beam Melting. *Mater. Charact.* **2015**, *102*, 47–61. [CrossRef]
31. Ziółkowski, G.; Chlebus, E.; Szymczyk, P.; Kurzac, J. Application of X-ray CT method for discontinuity and porosity detection in 316L stainless steel parts produced with SLM technology. *Arch. Civ. Mech. Eng.* **2014**, *14*, 608–614. [CrossRef]

Article

In-Situ Characterization of Pore Formation Dynamics in Pulsed Wave Laser Powder Bed Fusion

Seyed Mohammad H. Hojjatzadeh [1,2], Qilin Guo [1,2], Niranjan D. Parab [3], Minglei Qu [1,2], Luis I. Escano [1], Kamel Fezzaa [3], Wes Everhart [4], Tao Sun [5,*] and Lianyi Chen [1,2,*]

[1] Department of Mechanical Engineering, University of Wisconsin-Madison, Madison, WI 53706, USA; hojjatzadeh@wisc.edu (S.M.H.H.); qguo46@wisc.edu (Q.G.); mqu22@wisc.edu (M.Q.); escanovolque@wisc.edu (L.I.E.)
[2] Department of Materials Science and Engineering, University of Wisconsin-Madison, Madison, WI 53706, USA
[3] X-ray Science Division, Advanced Photon Source, Argonne National Laboratory, Lemont, IL 60439, USA; niranjanparab@gmail.com (N.D.P.); fezzaa@aps.anl.gov (K.F.)
[4] Department of Energy's Kansas City National Security Campus Managed by Honeywell FM&T, Kansas City, MO 64147, USA; weverhart@kcnsc.doe.gov
[5] Department of Materials Science and Engineering, University of Virginia, Charlottesville, VA 22904, USA
* Correspondence: ts7qw@virginia.edu (T.S.); lianyi.chen@wisc.edu (L.C.)

Citation: Hojjatzadeh, S.M.H.; Guo, Q.; Parab, N.D.; Qu, M.; Escano, L.I.; Fezzaa, K.; Everhart, W.; Sun, T.; Chen, L. In-Situ Characterization of Pore Formation Dynamics in Pulsed Wave Laser Powder Bed Fusion. *Materials* **2021**, *14*, 2936. https://doi.org/10.3390/ma14112936

Academic Editor: Tuhin Mukherjee

Received: 4 May 2021
Accepted: 25 May 2021
Published: 29 May 2021

Publisher's Note: MDPI stays neutral with regard to jurisdictional claims in published maps and institutional affiliations.

Abstract: Laser powder bed fusion (LPBF) is an additive manufacturing technology with the capability of printing complex metal parts directly from digital models. Between two available emission modes employed in LPBF printing systems, pulsed wave (PW) emission provides more control over the heat input compared to continuous wave (CW) emission, which is highly beneficial for printing parts with intricate features. However, parts printed with pulsed wave LPBF (PW-LPBF) commonly contain pores, which degrade their mechanical properties. In this study, we reveal pore formation mechanisms during PW-LPBF in real time by using an in-situ high-speed synchrotron x-ray imaging technique. We found that vapor depression collapse proceeds when the laser irradiation stops within one pulse, resulting in occasional pore formation during PW-LPBF. We also revealed that the melt ejection and rapid melt pool solidification during pulsed-wave laser melting resulted in cavity formation and subsequent formation of a pore pattern in the melted track. The pore formation dynamics revealed here may provide guidance on developing pore elimination approaches.

Keywords: laser powder bed fusion; additive manufacturing; pore; pulsed emission; X-ray imaging

1. Introduction

The laser powder bed fusion (LPBF) additive manufacturing (AM) process is a 3D printing technology, which selectively melts powders in successive thin layers to build three dimensional parts directly from digital models without the constraints of traditional manufacturing methods. Currently, the use of LPBF is rapidly growing with multiple industrial applications, such as in the medical, aerospace, defense, and automobile industries [1].

One of the primary distinctions between commercial LPBF systems is the type of laser emission mode employed [2]. In continuous wave LPBF (CW-LPBF) systems, the laser delivers energy continuously without interruption; while in pulsed wave LPBF (PW-LPBF) systems, the laser power is fast modulated to turn on and off repeatedly, delivering energy in pulses [3,4]. The short burst of energy with PW-LPBF creates a melt pool with more flexible control over the heat input, which is highly advantageous for printing finer features such as lattice structures [5]. However, parts printed with PW-LPBF exhibit pores because the pulsated laser can cause instability in the melt pool leading to the formation of pores [6]. Pores are the major defect in parts printed by LPBF AM, which adversely affect the mechanical properties [7], especially fatigue life [8].

While pore formation during the CW-LPBF process has been studied extensively with post-processing diagnostic techniques [9], in-situ X-ray imaging techniques [10–16] and high fidelity simulations [13], research on pore formation and its underlying mechanisms during PW-LPBF is limited. Therefore, it is important to implement in-situ diagnostic tools, such as state-of-the-art in situ x-ray imaging techniques, to perform fundamental studies on pore formation during the PW-LPBF process in real time.

In this study, we revealed the dynamics and mechanisms of pore formation during the PW-LPBF process by utilizing in-situ high-speed X-ray imaging with 100 ps temporal resolution and ~2 μm spatial resolution. The results of this study are vital for developing processing parameters to mitigate pore formation and therefore improve the mechanical performance and reliability of parts printed using PW-LPBF. In addition, the results of this research may have implications in other areas where pulsated laser is used [17–21].

2. Materials and Methods

High-speed high-resolution X-ray imaging (at the beamline 32-ID-B of the Advanced Photon Source, Argonne National Laboratory, Lemont, IL, USA) was utilized to probe pore formation dynamics during PW-LPBF in real time [22]. The schematic of the X-ray imaging system is displayed in Figure 1.

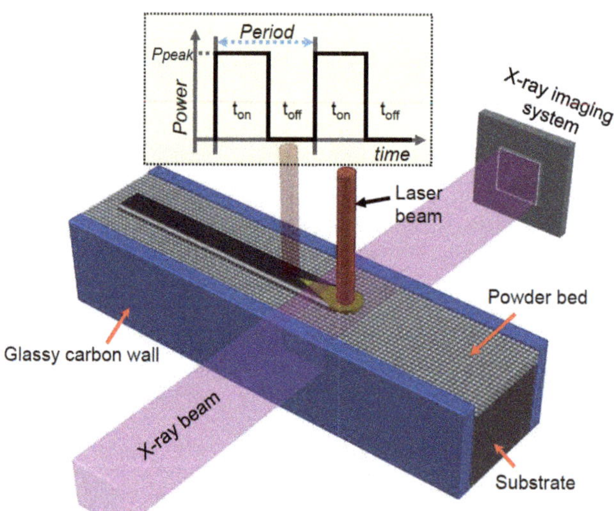

Figure 1. Schematic of the x-ray imaging experiment and the temporal characteristics of the PW-LPBF (pulsed wave laser powder bed fusion) process. The figure shows the temporal characteristic of square shaped pulses used in the experiment and the definition of laser on and off periods. Note that the actual output of the pulse shape may deviate from the perfect square shape specified in the laser control program.

The high-speed X-ray imaging system is composed of a miniature laser powder bed setup which is clamped between two glassy carbon container walls. A pseudo pink X-ray beam, with 1st harmonic energy at (24.7~25.3) keV penetrates through the metal and powder while a downstream detection system converts the transmitted x-ray beam into a visible light image using a scintillator. The converted signal is then recorded by a high-speed camera with a 10× magnification and spatial resolution of approximately 2 μm per pixel [22–25]. A recording frame rate of 50 kHz was used in this study. The experiments were performed inside a stainless-steel vacuum chamber, under 1 atm argon atmosphere. Ti-6Al-4V and Al6061 plates with the thicknesses of 0.4 and 0.7 mm, respectively, were used as the metal substrates. A layer of Al6061 powder with a thickness of ~100 μm was spread on the top of the Al6061 substrate metal to perform pulsed-LPBF AM experiments. In the experiments with Ti-6Al-4V substrate, no powder was added on the top of the substrate metal.

The key parameters to define the pulse in pulsed laser melting are frequency and laser duty cycle. The frequency (f) is defined as:

$$Frequency = \frac{1}{Period} = \frac{1}{t_{on} + t_{off}}$$ (1)

where t_{on} is defined as the time period when the laser is "on" in each pulse, called the laser-on period, and t_{off} denotes the time period when laser is "off" between the end of the pulse and the beginning of the consecutive pulse, called the laser-off period (see the inset of Figure 1). The laser duty cycle is the percentage of the laser-on time in the given modulated period and is defined as:

$$Duty\ cycle = \frac{t_{on}}{t_{off} + t_{on}} \times 100\%$$ (2)

An ytterbium fiber laser with the wavelength of 1070 nm, maximum output power of 520 W and a $D4\sigma$ diameter of ~100 μm was modulated by a square wave to emit with a given peak power at varying laser frequency (up to 50 kHz) and laser duty cycle (up to 99%) to perform single track laser melting on both powder bed and bare substrate samples. The laser scan velocity was varied from 0.3 to 1.5 m/s in the experiments.

The recorded x-ray images were processed using ImageJ to reduce the noise and enhance the contrast in each frame. The solid–liquid interface was identified in X-ray images by image processing where the image intensity at each pixel of Frame (i) was divided by the intensity of corresponding pixel in Frame ($i + 2$), such that the motionless part in the image was converted to blank background [25].

3. Results

Pore formation during the PW-LPBF process was studied by performing a series of X-ray imaging experiments at a frame rate of 50 kHz under varying laser frequency and laser duty cycle. Figure 2 and Supplementary Movies S1–S3 show pore formation during the PW-LPBF process of Al6061 under varying laser frequency (4, 7, and 10 kHz) and a constant laser duty cycle (50%). Pores are observed to form occasionally via the rapid collapse of the vapor depression at the end of the laser-on period in one pulse which is reminiscent of pore formation at the end of laser track during CW-LPBF AM. The mechanism of pore formation when the laser is turned off at the end of the track has been extensively studied before [11,13,15,22]. Under constant laser duty cycle (while laser power and scan speed are kept constant), the melt pool size is observed to be a function of laser frequency. As the laser frequency increased (from 4 to 10 kHz), a smaller melt pool and therefore a shallower depression zone formed. This caused the formation of pores from vapor depression collapse at the depth closer to the interface between the substrate and the powder layer, as shown in Figure 2.

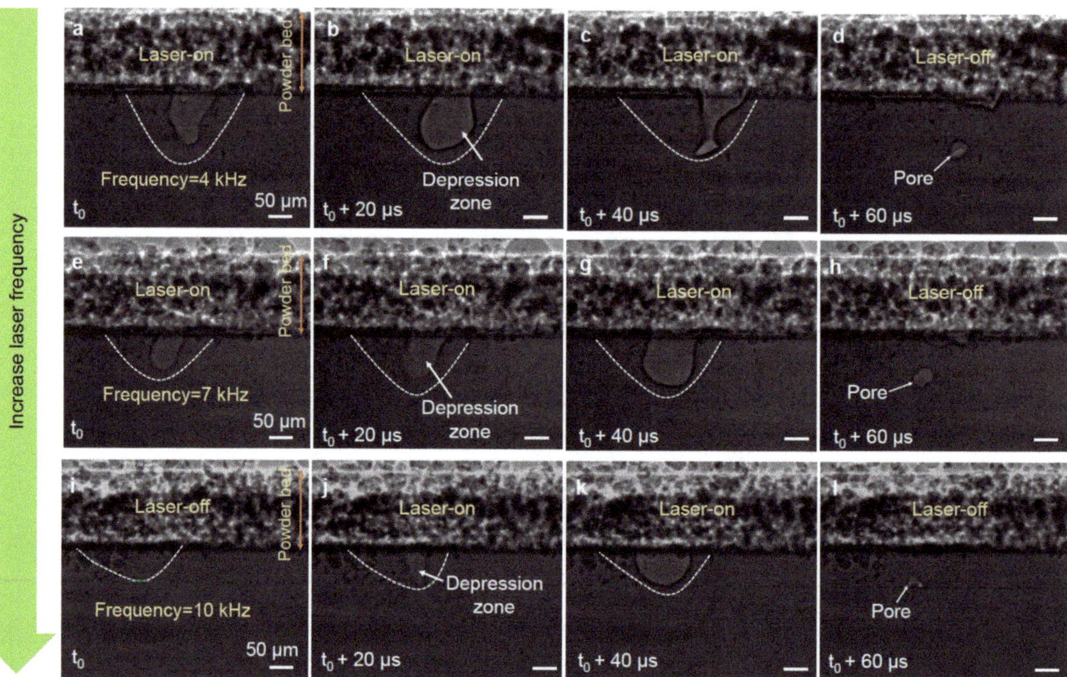

Figure 2. Pore formation under varying laser frequencies at a constant laser duty cycle: (**a–l**) Dynamic X-ray images showing pore formation during the PW-LPBF of Al6061 at laser frequencies of 4 kHz (**a–d**), 7 kHz (**e–h**), and 10 kHz (**i–l**) under a laser duty cycle of 50%, a laser power of 470 W and a scan speed of 0.4 m/s. In d, h and l, the melt pool boundary is not indicated to avoid blocking the view of the pores. Note that images do not display the complete duration of one pulse. The compiled movies showing the complete duration of two consecutive pulses are available in the Supplementary Material (Movies S1–S3).

Similarly, vapor depression collapse at the end of the laser-on period occasionally caused pore formation in the melt pool under a varying laser duty cycle and a constant laser frequency, as shown in Figure 3 and Supplementary Movies S4–S6. The increase in laser duty cycle and therefore longer laser exposure time in these experiments resulted in the formation of a larger melt pool and subsequently the formation of pores at the larger depth relative to the interface between the substrate and the powder layer. From these experimental observations, neither the size nor the number of pores were identified to be correlated with the laser frequency or duty cycle (Figures 2 and 3), which is ascribed to the random pore formation from vapor depression collapse during PW-LPBF.

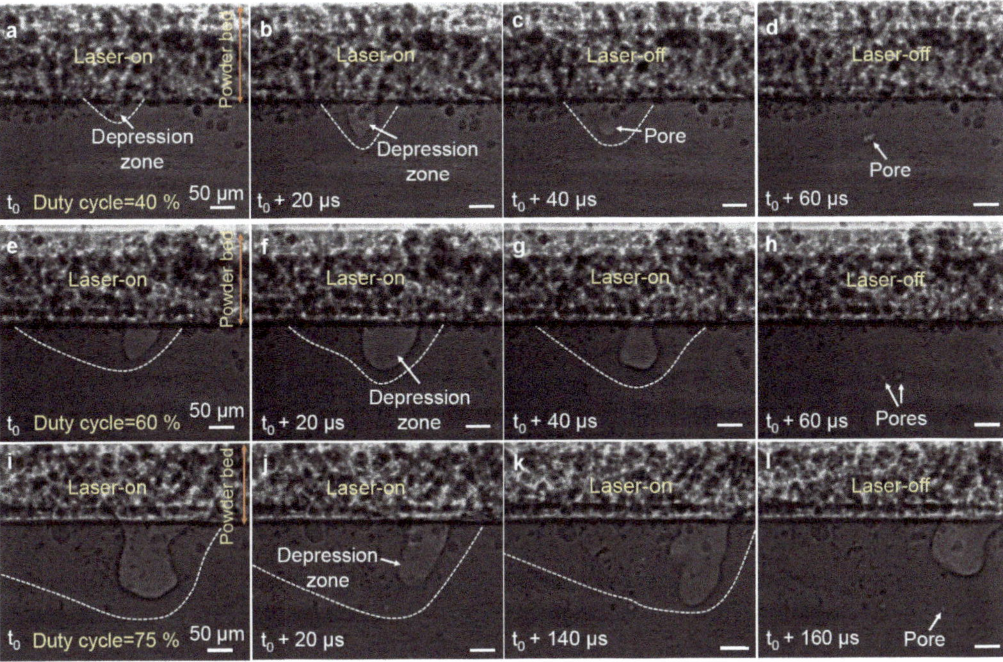

Figure 3. Pore formation under varying laser duty cycles at a constant laser frequency: (**a–l**) Dynamic X-ray images showing pore formation during the PW-LPBF process of Al6061 at the laser duty cycle of 40% (**a–d**), 60% (**e–h**), and 75% (**i–l**) at a laser frequency of 7 kHz, a laser power of 470 W and a scan speed of 0.4 m/s. In (**d,h,l**) the melt pool boundary is not indicated to avoid blocking the view of the pores. Note that images do not display the complete duration of one pulse. The compiled movies showing the complete duration of two consecutive pulses are available in the Supplementary Material (Movies S4–S6).

Low laser frequency is associated with longer pulse duration $\left(period = \frac{1}{frequency} \right)$ and therefore longer laser irradiation time. To further our understanding of pore formation at low laser frequency, we performed a series of X-ray imaging experiments during pulsed-wave laser melting of Ti-6Al-4V at 4 kHz under a varying laser duty cycle. Figure 4 and Supplementary Movie S7 display the X-ray image sequences during pulsed-wave laser melting of Ti-6Al-4V substrate at a laser frequency of 4 kHz and a duty cycle of 50%. The first frame (t_0, Figure 4a) shows the onset of the laser pulse when laser irradiation starts. First, the laser heating creates a vapor cavity by recoil pressure. After 120 μs, the laser irradiation stops, leading to a rapid freezing of the melt pool and formation of a cavity (Figure 4d). As the consecutive laser pulse begins, the depression zone emerges at a location of ~220 μm distance from the center of the first cavity (t_0 + 300 μs, Figure 4f). The laser irradiation stops again after 120 μs and results in the formation of the second cavity in the substrate (Figure 4h). As the laser moves forward, the cavity formation proceeds and a pattern of cavities is formed in the substrate material (t_0 + 900 μs, Figure 4p).

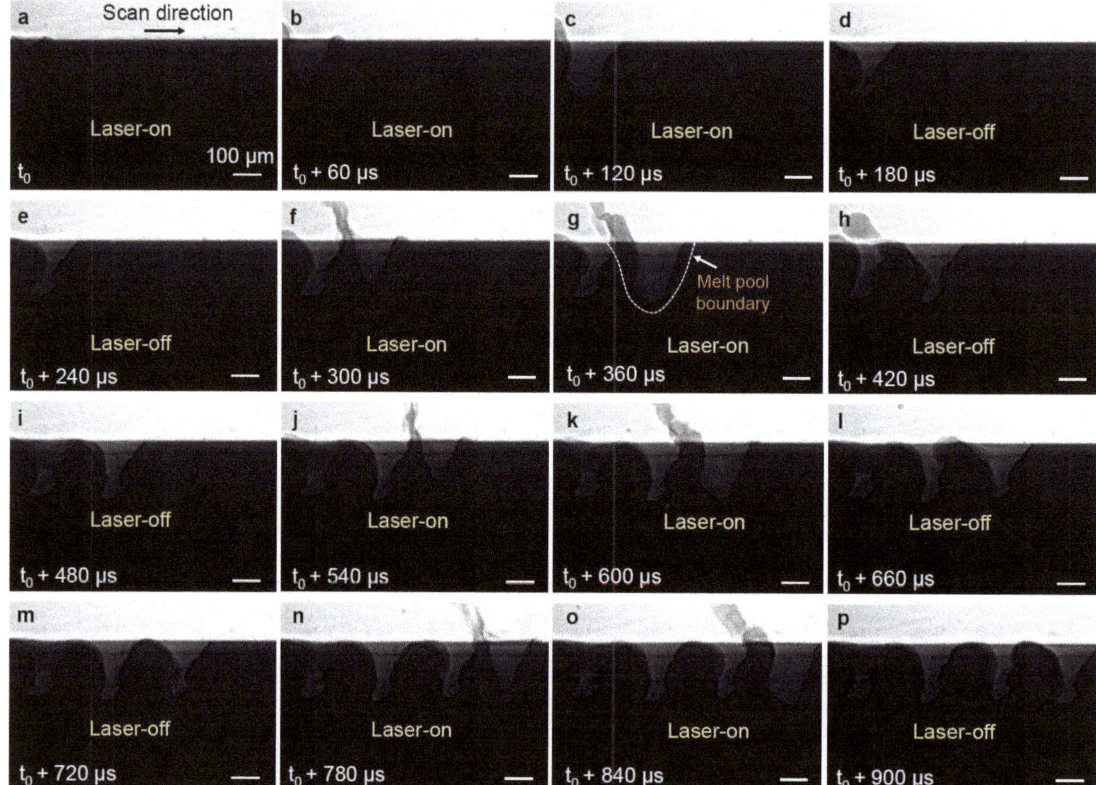

Figure 4. Cavity pattern formation at low laser frequency: (**a–p**) Dynamic X-ray images showing formation of cavity during pulsed-wave laser melting of Ti-6Al-4V substrate at a laser frequency of 4 kHz, a duty cycle of 50%, a laser power of 420 W, and a scan speed of 0.8 m/s. Note that some image frames during laser-on time and laser-off time have been skipped. The compiled movie showing the details is available in the Supplementary Material (Movie S7).

In the modulated laser, the distance that the laser travels at the time interval between pulses (commonly referred to as point distance) is decreased via the decrease in laser scan speed. This results in the formation of the overlap between melt pools of the consecutive pulses. Figure 5 and Supplementary Movie S8 show an X-ray imaging experiment with pulsed-wave laser melting of Ti-6Al-4V substrate at a laser frequency of 4 kHz, a duty cycle of 60% and a laser scan speed of 0.5 m/s. The cavity forms after the rapid freezing of the melt pool at the end of the laser-on period. As the consecutive pulse begins, the vapor depression emerges at a location where it interacts with the cavity, turning the cavity into a closed pore (Figure 5a–o). As the laser melting continues, a pattern of pores form in the substrate via this pore formation mechanism (Figure 5p).

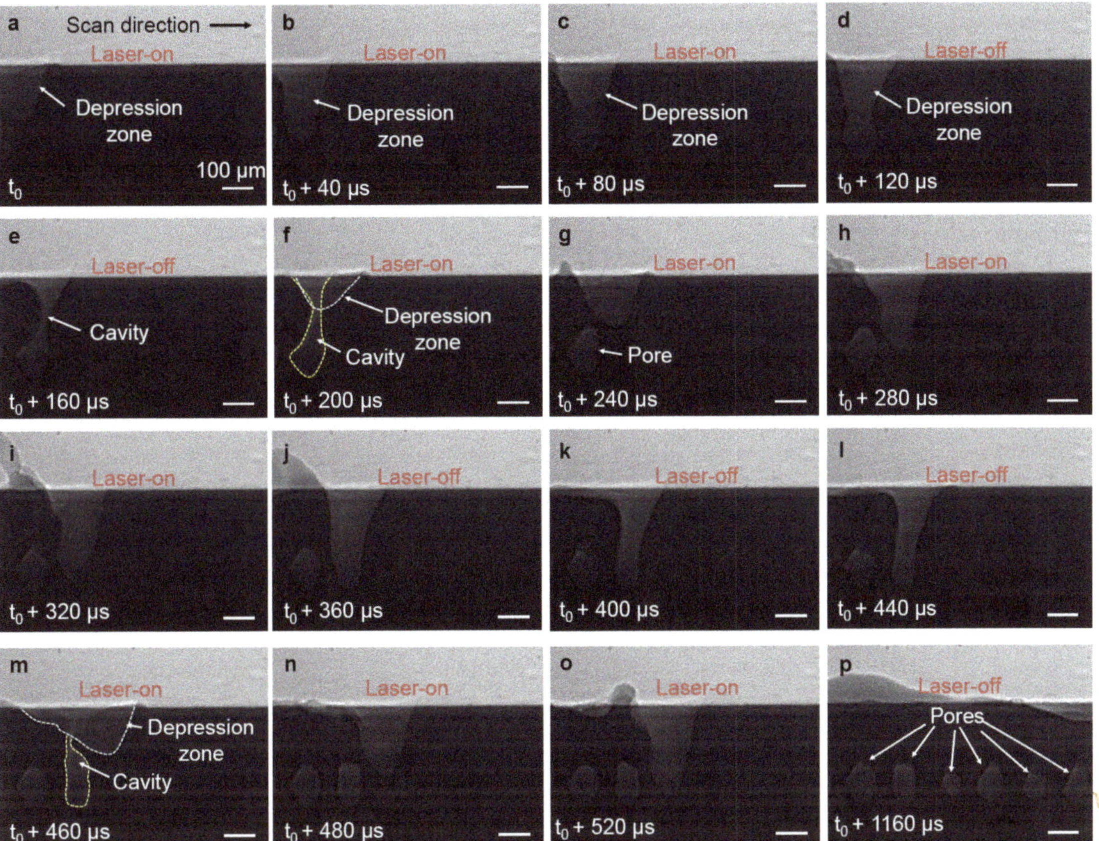

Figure 5. Pore pattern formation from cavities at low laser frequency: (**a–p**) Dynamic X-ray images showing the formation of pores from a cavity during pulsed-wave laser melting of Ti-6Al-4V at a laser frequency of 4 kHz, a duty cycle of 60%, a laser power of 470 W and a laser scan speed of 0.5 m/s. The pore formation from cavity in two consecutive pulses is shown in (**a–o**). Note that some image frames during the laser-on period and laser-off period have been skipped. The movie showing pore formation dynamics is available in the Supplementary Material (Movie S8).

4. Discussion

We constructed schematics to illustrate the formation mechanisms of the cavity pattern and pore pattern. The mechanism of cavity pattern formation is displayed in Figure 6a–f. During laser melting, vapor cavities are formed progressively in the substrate material by a strong vaporization-induced recoil pressure. The melt pool that forms in one pulse is observed to be only slightly larger than the vapor depression (as indicated in Figure 4g), appearing to form a layer of liquid around the vapor depression. A strong recoil pressure around the vapor depression pushes the molten metal to move rapidly along the vapor depression walls and ultimately eject away near the rim of the vapor depression in the form of a melt ligament and spatter (Figure 6b,c). As a result, a large amount of liquid metal is ejected away rapidly from the melted area around the vapor depression during laser melting. This phenomenon was recently simulated by a high-fidelity model [13]. As the laser turns off, the temperature around the depression zone decreases abruptly, which results in rapid solidification of the remaining liquid around the vapor depression, especially around the bottom part of the depression zone (Figures 4h and 6d–f).

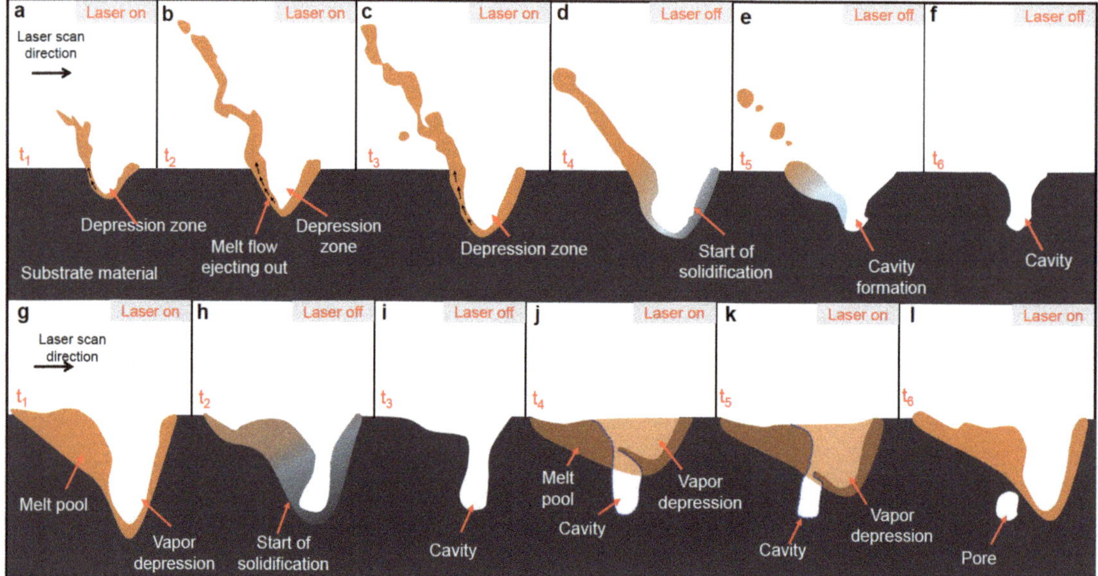

Figure 6. The mechanisms of pore and cavity pattern formation at low laser frequency: (**a–f**) Cavity formation mechanism; (**g–l**) Mechanism of pore formation from cavity at low laser frequency.

At the laser parameters setting used here, we did not observe the cavity pattering mechanism in Al6061 substrates. This can be ascribed to the higher thermal conductivity of Al6061 relative to Ti-6Al-4V (90 W/mK compared to 35 W/mK [26,27]). In Ti-6Al-4V, pulsed laser melting results in formation of a small melt pool in each pulse which starts to cool and solidify as soon as the pulse is complete. In Al6061 substrate material with higher thermal conductivity, on the other hand, a larger melt pool with longer solidification time forms in each pulse, which appears like a continuous melt pool by looking at consecutive pulses. As the laser irradiation stops in each pulse, there is sufficient melt around the vapor depression to reverse the direction towards the vapor depression side walls and fill the vapor cavity [28]. Similar cavity formation phenomena may be observed in Al6061 substrate by reducing the pulse duration and increasing the laser power. The mechanism of pore pattern formation from cavity is schematically shown in Figure 6g−l. The first cavity forms as a result of rapid solidification of the melt pool (especially around the bottom part of the depression zone) when the laser is switched off at the end of the laser-on period (Figure 6g−i). With the onset of the consecutive pulse, a new melt pool is created where it connects with the cavity formed in the previous pulse (Figure 6j). As the melted zone moves forward and grows in depth with the laser translating, the upper and middle portions of the cavity are filled with the liquid flowing from the melt pool, and the bottom portion of the cavity remains as a closed pore in the substrate material (Figure 6j−l). This process can continue until a pattern of pores is observed in the substrate material.

5. Conclusions

In summary, pore formation dynamics during pulsed wave LPBF AM process were directly observed by utilizing synchrotron x-ray imaging technique. The main conclusions are as follows:

1. The collapse of vapor depression, when laser irradiation stops at the end of the laser-on period in one pulse, was observed to occasionally induce pores during the PW-LPBF process under varying laser frequencies and duty cycles.

2. The melt pool and depression zone size changed with laser frequency and duty cycle. With the increase of the laser frequency or decrease of the duty cycle, the melt pool size and consequently the depression zone size decreased during PW-LPBF.
3. Our experimental observations did not reveal any correlation between the size nor the number of pores and the laser frequency or duty cycle.
4. In the depression/keyhole mode laser melting, at a low laser frequency with large point distance, cavity formation proceeds via the rapid solidification of the thin molten metal layer around the vapor cavity, which subsequently results in the formation of a cavity pattern in the substrate material.
5. In the depression/keyhole mode laser melting, at a low laser frequency with small point distance, the interaction of the cavity with the melt pool in the consecutive pulse results in the formation of closed pores and a pore pattern.
6. The results of this study will help the understanding of the PW-LPBF process and guide the development of processing approaches to mitigate pores.

Supplementary Materials: The following are available online at https://www.mdpi.com/article/10.3390/ma14112936/s1, Movie S1: The dynamics of pore formation during pulsed wave LPBF of Al6061 at a laser frequency of 4 kHz, duty cycle of 50%, laser power of 470 W, and laser scan speed of 0.4 m/s; Movie S2: The dynamics of pore formation during pulsed wave LPBF of Al6061 at a laser frequency of 7 kHz, duty cycle of 50%, laser power of 470 W, and laser scan speed of 0.4 m/s; Movie S3: The dynamics of pore formation during pulsed wave LPBF of Al6061 at a laser frequency of 10 kHz, duty cycle of 50%, laser power of 470 W, and laser scan speed of 0.4 m/s; Movie S4: The dynamics of pore formation during pulsed wave LPBF of Al6061 at a duty cycle of 40%, laser frequency of 7 kHz, laser power of 470 W, and laser scan speed of 0.4 m/s; Movie S5: The dynamics of pore formation during pulsed wave LPBF of Al6061 at a duty cycle of 60%, laser frequency of 7 kHz, laser power of 470 W, and laser scan speed of 0.4 m/s; Movie S6: The dynamics of pore formation during pulsed wave LPBF of Al6061 at a duty cycle of 75%, laser frequency of 7 kHz, laser power of 470 W, and laser scan speed of 0.4 m/s; Movie S7: The dynamics of cavity pattern formation during pulsed-wave laser melting of Ti-6Al-4V substrate at a laser frequency of 4 kHz, duty cycle of 50%, laser power of 420 W, and scan speed of 0.8 m/s; Movie S8: The dynamics of pore pattern formation during pulsed-wave laser melting of Ti-6Al-4V substrate at a laser frequency of 4 kHz, duty cycle of 60%, laser power of 470 W and laser scan speed of 0.5 m/s.

Author Contributions: Conceptualization, L.C.; methodology, Q.G., T.S., N.D.P., S.M.H.H., L.I.E., M.Q., K.F., W.E., and L.C.; formal analysis, S.M.H.H. and L.C.; investigation, S.M.H.H. and L.C.; resources, L.C., T.S.; data curation, Q.G., L.I.E., M.Q.; writing—original draft preparation, S.M.H.H. and L.C.; writing—review and editing, S.M.H.H. and L.C.; visualization, S.M.H.H. and L.C.; supervision, L.C.; project administration, L.C., W.E.; funding acquisition, L.C., T.S. and W.E. All authors have read and agreed to the published version of the manuscript.

Funding: This research was funded by the U.S. Department of Energy's Kansas City National Security Campus managed by Honeywell Federal Manufacturing & Technologies (FM&T) grant number [DE-NA0002839] and the U.S. National Science Foundation grant number [2002840] and the APC was funded by [DE-NA0002839].

Institutional Review Board Statement: Not applicable.

Informed Consent Statement: Not applicable.

Data Availability Statement: Not applicable.

Acknowledgments: The authors would like to thank Alex Deriy at Advanced Photon Source for his help with the beamline experiments. This research used resources of the Advanced Photon Source, a U.S. Department of Energy (DOE) Office of Science User Facility operated for the DOE Office of Science by Argonne National Laboratory under Contract No. DE-AC02-06CH11357. All data prepared, analyzed, and presented have been developed in a specific context of work and were prepared for internal evaluation and use pursuant to that work authorized under the referenced contract. Reference herein to any specific commercial product, process or service by trade name, trademark, manufacturer, or otherwise, does not necessarily constitute or imply its endorsement, recommendation, or favoring by the United States Government, any agency thereof or Honeywell

Federal Manufacturing & Technologies, LLC. This publication has been authored by Honeywell Federal Manufacturing & Technologies under Contract No. DE-NA0002839 with the U.S. Department of Energy. The United States Government retains and the publisher, by accepting the article for publication, acknowledges that the United States Government retains a nonexclusive, paid up, irrevocable, world-wide license to publish or reproduce the published form of this manuscript, or allow others to do so, for the United States Government purposes.

Conflicts of Interest: The authors declare no conflict of interest.

References

1. Herzog, D.; Seyda, V.; Wycisk, E.; Emmelmann, C. Additive manufacturing of metals. *Acta Mater.* **2016**, *117*, 371–392. [CrossRef]
2. Kim, J.; Ji, S.; Yun, Y.; Yeo, J. A review: Melt pool analysis for selective laser melting with continuous wave and pulse width modulated lasers. *Appl. Sci. Converg. Technol.* **2018**, *27*, 113–119. [CrossRef]
3. Demir, A.G.; Colombo, P.; Previtali, B. From pulsed to continuous wave emission in SLM with contemporary fiber laser sources: Effect of temporal and spatial pulse overlap in part quality. *Int. J. Adv. Manuf. Technol.* **2017**, *91*, 2701–2714. [CrossRef]
4. Modulation Matters—How to Build All Features Great and Small. Available online: https://www.linkedin.com/pulse/modulation-matters-how-build-all-features-great-small-marc-saunders/ (accessed on 27 May 2021).
5. Demir, A.G.; Previtali, B. Additive manufacturing of cardiovascular CoCr stents by selective laser melting. *Mater. Des.* **2017**, *119*, 338–350. [CrossRef]
6. Kuo, T.Y.; Lin, Y.D. Effects of different shielding gases and power waveforms on penetration characteristics and porosity formation in laser welding of inconel 690 alloy. *Mater. Trans.* **2007**, *48*, 219–226. [CrossRef]
7. DebRoy, T.; Wei, H.L.; Zuback, J.S.; Mukherjee, T.; Elmer, J.W.; Milewski, J.O.; Beese, A.M.; Wilson-Heid, A.; De, A.; Zhang, W. Additive manufacturing of metallic components—Process, structure and properties. *Prog. Mater. Sci.* **2018**, *92*, 112–224. [CrossRef]
8. Kasperovich, G.; Hausmann, J. Improvement of fatigue resistance and ductility of TiAl6V4 processed by selective laser melting. *J. Mater. Process. Technol.* **2015**, *220*, 202–214. [CrossRef]
9. Cunningham, R.; Nicolas, A.; Madsen, J.; Fodran, E.; Anagnostou, E.; Sangid, M.D.; Rollett, A.D. Analyzing the effects of powder and post-processing on porosity and properties of electron beam melted Ti-6Al-4V. *Mater. Res. Lett.* **2017**, *5*, 516–525. [CrossRef]
10. Leung, C.L.A.; Marussi, S.; Atwood, R.C.; Towrie, M.; Withers, P.J.; Lee, P.D. In situ X-ray imaging of defect and molten pool dynamics in laser additive manufacturing. *Nat. Commun.* **2018**, *9*, 1355. [CrossRef] [PubMed]
11. Martin, A.A.; Calta, N.P.; Hammons, J.A.; Khairallah, S.A.; Nielsen, M.H.; Shuttlesworth, R.M.; Sinclair, N.; Matthews, M.J.; Jeffries, J.R.; Willey, T.M.; et al. Ultrafast dynamics of laser-metal interactions in additive manufacturing alloys captured by in situ X-ray imaging. *Mater. Today Adv.* **2019**, *1*, 100002. [CrossRef]
12. Gould, B.; Wolff, S.; Parab, N.; Zhao, C.; Lorenzo-Martin, M.C.; Fezzaa, K.; Greco, A.; Sun, T. In situ analysis of laser powder bed fusion using simultaneous high-speed infrared and x-ray imaging. *JOM* **2021**, *73*, 201–211. [CrossRef]
13. Khairallah, S.A.; Martin, A.A.; Lee, J.R.I.; Guss, G.; Calta, N.P.; Hammons, J.A.; Nielsen, M.H.; Chaput, K.; Schwalbach, E.; Shah, M.N.; et al. Controlling interdependent meso-nanosecond dynamics and defect generation in metal 3D printing. *Science* **2020**, *368*, 660–665. [CrossRef]
14. Hojjatzadeh, S.M.H.; Parab, N.D.; Guo, Q.; Qu, M.; Xiong, L.; Zhao, C.; Escano, L.I.; Fezzaa, K.; Everhart, W.; Sun, T.; et al. Direct observation of pore formation mechanisms during LPBF additive manufacturing process and high energy density laser welding. *Int. J. Mach. Tools Manuf.* **2020**, *153*, 103555. [CrossRef]
15. Martin, A.A.; Calta, N.P.; Khairallah, S.A.; Wang, J.; Depond, P.J.; Fong, A.Y.; Thampy, V.; Guss, G.M.; Kiss, A.M.; Stone, K.H.; et al. Dynamics of pore formation during laser powder bed fusion additive manufacturing. *Nat. Commun.* **2019**, *10*, 1–10. [CrossRef] [PubMed]
16. Cunningham, R.; Zhao, C.; Parab, N.; Kantzos, C.; Pauza, J.; Fezzaa, K.; Sun, T.; Rollett, A.D. Keyhole threshold and morphology in laser melting revealed by ultrahigh-speed x-ray imaging. *Science* **2019**, *363*, 849–852. [CrossRef] [PubMed]
17. Gao, X.-L.; Zhang, L.-J.; Liu, J.; Zhang, J.-X. Porosity and microstructure in pulsed Nd:YAG laser welded Ti6Al4V sheet. *J. Mater. Process. Technol.* **2014**, *214*, 1316–1325. [CrossRef]
18. AlShaer, A.W.; Li, L.; Mistry, A. The effects of short pulse laser surface cleaning on porosity formation and reduction in laser welding of aluminium alloy for automotive component manufacture. *Opt. Laser Technol.* **2014**, *64*, 162–171. [CrossRef]
19. Perry, T.L.; Werschmoeller, D.; Li, X.; Pfefferkorn, F.E.; Duffie, N.A. Pulsed laser polishing of micro-milled Ti6Al4V samples. *J. Manuf. Process.* **2009**, *11*, 74–81. [CrossRef]
20. Wu, Z.; Yin, K.; Wu, J.; Zhu, Z.; Duan, J.-A.; He, J. Recent advances in femtosecond laser-structured Janus membranes with asymmetric surface wettability. *Nanoscale* **2021**, *13*, 2209–2226. [CrossRef]
21. Gao, X.; Feng, W.; Zhu, Z.; Wu, Z.; Li, S.; Kan, S.; Qiu, X.; Guo, A.; Chen, W.; Yin, K. Rapid Fabrication of Superhydrophilic Micro/Nanostructured Nickel Foam Toward High-Performance Glucose Sensor. *Adv. Mater. Interfaces* **2021**, *8*, 2002133. [CrossRef]
22. Hojjatzadeh, S.M.H.; Parab, N.D.; Yan, W.; Guo, Q.; Xiong, L.; Zhao, C.; Qu, M.; Escano, L.I.; Xiao, X.; Fezzaa, K.; et al. Pore elimination mechanisms during 3D printing of metals. *Nat. Commun.* **2019**, *10*, 3088. [CrossRef] [PubMed]
23. Zhao, C.; Fezzaa, K.; Cunningham, R.W.; Wen, H.; De Carlo, F.; Chen, L.; Rollett, A.D.; Sun, T. Real-time monitoring of laser powder bed fusion process using high-speed X-ray imaging and diffraction. *Sci. Rep.* **2017**, *7*, 3602. [CrossRef]

24. Guo, Q.; Zhao, C.; Escano, L.I.; Young, Z.; Xiong, L.; Fezzaa, K.; Everhart, W.; Brown, B.; Sun, T.; Chen, L. Transient dynamics of powder spattering in laser powder bed fusion additive manufacturing process revealed by in-situ high-speed high-energy x-ray imaging. *Acta Mater.* **2018**, *151*, 169–180. [CrossRef]
25. Guo, Q.; Zhao, C.; Qu, M.; Xiong, L.; Escano, L.I.; Hojjatzadeh, S.M.H.; Parab, N.D.; Fezzaa, K.; Everhart, W.; Sun, T.; et al. In-situ characterization and quantification of melt pool variation under constant input energy density in laser powder bed fusion additive manufacturing process. *Addit. Manuf.* **2019**, *28*, 600–609. [CrossRef]
26. Ning, J.; Mirkoohi, E.; Dong, Y.; Sievers, D.E.; Garmestani, H.; Liang, S.Y. Analytical modeling of 3D temperature distribution in selective laser melting of Ti-6Al-4V considering part boundary conditions. *J. Manuf. Process.* **2019**, *44*, 319–326. [CrossRef]
27. Jagadeesh, S.K.; Ramesh, C.S.; Mallikarjuna, J.M.; Keshavamurthy, R. Prediction of cooling curves during solidification of Al 6061-SiCp based metal matrix composites using finite element analysis. *J. Mater. Process. Technol.* **2010**, *210*, 618–623. [CrossRef]
28. Khairallah, S.A.; Anderson, A.T.; Rubenchik, A.; King, W.E. Laser powder-bed fusion additive manufacturing: Physics of complex melt flow and formation mechanisms of pores, spatter, and denudation zones. *Acta Mater.* **2016**, *108*, 36–45. [CrossRef]

materials

Article

Contrasting the Role of Pores on the Stress State Dependent Fracture Behavior of Additively Manufactured Low and High Ductility Metals

Alexander E. Wilson-Heid [1], Erik T. Furton [1] and Allison M. Beese [1,2,*]

[1] Department of Materials Science and Engineering, Pennsylvania State University, University Park, PA 16802, USA; axw66@psu.edu (A.E.W.-H.); emf5640@psu.edu (E.T.F.)
[2] Department of Mechanical Engineering, Pennsylvania State University, University Park, PA 16802, USA
[*] Correspondence: beese@matse.psu.edu

Abstract: This study investigates the disparate impact of internal pores on the fracture behavior of two metal alloys fabricated via laser powder bed fusion (L-PBF) additive manufacturing (AM)—316L stainless steel and Ti-6Al-4V. Data from mechanical tests over a range of stress states for dense samples and those with intentionally introduced penny-shaped pores of various diameters were used to contrast the combined impact of pore size and stress state on the fracture behavior of these two materials. The fracture data were used to calibrate and compare multiple fracture models (Mohr-Coulomb, Hosford-Coulomb, and maximum stress criteria), with results compared in equivalent stress (versus stress triaxiality and Lode angle) space, as well as in their conversions to equivalent strain space. For L-PBF 316L, the strain-based fracture models captured the stress state dependent failure behavior up to the largest pore size studied (2400 μm diameter, 16% cross-sectional area of gauge region), while for L-PBF Ti-6Al-4V, the stress-based fracture models better captured the change in failure behavior with pore size up to the largest pore size studied. This difference can be attributed to the relatively high ductility of 316L stainless steel, for which all samples underwent significant plastic deformation prior to failure, contrasted with the relatively low ductility of Ti-6Al-4V, for which, with increasing pore size, the displacement to failure was dominated by elastic deformation.

Keywords: ductile fracture; stress state; Ti-6Al-4V; 316L stainless steel; laser powder bed fusion

Citation: Wilson-Heid, A.E.; Furton, E.T.; Beese, A.M. Contrasting the Role of Pores on the Stress State Dependent Fracture Behavior of Additively Manufactured Low and High Ductility Metals. *Materials* **2021**, *14*, 3657. https://doi.org/10.3390/ma14133657

Academic Editor: Federico Mazzucato

Received: 28 May 2021
Accepted: 26 June 2021
Published: 30 June 2021

Publisher's Note: MDPI stays neutral with regard to jurisdictional claims in published maps and institutional affiliations.

1. Introduction

A major area of focus in the field of additive manufacturing (AM) is understanding pore formation, and process optimization with the goal of creating fully dense, defect-free components [1–3]. Studying the effect of pores on metal failure, along with the development of failure models based on theory and experiments has a long and continued history for conventionally processed ductile metals due to the importance of avoiding failure in load bearing components or during metal forming [4–9]. For example, in 1977 Gurson developed a model in terms of a yield function and microstructurally informed void volume fraction to understand void growth and ductile failure using simplified spherical and cylindrical void models [6]. By combining the existing frameworks for fracture modeling with the unique manufacturing capabilities of AM, new insight into the effect of pores on ductile failure is possible.

Laser powder bed fusion (L-PBF) AM is a process for building metallic components layer-by-layer using a focused laser heat source to melt a selected 2-dimensional pattern in a thin layer (10–100 μm) of powder feedstock to a baseplate, lowering the baseplate by a prescribed layer thickness, spreading on a new layer, scanning the next layer pattern, which fuses to the previous layer, and repeating until a final 3-dimensional (3D) component is completed. In AM, there are numerous processing parameters that dictate a completed component's quality. One primary evaluation metric for optimizing a parameter set for a given alloy is component density, or the reduction of pores. Pores in AM parts can be formed via different mechanisms [10]; two of the primary mechanisms are: (1) gas entrapment during melting and solidification of the meltpools, analogous to that seen in casting and welding, which generates spherical pores and (2) irregular morphology lack-of-fusion (LoF) pores, which are formed due to incomplete fusion between layers or adjacent meltpools on the same layer along the heat source scanning path, and which can have sharp features. LoF pores are of primary interest in AM because of their significantly more detrimental influence on material ductility compared to spherical pores [11].

In addition to pores having a harmful effect on the fracture behavior of ductile metals, stress state is also known to play an important role in fracture [12–14]. Stress state can be defined using the two parameters stress triaxiality (η) and Lode angle parameter ($\bar{\theta}$). The stress triaxiality is the ratio of the mean stress (σ_m) and von Mises equivalent stress ($\bar{\sigma}_{VM}$):

$$\eta = \frac{\sigma_m}{\bar{\sigma}_{VM}} \text{ with } \sigma_m = \frac{1}{3}I_1 \text{ and } \bar{\sigma}_{VM} = \sqrt{3J_2} \tag{1}$$

where $I_1 = \sigma_{kk}$ is the first invariant of the stress tensor, σ, and $J_2 = \frac{1}{2}s_{ij}s_{ij}$ is the second invariant of the deviatoric stress tensor, s. The normalized Lode angle parameter is a function of the third invariant of the deviatoric stress tensor, $J_3 = det(s_{ij})$, and is defined as:

$$\bar{\theta} = 1 - \frac{2}{\pi}arccos\left[\frac{27}{2}\frac{J_3}{\bar{\sigma}_{VM}^3}\right]. \tag{2}$$

Increased stress triaxiality is known to accelerate the void nucleation and growth process in ductile metals, resulting in reduced ductility even in fully dense samples [5,15]. Designing load bearing components to be safe under the spatially varying stress state is an important consideration for engineers in the design against failure.

Fracture models that describe the effects of stress state on ductile failure have primarily been presented in mixed stress-strain space, meaning equivalent plastic strain to failure versus η and $\bar{\theta}$ (referred to here as equivalent strain space) because in ductile fracture, the resolution in strain is typically much larger than the resolution in stress, that is, large differences in strain result in relatively minor differences in stress due to the fact that the elastic contribution to failure is negligible compared to the plastic contribution. However, for alloys with little to no plastic deformation to failure, describing failure in terms of strain to failure becomes more challenging. This is shown schematically in Figure 1, which compares the engineering stress-strain curves for the two materials in this study—Ti-6Al-4V and 316L stainless steel (316L).

In this study, the effect of pores relative to the behavior of dense samples was assessed in two different alloys using well-known fracture models calibrated in both equivalent stress and strain space. Data over a wide range of stress states from previous studies by the authors on L-PBF stainless steel 316L [16,17] and L-PBF Ti-6Al-4V [18–20] that included intentionally manufactured, penny-shaped pores of varying diameter were used to calibrate fracture models for each pore size in stress triaxiality versus Lode angle parameter versus equivalent stress space (Haigh-Westergaard space, referred to here as equivalent stress space) and equivalent strain space. By comparing the ductile 316L alloy (>60% engineering strain to failure under uniaxial tension) to the less ductile Ti-6Al-4V (<10% engineering strain to failure under uniaxial tension) in both equivalent stress and strain space as a function of pore size, an assessment of dominant fracture mechanism changes, if any, can be made for each material. Additionally, the appropriateness of fracture models in equivalent stress space versus equivalent plastic strain space for capturing the effect of pore size on the failure behavior of both materials is discussed.

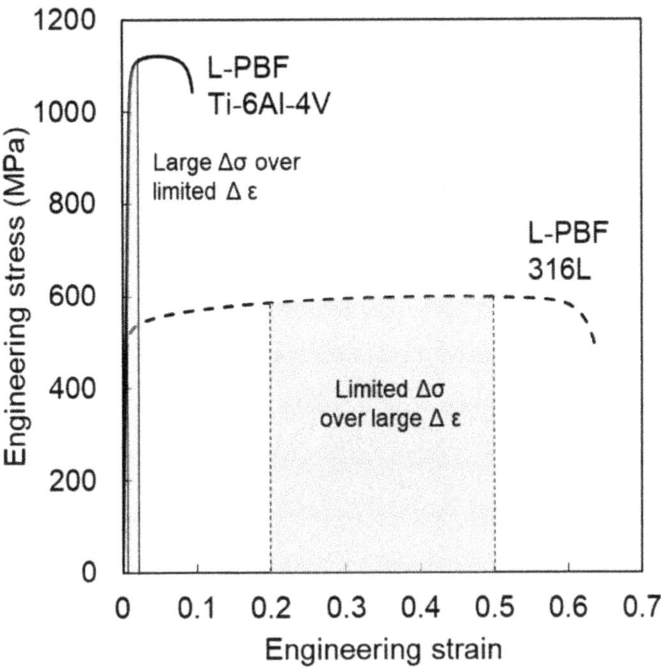

Figure 1. Uniaxial tension engineering stress vs. strain curves for L-PBF Ti-Al-4V and 316L that schematically highlights the differences of resolution in stress and strain for the two alloys.

2. Materials and Methods

The complete description of the manufacturing process, samples, experimental set-up, and simulations are described in extensive detail for the 316L builds in Refs. [16,17] and for the Ti-6Al-4V builds in Refs. [18,20]. The following is an overview of information that is most pertinent to the current study.

2.1. Experimental Methods

Samples used in this study for both alloys were manufactured on a 3DSystems, Inc. ProX 320 L-PBF AM machine (3DSystems, Rock Hill, SC, USA). No post-processing heat treatment was used for the 316L builds, while a post-processing stress relief heat treatment of 650 °C for 3 h in an argon environment was used for the Ti-6Al-4V builds. Data obtained using five different sample geometries, selected as they have a pre-determined failure initiation location, and corresponding to five unique stress states, were used in the current study: pure shear, equibiaxial tension (punch tests), and round notched tension with three different notch radii (3 mm (R3), 5 mm (R5), and 12 mm (R12)), as shown in Figure 2. The three notched tension geometries each had a minimum cross-sectional diameter of 6 mm. The gauge regions of all samples were fabricated using computer numeric control (CNC) machining (Lynx 220L, Doosan Machine Tools Co., Ltd., Pine Brook, NJ, USA).

The five sample geometries for both materials were evaluated in the dense condition, that is using process parameters optimized to build each material. Only the R3, R5, and R12 tests selected in the current study had single, penny shaped pores of varying diameters that were directly designed into the CAD files for the samples, and therefore, included at the center of each sample during the AM fabrication. The pore diameters in µm (and their percentage of the cross-sectional area of the notched tension samples) evaluated in this study were: 300 (0.3%), 600 (1%), 1200 (4%), and 2400 (16%). The four pore sizes were chosen in the current analysis because they interrogated a wide range of percentage of the cross-sectional area that led to varied mechanical response relative to dense samples; additionally, all four pore sizes were studied in each alloy, allowing for a direct comparison. In the 316L material, each of the pores were designed to be 180 μm or 3 layers tall cylinders with the varying diameters, while in the Ti-6Al-4V samples, the internal pores were designed with a conical feature on the top surface after 180 μm vertical walls to prevent pore closure due to dross formation in this alloy system. As discussed in Ref. [20], a study on the effect of pore height performed for Ti-6Al-4V indicated that the pore height did not impact the measured strength or ductility in uniaxial tension specimens.

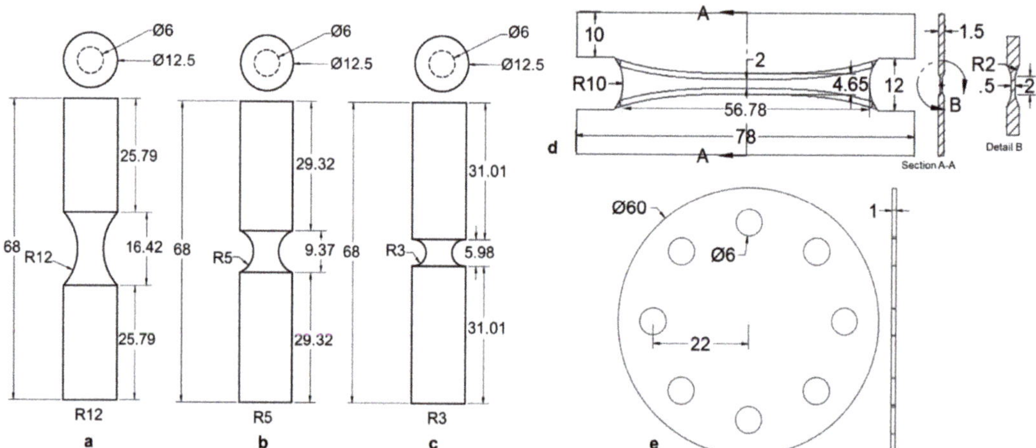

Figure 2. Drawings of mechanical test specimens used to calibrate fracture criterion. (**a–c**) The cylindrical notched tension geometries that were tested in the dense condition and with single, penny-shaped pores of varying diameter at the center. (**d**) The butterfly test geometry used to evaluate material properties under pure shear and (**e**) the punch test geometry used to evaluate equibiaxial tension. Both (**d,e**) were only tested in the dense condition. All dimensions in mm.

For all tests at least two repetitions were completed for each combination of stress state and pore size, including the dense samples. Tests were performed to failure on their respective load frames. Displacement and strain, using a virtual extensometer, were measured using digital image correlation (DIC), a non-contact surface strain field measurement technique. For each test, two different definitions of failure were examined, displacement to maximum force and displacement to material separation, and were used to inform the simulation data in Section 2.1 and for calibration of the equivalent stress and strain fracture models, respectively, in Section 2.3.

2.2. Finite Element Analysis Simulations

Calibrated plasticity models for 316L [17] and Ti-6Al-4V [19] were developed previously by the authors and used in simulations of each dense geometry using the finite element method in the commercial software Abaqus [21]; the complete model details for each fracture geometry are provided in Refs. [16–18,20]. Finite element analysis (FEA) simulations of each dense geometry were used to probe, as a function of applied displacement, the components of the Cauchy stress tensor (σ), stress state (η and $\bar{\theta}$), and von Mises equivalent stress ($\bar{\sigma}_{VM}$) and equivalent plastic strain ($\bar{\varepsilon}$) in the centermost element of each sample geometry, which is where failure is assumed to initiate. Simulations were performed to mimic the experimental displacement to catastrophic failure of the dense specimens, as measured via the virtual extensometer in experiments and an equivalent extensometer in each of the FEA models. The average stress triaxiality and Lode angle parameter up to the experimentally measured displacement to failure for each test condition (i.e., pore size and sample geometry) were calculated using:

$$\eta_{av} = \frac{1}{\bar{\varepsilon}_f^p} \int_0^{\bar{\varepsilon}_f^p} \eta \, d\bar{\varepsilon} \tag{3}$$

and:

$$\bar{\theta}_{av} = \frac{1}{\bar{\varepsilon}_f^p} \int_0^{\bar{\varepsilon}_f^p} \bar{\theta} \, d\bar{\varepsilon}. \tag{4}$$

The equivalent stress from the FEA model was recorded at the displacement corresponding to the average experimental displacement to maximum force for every test condition. The equivalent plastic strain from the FEA model was recorded at the displacement corresponding to the average experimental displacement to catastrophic failure, or material separation, for every test condition.

To summarize, for both materials the data sets were analyzed in two different 3D spaces (η_{av}, $\bar{\theta}_{av}$, $\bar{\sigma}_{VM}$), or equivalent stress space, and ($\eta(\bar{\varepsilon})$, $\bar{\theta}(\bar{\varepsilon})$, $\bar{\varepsilon}_f$), or equivalent strain space. For each 3D space there were five data points for the dense condition (pure shear, punch, R3, R5, and R12) and three additional data points for each pore size (R3, R5, and R12). These data will be referred to in subsequent sections in the calibration of the fracture models in equivalent stress and strain spaces.

2.3. Fracture Models

In this study, three fracture models were calibrated in the 3D equivalent stress space of (η_{av}, $\bar{\theta}_{av}$, $\bar{\sigma}_{VM}$) and two fracture models in the 3D equivalent strain space of ($\eta(\bar{\varepsilon})$, $\bar{\theta}(\bar{\varepsilon})$, $\bar{\varepsilon}_f$), which are transformed versions of two of the stress-based models using assumed plasticity frameworks as described in Refs. [22,23]. Each model was calibrated for the dense material and each pore size. For the fracture surface calibrations for dense material, the dense pure shear, punch, R3, R5, and R12 data were used. For each subsequent pore size, five tests were also used in calibration: the dense pure shear and punch data were used as "anchor points" as the pore configurations being considered in this study (with the cylinder axis oriented parallel to the vertical build direction) are assumed to have relatively negligible impact on ductility in those stress states, and the three notched tension data for the corresponding pore size.

2.3.1. Equivalent Stress versus Stress Triaxiality and Lode Angle Space
Maximum Stress Failure Criterion

The maximum principal stress fracture criterion is based on the premise that a crack or defect will grow in a direction perpendicular to the maximum principal stress, resulting in failure when the maximum principal stress reaches a critical value. This model is given as:

$$\sigma_c = \sigma_{max} = \max(\sigma_1, \sigma_2, \sigma_3) \tag{5}$$

where σ_1, σ_2, and σ_3 are the principal stress components, and σ_c is the critical maximum principal stress resulting in fracture. For each notched tension geometry, the average σ_c values from the three notched tensions tests for the dense data, as well as each pore size, were calculated and are given in Tables 1 and 2 for both materials. To plot the maximum stress fracture loci for dense material and each pore size, von Mises equivalent stress, η, and $\bar{\theta}$ were calculated and plotted as a function of the three principal stress components with σ_c held constant. The discrete points were then interpolated to generate a fracture locus using only a single input, σ_c.

Table 1. Calibrated stress-based fracture model parameters for 316L using data from dense pure shear and equibiaxial tension tests combined with dense, 300 μm, 600 μm, 1200 μm, and 2400 μm pore round notched tension tests. Error is the mean absolute percentage error for the model predicted values of equivalent stress on the fracture surface compared to the experimental values of equivalent stress across each stress state used in calibration.

L-PBF 316L		Pore Diameter (μm)				
		Dense	300	600	1200	2400
Max Stress	σ_{max}	1135	1151	1128	1073	1017
Mohr-Coulomb	c_1	0.516	0.507	0.468	0.383	0.337
	c_2	800	788	736	624	564
	Error	5.2%	5.0%	5.6%	6.7%	9.7%
Hosford-Coulomb	a	1.11	1.32	1.42	1.44	1.50
	b	1519	1452	1431	1427	1414
	c	0.422	0.392	0.414	0.451	0.500
	Error	4.9%	4.6%	5.0%	6.3%	8.7%

Table 2. Calibrated stress-based fracture model parameters for Ti-6Al-4V using data from dense pure shear and equibiaxial tension tests combined with dense, 300 μm, 600 μm, 1200 μm, and 2400 μm pore round notched tension tests. Error is the mean absolute percentage error for the model predicted values of equivalent stress on the fracture surface compared to the experimental values of equivalent stress across each stress state used in calibration.

L-PBF Ti-6Al-4V		Pore Diameter (μm)				
		Dense	300	600	1200	2400
Max Stress	σ_{max}	1817	1774.33	1647	1507	1128
Mohr-Coulomb	c_1	0.959	0.956	0.760	0.675	0.521
	c_2	1456	1429	1150	1030	814
	Error	13.4%	9.9%	7.5%	4.7%	9.6%
Hosford-Coulomb	a	1.23	0.944	1.06	0.862	0.438
	b	1376	1498	1439	1557	2428
	c	0.055	0.133	0.133	0.230	1.20
	Error	1.2%	1.2%	1.7%	2.3%	9.6%

Mohr-Coulomb Failure Criterion

The Mohr-Coulomb (MC) fracture criterion is a classical stress-based criterion that has its origins in describing brittle failure (e.g., rocks and ceramics) [24]. The model is phenomenological in that it describes fracture as occurring when a combination of normal stress (the intermediate principal stress is ignored) and shear stress reach a critical value. The model in the 3D space of $(\eta, \bar{\theta}, \sigma_{VM})$ was initially presented by Bai et al. [22] and is given as:

$$\bar{\sigma}_{vM}^{MC}[\eta,\bar{\theta}] = c_2 \left[\sqrt{\frac{1+c_1^2}{3}} cos\left(\frac{\pi}{6}-\bar{\theta}\right) + c_1 \left(\eta + \frac{1}{3}sin\left(\frac{\pi}{6}-\bar{\theta}\right)\right) \right]^{-1} \tag{6}$$

where c_1 (friction coefficient) and c_2 (shear resistance) are model parameters that are calibrated. The ranges of the model parameters are $c_1 \geq 0$ and $c_2 > 0$. The optimized model parameters for the dense material and the different pores sizes were determined using a Matlab function (fmincon) (R2021a, 2021, Mathworks, Natick, MA, USA) that finds the minimum of a constrained nonlinear multivariable function; in the current study, this function probed fracture model parameters in a defined range and calculated the mean absolute percentage error (MAPE) between the predicted equivalent stress on the fracture surface for each evaluated parameter set and the fixed combined experimental/simulation scatter data at the same $(\eta, \bar{\theta})$ coordinates. The optimized parameter sets for each pore size are shown in Tables 1 and 2.

Hosford-Coulomb Failure Criterion

The Hosford-Coulomb (HC) model is a stress-based failure criterion that describes failure in terms of a material's deviatoric strength, as originally postulated by Coulomb in 1776 [25]. Unlike the MC model, the HC model considers the contribution from the intermediate principal stress by replacing the maximum shear stress contribution in the MC model with the Hosford equivalent stress [26]. The model in the 3D space of $(\eta, \bar{\theta}, \bar{\sigma}_{VM})$ was initially presented by Mohr and Marcadet [14] and is given as:

$$\bar{\sigma}_{vM}^{HC}\left[\eta, \bar{\theta}\right] = \frac{b}{\left\{\frac{1}{2}\left[(f_1 - f_2)^a + (f_2 - f_3)^a + (f_1 - f_3)^a\right]\right\}^{\frac{1}{a}} + c(2\eta + f_1 + f_3)} \tag{7}$$

with:

$$f_1 = \frac{2}{3}\cos\left[\frac{\pi}{6}(1 - \bar{\theta})\right] f_2 = \frac{2}{3}\cos\left[\frac{\pi}{6}(3 + \bar{\theta})\right] f_3 = -\frac{2}{3}\cos\left[\frac{\pi}{6}(1 + \bar{\theta})\right] \tag{8}$$

where model parameters a (Hosford exponent—controls the Lode angle parameter dependence), b (controls the height of the fracture surface), and c (controls the stress triaxiality dependence) were calibrated for each pore size in the current study. The Matlab function (fmincon) used in the calibration of the MC model parameters was adopted for the HC optimization using the same experimental/simulation data for each pore size. The optimized parameter sets for each pore size are shown in Tables 1 and 2.

2.3.2. Equivalent Plastic Strain versus Stress Triaxiality and Lode Angle Space

Taking into account the higher resolution in strain to failure than stress to failure generally observed in ductile metals, the stress-based fracture criteria can be transformed to strain space through a transformation based on an appropriate plasticity model framework as described in Refs. [14,22]. For both materials discussed here, isotropic plasticity models were assumed and used in simulations.

The first strain-based fracture model investigated was the modified Mohr-Coulomb (MMC) fracture criterion [22]. This model is based on transforming the MC model from stress space to strain space by assuming a plasticity framework as described in [22]. This results in a definition of strain to failure, under the constraints of proportional loading, in $(\eta_{av}, \bar{\theta}_{av}, \bar{\varepsilon}_f)$ space of:

$$\bar{\varepsilon}_f^{MMC} = \left\{\frac{A}{c_2}\left[c_\theta^s + \frac{\sqrt{3}}{2-\sqrt{3}}\left(c_\theta^{ax} - c_\theta^s\right)\left(\sec\left(\frac{\bar{\theta}\pi}{6}\right) - 1\right)\right]\left[\sqrt{\frac{1+c_1^2}{3}}\cos\left(\frac{\bar{\theta}\pi}{6}\right)\right. \right.$$
$$\left.\left. + c_1\left(\eta + \frac{1}{3}\sin\left(\frac{\bar{\theta}\pi}{6}\right)\right)\right]\right\}^{-\frac{1}{n}} \tag{9}$$

with:

$$c_\theta^{ax} = \begin{cases} 1 & \bar{\theta} \geq 0 \\ c_\theta^c & \bar{\theta} < 0 \end{cases}. \tag{10}$$

The parameters A and n are parameters from the equations used to describe the rate of strain hardening in the plasticity models given in [17,19], and these values were held constant for each pore size, while c_1, c_2, c_θ^s, and c_θ^c were calibrated for each pore size in the current study. The parameters c_1 and c_2 have the same effect as in the stress-based MC model, and c_θ^s and c_θ^c control the Lode angle parameter dependence and asymmetry, respectively, of the calibrated fracture surfaces.

The Hosford-Coulomb fracture criteria in equivalent plastic strain space is a phenomenological fracture model that was developed on the hypothesis that the fracture initiation in a ductile metal coincides with the formation of a primary or secondary band of localization, where the moment of this localization can be described by a critical combination of the Hosford equivalent stress and the normal stress on the plane of maximum shear. Marcadet and Mohr [14] and Gu and Mohr proposed strain-based fracture models based on this critical combination of equivalent and normal stress by transforming the stress-based failure criterion to strain space through assumed plasticity models. The model used here is the one proposed in Gu and Mohr [23], which, under the constraints of proportional loading, is given as:

$$\bar{\varepsilon}_f^{HC}(\sigma/\bar{\sigma}) = b\left(\frac{1+c}{g_{HC}\left(\frac{\sigma}{\bar{\sigma}}\right)}\right)^{\frac{1}{d}} \tag{11}$$

with:

$$g_{HC}\left(\frac{\sigma}{\bar{\sigma}}\right) = \left\{\frac{1}{2}\left[(f_1 - f_2)^a + (f_2 - f_3)^a + (f_1 - f_3)^a\right]\right\}^{\frac{1}{a}} + c(2\eta + f_1 + f_3). \tag{12}$$

In this model, a, b, and c all retain their meaning from the HC stress-based failure criterion and d increases or decreases the curvature of the fracture locus, where a larger value of d results in less curvature and therefore a flatter surface.

Parameters for both models were calibrated using a Matlab code (fmincon) that optimized parameter values such that mean absolute percentage error for the damage indicator (D), calculated as:

$$D = \int_0^{\bar{\varepsilon}_f} \frac{1}{\bar{\varepsilon}_f^{MMC,}} d\bar{\varepsilon} \text{ and } D = \int_0^{\bar{\varepsilon}_f} \frac{1}{\bar{\varepsilon}_f^{HC}} d\bar{\varepsilon}, \tag{13}$$

was minimized. A target value of D = 1, corresponding to material failure, was used in the optimization code for calculating error.

3. Results and Discussion

3.1. Effect of Pores in Equivalent Strain versus Stress Triaxiality and Lode Angle Parameter Space

The results from the models in the equivalent plastic strain space will first be discussed because these models are most often used when describing ductile failure behavior and provided a baseline performance with which the equivalent stress-based models were contrasted. The results of the strain-based fracture model calibration for both the MMC and strain-based HC model are shown in Figure 3 and the resulting model parameters are given in Tables 3 and 4.

Table 3. Calibrated strain-based ductile fracture model parameters for 316L using data from dense pure shear and equibiaxial tension tests combined with dense, 300 μm, 600 μm, 1200 μm, and 2400 μm pore round notched tension Table 1. which represents perfect model agreement with experiments.

L-PBF 316L		Pore Diameter (μm)				
		Dense	300	600	1200	2400
Modified Mohr-Coulomb	c_1	0.724	0.627	0.672	0.804	1.026
	c_2	1665	1292	1256	1225	1190
	c_θ^s	1.99	1.61	1.53	1.41	1.23
	c_θ^c	0.995	0.981	0.917	0.8	0.623
	Error	5.8%	4.3%	1.7%	4.3%	7.3%
Hosford-Coulomb	a	0.551	0.562	0.705	1.04	1.16
	b	1.34	1.34	1.09	0.688	0.553
	c	0.274	0.32	0.249	0.24	0.267
	d	0.473	0.352	0.309	0.377	0.321
	Error	7.1%	3.2%	0.7%	5.5%	17%

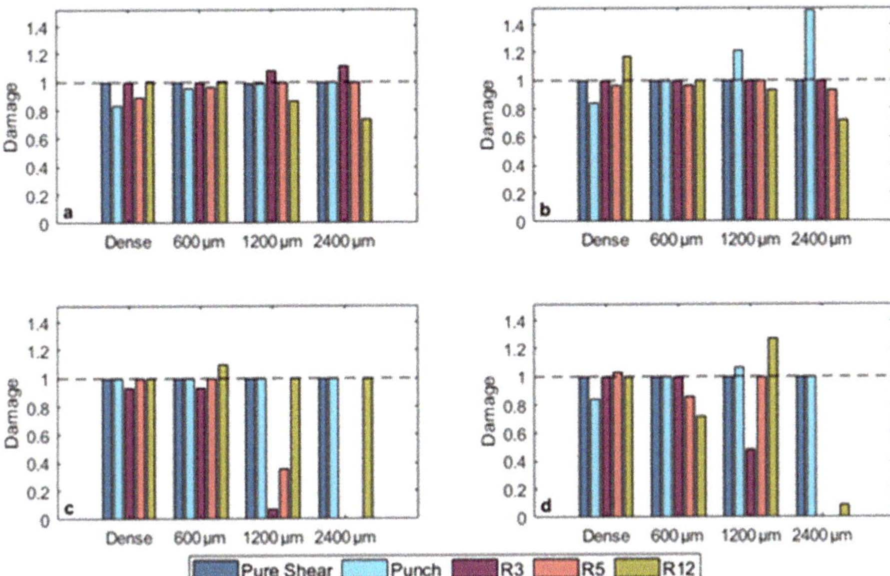

Figure 3. Damage accumulation at fracture for (**a,b**) 316L and (**c,d**) Ti-6Al-4V using the calibrated (**a,c**) modified Mohr-Coulomb and (**b,d**) Hosford-Coulomb models. The dashed line at a value of damage equal to one represents perfect agreement between the model and the experimental/simulation data for each test. Numbers on the horizontal axes denote the pore diameter at the center of the samples.

Table 4. Calibrated strain-based ductile fracture model parameters for Ti-6Al-4V using data from dense pure shear and equibiaxial tension tests combined with dense, 300 μm, 600 μm, 1200 μm, and 2400 μm pore round notched tension tests. Error is mean absolute percentage error for the model-calibrated damage accumulation in all five tests compared to a damage value of 1, which represents perfect model agreement with experiments.

L-PBF Ti-6Al-4V		Pore Diameter (μm)				
		Dense	**300**	**600**	**1200**	**2400**
	c_1	0.069	0.109	0.139	0.156	0.219
Modified	c_2	694	697	697	697	690
Mohr-	c_θ^s	0.981	0.982	0.978	0.976	0.954
Coulomb	c_θ^c	1.037	0.995	0.961	0.944	0.87
	Error	1.3%	11%	3.3%	32%	40%
	a	0.46	0.638	1.22	1.33	1.33
Hosford-	b	0.451	0.478	0.193	0.236	0.235
Coulomb	c	0.415	0.296	0.075	0.058	0.057
	d	0.154	0.059	0.018	0.008	0.008
	Error	3.7%	1.3%	8.4%	17%	58%

3.1.1. L-PBF 316L

The damage prediction shows that for 316L, both the MMC and strain-based HC models do a relatively good job at accurately capturing the failure behavior of the dense samples and the samples with 600 μm and 1200 μm diameter intentional pores. The mean absolute percentage error across all test conditions was 4.7% for the MMC model and 6.7% for the HC model. However, for the HC model the largest error was for the 2400 μm pore diameter (16% of the cross-sectional area), where the model did not accurately capture the punch, R5, and R12 behavior simultaneously, resulting in a MAPE of 17%. The low error in the 316L calibrated parameters resulted in good fitting of the fracture surfaces relative to the experimental/simulated scatter data, as shown in Figure 4.

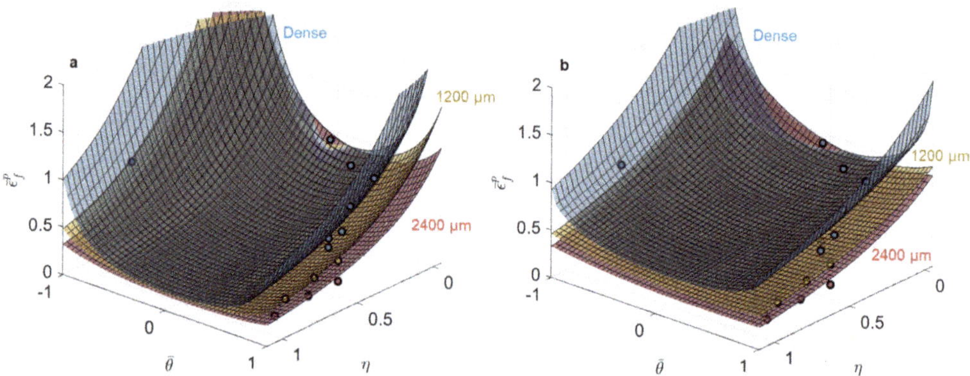

Figure 4. Calibrated (**a**) modified Mohr–Coulomb and (**b**) Hosford–Coulomb fracture loci in stress triaxiality vs. Lode angle parameter vs. equivalent plastic strain space for 316L using the dense, 1200 μm, and 2400 μm pore data. Note that the horizontal positions of each symbol correspond to the average stress triaxiality and Lode angle parameter during loading, while the vertical position corresponds to the experimental strain to failure of the average stress state under the assumption of proportional loading; therefore, the symbols are not expected to lie on the fracture surfaces, which took into account loading history in the accumulation of damage.

With increased pore size in the 316L, there was a noticeable flattening behavior of the surfaces using both the MMC and strain-based HC models; in other words, failure behavior became less stress state dependent as a function of increased pore size. The strain-based models captured the fracture behavior in 316L with the inclusion of pores well because even with the introduction of the 2400 μm diameter pore (16% of the cross-sectional area), the equivalent plastic strain to failure was >20% at failure, which significantly exceeds elastic deformation; thus, these data still lie in the region of high strain resolution. Overall, the MMC model did a better job capturing the fracture behavior as a function of pore size using the five tests in this study for calibration of the model parameters.

3.1.2. L-PBF Ti-6Al-4V

The calibrated model parameters for Ti-6Al-4V had relatively low average error in the damage prediction for the dense, 300 μm, and 600 μm diameter pore experimental/simulation test data using both the calibrated MMC (5.2% MAPE) and the strain-based HC (4.5% MAPE) models. However, the ability of the models to accurately describe the evolution of damage to material failure drastically declined for the samples with the 1200 μm diameter (4% of the cross-sectional area) and 2400 μm diameter (16% of the cross-sectional area) pores. These large errors are primarily driven by errors in fitting the notched tension tests, where the inclusion of the large diameter pores results in failure at little to no plastic strain [20].

Plotting the fracture surfaces for 1200 μm diameter (4% of the cross-sectional area) and 2400 μm diameter (16% of the cross-sectional area) pores, as shown in Figure 5 for both models, resulted in surfaces that were hard to distinguish from each other and did not perfectly capture the data. The breakdown of the models' ability to capture the data accurately in the equivalent plastic strain space for the test conditions with the largest pores was due to the fact that, with increased pore size and reduced displacement to failure, the elastic strain contribution became non-negligible or even dominant, compared to the plastic contribution, for failure in the Ti-6Al-4V.

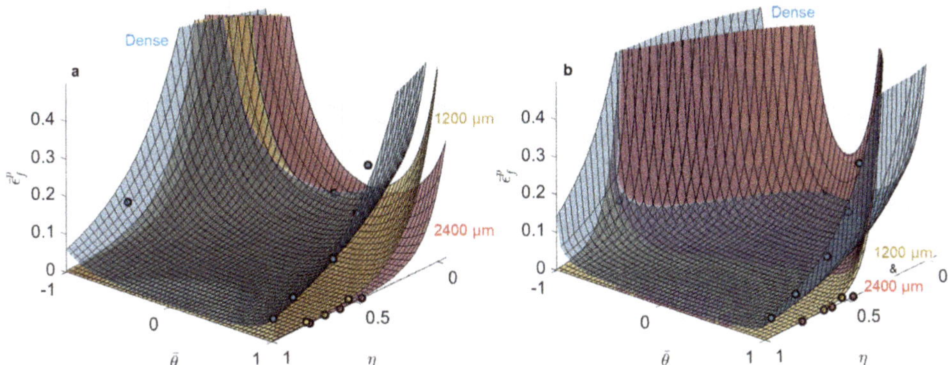

Figure 5. Calibrated (**a**) modified Mohr–Coulomb and (**b**) Hosford–Coulomb fracture loci in stress triaxiality vs. Lode angle parameter vs. equivalent plastic strain space for Ti-6Al-4V using dense, 1200 μm, and 2400 μm pore data. Note that the horizontal positions of each symbol correspond to the average stress triaxiality and Lode angle parameter during loading, while the vertical position corresponds to the experimental strain to failure of the average stress state under the assumption of proportional loading; therefore the symbols are not expected to lie on the fracture surfaces, which took into account loading history in the accumulation of damage.

3.2. Effect of Pores in Equivalent Stress versus Stress Triaxiality and Lode Angle Parameter Space

3.2.1. L-PBF 316L

The equivalent strain-based models captured the change in fracture behavior with increasing pore diameters well, and based on the limited strain hardening in the 316L as shown in Figure 1, the ability to resolve the change in equivalent failure stress, with increasing pore size, should be restricted. Both the MC and HC stress-based models captured the experimental/simulated data well with all calculated MAPE below 10%.

The fracture surface shapes remained similar with increasing pore size, but the magnitude of equivalent stress across the entire surface was reduced as pore size increased. The MC model exhibits the most change in shape with increased pore size, where the calibrated models for the samples with 1200 μm (4% of the cross-sectional area) and 2400 μm (16% of the cross-sectional area) diameter pores resulted in a flattening of the surface along the edge where Lode angle parameter equals 1, which is where the three notched tension tests lie, as shown in Figure 6b. The maximum stress model captured the notched tension tests well, but all surfaces predict much higher stress to failure than observed for the pure shear test. However, it should be noted that experimentally measured failure under shear should be taken as a lower bound (see, e.g., Ref. [27]). In general, the models capture the data well in equivalent stress spaces for the 316L, however the limited loss in strength with increased pore size makes it challenging to differentiate the calibrated fracture surfaces from one another compared to the equivalent strain space.

3.2.2. L-PBF Ti-6Al-4V

As the stress-based fracture models investigated here were developed to describe fracture behavior of brittle materials, it follows that a stress space criterion would better capture the detriment to mechanical behavior, due to the introduction of large pores, in already limited-ductility Ti-6Al-4V, better than a strain space representation based on plastic deformation. For Ti-6Al-4V, there were larger differences in the two models' abilities to capture failure behavior with increased pore size; the MC model average MAPE was 9%, and the model had the most difficulty capturing the dense behavior (13.4% MAPE), but for the HC model the average MAPE was only 3%. The maximum stress model was able to capture the drop in equivalent stress to failure for samples with increased pore size; however, it did a poor job capturing the pure shear and punch tests, as shown in Figure 7a.

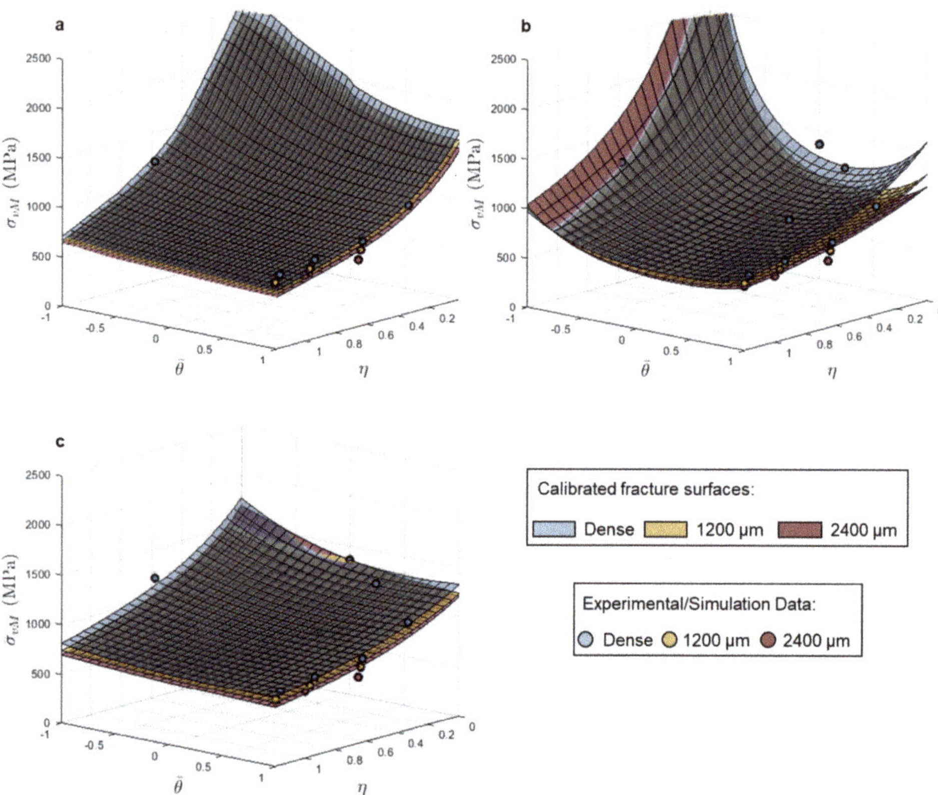

Figure 6. Calibrated (**a**) maximum stress, (**b**) Mohr–Coulomb and (**c**) Hosford–Coulomb fracture loci in stress triaxiality vs. Lode angle parameter vs. equivalent stress space for 316L using dense, 1200 μm, and 2400 μm pore data.

Between the MC and HC models, the most evident change in calibrated fracture surface shape in stress space was between the surfaces for samples with 1200 μm diameter (4% of the cross-sectional area) and 2400 μm diameter (16% of the cross-sectional area) pores, as shown in Figure 7c. The calibrated HC surfaces showed a clear change from relatively stress state independent failure (flat surface) to more stress state dependent failure in the equivalent stress space for samples with the 2400 μm diameter (16% of the cross-sectional area) pore. For samples with a pore diameter of 1200 μm (4% of the cross-sectional area), the contributions of the elastic and plastic strain components were similar in magnitude; at failure the equivalent plastic strain was less than 1.5% for all three notched tension geometries. All samples with a pore diameter of 2400 μm (16% of the cross-sectional area), failed in an elastic deformation-dominated regime, for which there was significantly greater resolution in stress compared to plastic strain. Therefore, the brittle fracture derived models, based on strength limits, better captured the fracture behavior with the 1200 μm (4% of the cross-sectional area) and the 2400 μm (16% of the cross-sectional area) diameter pores.

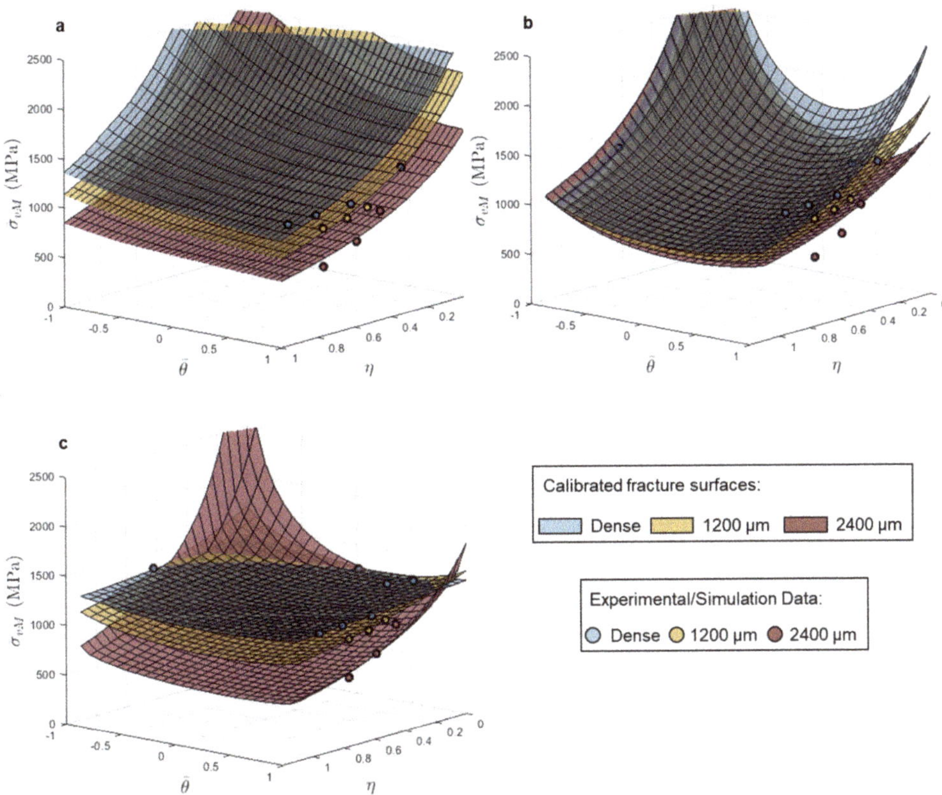

Figure 7. Calibrated (**a**) maximum stress, (**b**) Mohr–Coulomb, and (**c**) Hosford–Coulomb fracture loci in stress triaxiality vs. Lode angle parameter vs. equivalent stress space for Ti-6Al-4V using dense, 1200 μm pore, and 2400 μm pore data.

4. Conclusions

This study takes advantage of the layer-by-layer manufacturing capabilities of AM to study the impact of controlled internal pores in two different metal alloys on stress-state and flaw-size dependent failure behavior. Fracture models presented in both equivalent stress (versus stress state) and strain (versus stress state) space were calibrated and their efficacy for describing the experimental/simulation results were discussed. The primary conclusions of this study are:

- L-PBF 316L and Ti-6Al-4V were shown to have drastically different stress state dependent fracture behavior in the dense condition, and these differences were exacerbated with the introduction of internal pores. Ultimately, the fracture behavior of relatively high ductility, and therefore defect tolerant, 316L was better captured by ductile fracture models based on an accumulation of damage with plastic deformation due to the significant plastic deformation to fracture observed in all samples, including those with pores. Conversely, the fracture behavior of relatively low ductility, and defect intolerant, Ti-6Al-4V was better captured by the fracture models derived based on critical strength values due to the limited or negligible plastic deformation preceding failure, particularly in samples with pores.
- For L-PBF 316L, the inclusion of the 1200 μm (4% of the cross-sectional area) and the 2400 μm (16% of the cross-sectional area) diameter pores in samples resulted in calibrated fracture surfaces in equivalent plastic strain space that had reduced stress

state dependent failure, or flatter fracture surfaces, with increased pore size as failure in these samples became dominated by pore size rather than stress triaxiality.

- The effect of pores on the fracture behavior of L-PBF 316L was best captured in equivalent plastic strain space as significant equivalent plastic strain to failure was retained even with the samples that had the largest diameter pores (2400 μm or 16% of the cross-sectional area). Specifically, the modified Mohr-Coulomb model calibrated with pure shear, equibiaxial tension, and three unique round notched tension geometries (with and without intentional penny-shaped pores of varying diameters) most accurately captured the failure behavior of L-PBF 316L.
- For L-PBF Ti-6Al-4V, the use of equivalent stress-based fracture models, initially proposed for brittle materials, to evaluate the effect of internal pores of varying diameter was shown to be most appropriate.
- The equivalent stress-based Hosford-Coulomb failure criterion most accurately captured the failure behavior of L-PBF Ti-6Al-4V samples as a function of pore size. For samples with the largest diameter pores (2400 μm or 16% of the cross-sectional area), the fracture behavior, as visualized with the HC fracture surfaces in stress space, became more stress state dependent compared to the calibrated model for dense Ti-6Al-4V.

Author Contributions: Conceptualization, A.M.B.; methodology, A.E.W.-H. and A.M.B.; software, A.E.W.-H. and A.M.B.; validation A.E.W.-H., E.T.F. and A.M.B.; formal analysis, A.E.W.-H. and A.M.B.; investigation, A.E.W.-H. and E.T.F.; resources, A.M.B.; data curation, A.E.W.-H. and E.T.F.; writing—original draft preparation, A.E.W.-H.; writing—review and editing, A.M.B.; visualization, A.E.W.-H.; supervision, A.M.B.; project administration, A.M.B.; funding acquisition, A.M.B. All authors have read and agreed to the published version of the manuscript.

Funding: The financial support provided by the National Science Foundation through award number CMMI-1652575 is gratefully acknowledged. Any opinions, findings, and conclusions or recommendations expressed in this material are those of the authors and do not necessarily reflect the views of the National Science Foundation.

Institutional Review Board Statement: Not applicable.

Informed Consent Statement: Not applicable.

Data Availability Statement: Data available on request. The data presented in this study are available on request from the corresponding author.

Conflicts of Interest: The authors declare no conflict of interest.

References

1. Tang, M.; Pistorius, P.C.; Beuth, J.L. Prediction of lack-of-fusion porosity for powder bed fusion. *Addit. Manuf.* **2017**, *14*, 39–48. [CrossRef]
2. Coeck, S.; Bisht, M.; Plas, J.; Verbist, F. Prediction of lack of fusion porosity in selective laser melting based on melt pool monitoring data. *Addit. Manuf.* **2019**, *25*, 347–356. [CrossRef]
3. Popov, V.; Grilli, M.; Koptyug, A.; Jaworska, L.; Katz-Demyanetz, A.; Klobčar, D.; Balos, S.; Postolnyi, B.O.; Goel, S. Powder Bed Fusion Additive Manufacturing Using Critical Raw Materials: A Review. *Materials* **2021**, *14*, 909. [CrossRef]
4. McClintock, F.A. A Criterion for Ductile Fracture by the Growth of Holes. *J. Appl. Mech.* **1968**, *35*, 363–371. [CrossRef]
5. Rice, J.R.; Tracey, D.M. On the ductile enlargement of voids in triaxial stress fields. *J. Mech. Phys. Solids* **1969**, *17*, 201–217. [CrossRef]
6. Gurson, A.L. Continuum Theory of Ductile Rupture by Void Nucleation and Growth: Part I—Yield Criteria and Flow Rules for Porous Ductile Media. *J. Eng. Mater. Technol.* **1977**, *99*, 2–15. [CrossRef]
7. Pardoen, T.; Doghri, I.; Delannay, F. Experimental and numerical comparison of void growth models and void coalescence criteria for the prediction of ductile fracture in copper bars. *Acta Mater.* **1998**, *46*, 541–552. [CrossRef]
8. Tvergaard, V.; Hutchinson, J.W. Two mechanisms of ductile fracture: Void by void growth versus multiple void interaction. *Int. J. Solids Struct.* **2002**, *39*, 3581–3597. [CrossRef]
9. Benzerga, A.A.; Leblond, J.-B. Ductile Fracture by Void Growth to Coalescence. *Adv. Appl. Mech.* **2010**, *44*, 169–305. [CrossRef]
10. Snow, Z.; Nassar, A.R.; Reutzel, E.W. Invited Review Article: Review of the formation and impact of flaws in powder bed fusion additive manufacturing. *Addit. Manuf.* **2020**, *36*, 101457. [CrossRef]

11. Ronneberg, T.; Davies, C.M.; Hooper, P.A. Revealing relationships between porosity, microstructure and mechanical properties of laser powder bed fusion 316L stainless steel through heat treatment. *Mater. Des.* **2020**, *189*, 108481. [CrossRef]
12. Needleman, A.; Tvergaard, V. An analysis of ductile rupture in notched bars. *J. Mech. Phys. Solids* **1984**, *32*, 461–490. [CrossRef]
13. Bai, Y.; Wierzbicki, T. A new model of metal plasticity and fracture with pressure and Lode dependence. *Int. J. Plast.* **2008**, *24*, 1071–1096. [CrossRef]
14. Mohr, D.; Marcadet, S.J. Micromechanically-motivated phenomenological Hosford–Coulomb model for predicting ductile fracture initiation at low stress triaxialities. *Int. J. Solids Struct.* **2015**, *67–68*, 40–55. [CrossRef]
15. Budiansky, B.; Hutchinson, J.; Slutsky, S. Void Growth and Collapse in Viscous Solids. In *Mechanics of Solids*; Pergamon Press Ltd.: Oxford, UK, 1982; pp. 13–45. [CrossRef]
16. Wilson-Heid, A.E.; Beese, A.M. Combined effects of porosity and stress state on the failure behavior of laser powder bed fusion stainless steel 316L. *Addit. Manuf.* **2021**, *39*, 101862. [CrossRef]
17. Wilson-Heid, A.E.; Qin, S.; Beese, A.M. Multiaxial plasticity and fracture behavior of stainless steel 316L by laser powder bed fusion: Experiments and computational modeling. *Acta Mater.* **2020**, *199*, 578–592. [CrossRef]
18. Wilson-Heid, A.E.; Beese, A.M. Fracture of laser powder bed fusion additively manufactured Ti–6Al–4V under multiaxial loading: Calibration and comparison of fracture models. *Mater. Sci. Eng. A* **2019**, *761*, 137967. [CrossRef]
19. Wilson-Heid, A.E.; Qin, S.; Beese, A.M. Anisotropic multiaxial plasticity model for laser powder bed fusion additively manufactured Ti–6Al–4V. *Mater. Sci. Eng. A* **2018**, *738*, 90–97. [CrossRef]
20. Furton, E.T.; Wilson-Heid, A.E.; Beese, A.M. Effect of stress triaxiality and penny-shaped pores on tensile properties of laser powder bed fusion Ti-6Al-4V. 2021; in press.
21. Simulia, Abaqus User Manual v2016. 2016. Available online: https://www.3ds.com/products-services/simulia/ (accessed on 1 November 2020).
22. Bai, Y.; Wierzbicki, T. Application of extended Mohr–Coulomb criterion to ductile fracture. *Int. J. Fract.* **2010**, *161*, 1–20. [CrossRef]
23. Gu, G.; Mohr, D. Anisotropic Hosford–Coulomb fracture initiation model: Theory and application. *Eng. Fract. Mech.* **2015**, *147*, 480–497. [CrossRef]
24. Labuz, J.F.; Zang, A. Mohr–Coulomb Failure Criterion. *Rock Mech. Rock Eng.* **2012**, *45*, 975–979. [CrossRef]
25. Coulomb, C. Essai sur une application des règles des maximis et minimis à quelques problèmes de statique relatifs à l'architecture. *Mem. Acad. Roy. Div. Sav.* **1776**, *7*, 343–387.
26. Hosford, W.F. A Generalized Isotropic Yield Criterion. *J. Appl. Mech.* **1972**, *39*, 607–609. [CrossRef]
27. Ghahremaninezhad, A.; Ravi-Chandar, K. Ductile failure behavior of polycrystalline Al 6061-T6 under shear dominant loading. *Int. J. Fract.* **2013**, *180*, 23–39. [CrossRef]

materials

Article

Modelling the Variability and the Anisotropic Behaviour of Crack Growth in SLM Ti-6Al-4V

Rhys Jones [1,2,*], Calvin Rans [3], Athanasios P. Iliopoulos [4], John G. Michopoulos [4], Nam Phan [5] and Daren Peng [1]

1   Centre of Expertise for Structural Mechanics, Department of Mechanical and Aerospace Engineering, Monash University, Clayton, VIC 3800, Australia; daren.peng@monash.edu
2   ARC Industrial Transformation Training Centre on Surface Engineering for Advanced Materials, Faculty of Science, Engineering and Technology, Swinburne University of Technology, John Street, Hawthorn, VIC 3122, Australia
3   Faculty of Aerospace Engineering, Delft University of Technology, Kluyverweg 1, 2629 HS Delft, The Netherlands; C.D.Rans@tudelft.nl
4   Computational Multiphysics Systems Laboratory, Code 6394, Center for Materials Physics and Technology, US Naval Research Laboratory, Washington, DC 20375, USA; athanasios.iliopoulos@nrl.navy.mil (A.P.I.); john.michopoulos@nrl.navy.mil (J.G.M.)
5   Structures Division, Naval Air Systems Command, Patuxent River, MD 20670, USA; nam.phan@navy.mil
*   Correspondence: rhys.jones@monash.edu

Abstract: The United States Air Force (USAF) Guidelines for the Durability and Damage Tolerance (DADT) certification of Additive Manufactured (AM) parts states that the most difficult challenge for the certification of an AM part is to establish an accurate prediction of its DADT. How to address this challenge is the focus of the present paper. To this end this paper examines the variability in crack growth in tests on additively manufactured (AM) Ti-6Al-4V specimens built using selective layer melting (SLM). One series of tests analysed involves thirty single edge notch tension specimens with five build orientations and two different post heat treatments. The other test program analysed involved ASTM standard single edge notch specimens with three different build directions. The results of this study highlight the ability of the Hartman–Schijve crack growth equation to capture the variability and the anisotropic behaviour of crack growth in SLM Ti-6Al-4V. It is thus shown that, despite the large variability in crack growth, the intrinsic crack growth equation remains unchanged and that the variability and the anisotropic nature of crack growth in this test program is captured by allowing for changes in both the fatigue threshold and the cyclic fracture toughness.

Keywords: additive manufacture; SLM Ti-6Al-4V; variability; anisotropy; fatigue crack growth

Citation: Jones, R.; Rans, C.; Iliopoulos, A.P.; Michopoulos, J.G.; Phan, N.; Peng, D. Modelling the Variability and the Anisotropic Behaviour of Crack Growth in SLM Ti-6Al-4V. *Materials* **2021**, *14*, 1400. https://doi.org/10.3390/ma14061400

Academic Editor: Tuhin Mukherjee

Received: 16 February 2021
Accepted: 11 March 2021
Published: 13 March 2021

Publisher's Note: MDPI stays neutral with regard to jurisdictional claims in published maps and institutional affiliations.

1. Introduction

The regulatory requirements associated with additively manufactured (AM) parts for both civil and military aircraft are summarised in [1–3]. As noted in [3,4], and in the United Staes Air Force (USAF) airworthiness certification standard MIL-STD-1530D [5], it is essential that the variability in the crack growth rates be understood. This requirement is also highlighted in USAF Structures Bulletin EZ-19-01 [4], which specifically addresses the USAF guidelines for the durability and damage tolerance (DADT) certification of additively manufactured aircraft structural parts. Indeed, the ability to accurately assess the variability in crack growth is particularly important when performing the risk of failure analysis mandated in the US Joint Services Structural Guideline JSSG2006 [6]. As explained in Section 5.3 of MIL-STD-1530D [5] analysis is central to airworthiness certification, and the purpose of experimental tests is "to validate or correct analysis methods and results, and to demonstrate that requirements are achieved".

The study by Virkler and Hillberry [7] is acknowledged to be the first paper to highlight the variability that can arise in the measured fatigue (long) crack growth rates

(da/dN, where a is the crack length and N is the number of cycles) in conventionally manufactured metals. Whilst the paper by Iliopoulos et al. [8] highlighted the extensive variability that is associated with long cracks in Ti-6Al-4V built using a number of different AM processes, the paper by Rans et al. [9], which presented the da/dN versus ΔK curves associated with thirty Ti-6Al-4V single edge notch tension (SENT) specimens built using selective laser melt (SLM), is arguably the first to present a similar in depth study to that of Virkler where attention was focused on a single AM process. Unfortunately, as explained in [8], the expression used in [9] to determine the range in the stress intensity factor in a load cycle ($\Delta K = K_{\max} - K_{\min}$, where $K_{\max}$ and $K_{\min}$, are the maximum and minimum values of the stress intensity factor in a cycle) was inaccurate. Whilst correct expression for ΔK was given in [8] the corresponding da/dN versus ΔK were not. Consequently, one of the primary purposes of this paper is to present the accurate curves associated with these thirty SLM Ti-6Al-4V single edge notch tension (SENT) specimens and thereby highlight the extent of the variability associated with crack growth in SLM Ti-6Al-4V.

The paper by Molent and Jones [10] was the first to reveal that the variability in the da/dN versus ΔK curves given in [7] could be captured by allowing for the variability in the fatigue threshold term ΔK_{thr} in Equation (2) the Hartman–Schijve equation [11], viz.

$$da/dN = D(\Delta\kappa)^p \tag{1}$$

The terms D and p are constants, and $\Delta\kappa$ is the crack driving force as given by Schwalbe in [12], viz.

$$\Delta\kappa = (\Delta K - \Delta K_{thr})/(1 - (K_{\max}/A))^{1/2} \tag{2}$$

where the term A is the cyclic fracture toughness. It has subsequently been shown [8,13–18] that the variability in crack growth in AM materials can often be accounted for by allowing for the variability in the fatigue threshold term ΔK_{thr} and the cyclic fracture toughness term (A).

To address the main issue of accurately predicting the DADT, this paper also focuses in evaluating if this formulation can also be used to account for the variability in the da/dN versus ΔK curves presented by Ran's et al. in [9]. The outcome of this initial study is that when da/dN is plotted as a function of $\Delta\kappa$ then each of these thirty curves essentially collapse onto the same master curve obtained for the growth of both long and small cracks in conventionally manufacture Ti-6Al-4V. It should be stressed that, this seminal finding represents the first time that any fracture mechanics-based study has been shown to be able to capture the underlying response in such a large cross section of tests on AM specimens built using a single AM facility. This example is particularly important given that MIL-STD-1530D mandates the use of fracture mechanics-based analyses in the certification process and that USAF Structures Bulletin EZ-19-01 states that the most difficult challenge for AM structural is to establish an "accurate prediction of structural performance" specific to DADT.

This study is complemented by a subsequent investigation into the ability of Equations (1) and (2) to capture the anisotropic behaviour of crack growth in ASTM compact tension (CT) SLM Ti-6Al-4V specimens. As such the studies presented in this paper illustrate how to address the challenge delineated in Structures Bulletin EZ-19-01, namely how to allow for the variability seen in crack growth in AM parts.

2. Materials and Methods

The data analysed in the present paper are taken either journals that are both peer reviewed and publicly available, refereed Conferences and texts that are publicly available (ISBN numbers are given in the associated reference), or from Google searches. Of these references ten are SCOPUS listed Journals, and five are available on various US government websites. The Book Chapters and Books referenced can all be found listed in SCOPUS, one reference can be found on the FAA website. The keywords used to find the references: Additive manufacturing, durability, damage tolerance, variability, and Hartman–Schijve. The exception to this is [2] which was presented at the Proceedings Indian Structural

Integrity Society, 3rd Structural Integrity Conference and Exhibition (SICE), IIT, Mumbai, India, 11 December 2020 and which is not as yet available online.

3. Modelling the Variability in SLM TI-6AL-4V

Let us first examine the variability in the *da/dN* versus ΔK curves given in [9] for crack growth in SLMTi-6Al-4V. In this study thirty $R = 0.1$ tests were performed on single edge notch tension (SENT). The specimens tested had five different build orientations (0, 30, 45, 60 and, 90 degrees), and two different post heat treatments, namely: (a) specimens annealed at 735 °C and (b) specimens annealed at 735 °C and then subjected to hot isostatic pressing (HIP) for two hours. In [9] the build direction was defined relative to the crack in the SENT specimen. (By this it is meant that a build direction of 90 degrees corresponds to the case when the crack was at nright angle to the build direction.) Details of the processes, and the specimen identifiers are given in Table 1.

Table 1. Notation associated with the specimen tests reported in [9].

Build Angle	Treatment	Descriptor
0°	annealed at 735 °C	00-2
ibid	ibid	00-3
ibid	ibid	00-4
ibid	annealed at 735 °C and then HIPed for	00-6
ibid	2 h at 920 °C and 1000 bar	00-7
ibid	ibid	00-8
30°	annealed at 735 °C	30-2
ibid	ibid	30-3
ibid	ibid	30-4
ibid	annealed at 735 °C and then HIPed for 2 h at 920 °C 1000 bar	30-6
ibid	ibid	30-7
ibid	ibid	30-8
45°	annealed at 735 °C	45-2
ibid	ibid	45-3
ibid	ibid	45-4
ibid	annealed at 735 °C and then HIPed for	45-5
ibid	2 hrs 920 °C 1000 bar	45-6
ibid	ibid	45-8
60°	annealed at 735 °C	60-2
ibid	ibid	60-3
ibid	annealed at 735 °C and then HIPed for	60-6
ibid	2 hrs at 920 °C and 1000 bar	60-7
ibid	ibid	60-8
90°	annealed at 735 °C	90-2
ibid	ibid	90-3
ibid	ibid	90-4
ibid	annealed at 735 °C and then HIPed for 2 h 920 °C 1000 bar	90-5
ibid	ibid	90-6
ibid	ibid	90-8

The variability in the thirty *da/dN* versus ΔK curves is shown in Figure 1. (As noted in [8] the *da/dN* versus ΔK curves given in [9] were incorrect since the expression used to determine ΔK was incorrect. Whilst this error was corrected in [8], only a few selected *da/dN* versus ΔK curves were given.) Figure 2 reveals that the *da/dN* versus ΔK curves are largely bounded above by that of specimen 30-3, and below by specimen 90-8, which is HIPed. For comparison Figure 1 also contains the $R = -1$ *da/dN* versus ΔK curve determined in [17] for the growth of a short surface crack in an AM Ti-6Al-4V cylindrical specimen, that was fabricated using an M290 Laser Beam Powder Bed Fusion (LB-PBF) facility, subjected to constant amplitude loading with a maximum stress of 910 MPa. This curve is labelled LB-PBF1. The cyclic fracture toughness (A) of the material in this test was approximately 85 MPa $\sqrt{m}$, and ΔK_{thr} was approximately 0.1 MPa $\sqrt{m}$, see [17]. Figure 1 reveals that the crack growth rate in specimen 30-3 is similar to the growth rate seen by the surface crack in specimen LB-PBF1.

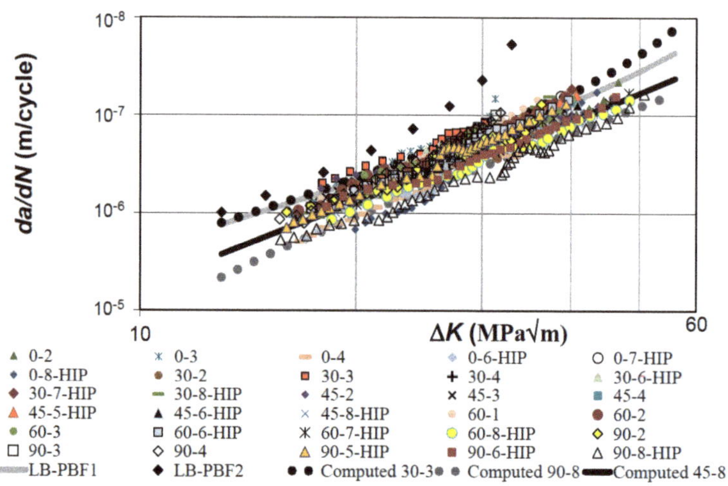

Figure 1. Variability in the thirty SLM Ti-6Al-4V tests reported in [9].

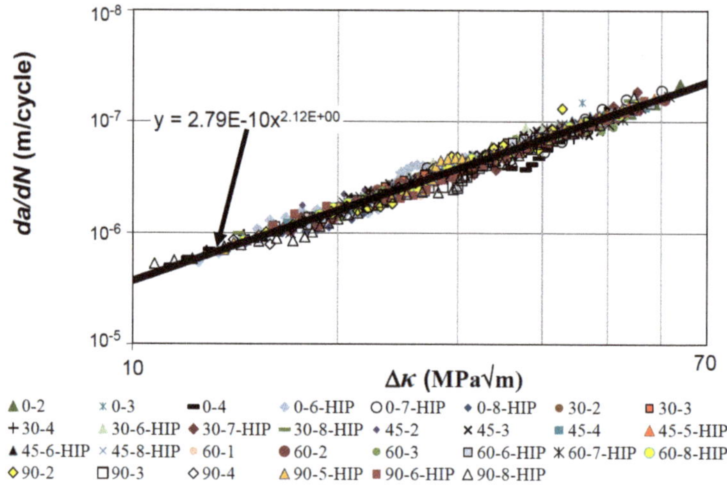

Figure 2. Hartman–Schijve representation of the thirty SLM Ti-6Al-4V tests.

To further illustrate the variability associated with AM Ti-6Al-4V Figure 1 also contains the $R = -1$ da/dN versus ΔK curve determined in [17] for a short surface crack in a specimen built using a Renishaw AM250 LB-PBF machine. In this instance the maximum applied stress was 268 MPa. This curve is labelled LB-PBF2. The cyclic fracture toughness (A) of the material in this test was found to be approximately 37 MPa $\sqrt{m}$ and the fatigue threshold term (ΔK_{thr}) was approximately 0.1 MPa $\sqrt{m}$, see [17]. Here, it should be noted that [17] explained that the difference in the crack growth rates in these two LB-PBF tests was primarily due to the differences in the cyclic fracture toughness's.

Figure 2 reveals that if the curves shown in Figure 1 are plotted with *da/dN* as a function of $\Delta\kappa$, then (allowing experimental error) the scatter in these thirty curves essentially vanishes. Figure 2 also reveals that these thirty tests lie on the same *da/dN* versus $\Delta\kappa$ curve determined for both long and short cracks in conventionally and AM Ti-6Al-4V, viz.

$$da/dN = 2.79 \times 10^{-10} \left[(\Delta K - \Delta K_{\text{thr}})/(1 - K_{\text{max}}/A)^{1/2} \right]^{1.99} \tag{3}$$

The values of ΔK_{thr} and A used in Figure 2, and the corresponding values of the coefficient of determination (R^2) are given in Table 2. (The mean value of the coefficients of determination given in Table 2 is approximately 0.96.) Here, it should be recalled that as shown in [13] the ASTM definition of the fracture toughness (ΔK_{th}), which is arbitrarily chosen to be the value of ΔK at a crack growth rate *da/dN* of 10^{-10} m/cycle [19], is related to ΔK_{thr} is via the expression:

$$\Delta K_{\text{th}} = \Delta K_{\text{thr}} + 0.62 \tag{4}$$

Table 2. Values used in Figure 2.

Build Angle	Descriptor	ΔK_{thr} (MPa $\sqrt{\text{m}}$)	A (MPa $\sqrt{\text{m}}$)	Coefficient of Determination (R^2)
0°	00-2	5.10	88.0	0.97
ibid	00-3	1.50	54.5	0.95
ibid	00-4	7.80	73.0	0.91
ibid	00-6 (HIPed)	5.40	107.0	0.96
ibid	00-7 (HIPed)	4.92	67.0	0.95
ibid	00-8 (HIPed)	8.20	70.0	0.99
30°	30-2	5.90	105.0	0.95
ibid	30-3	0.10	63.5	0.97
ibid	30-4	4.10	73.0	0.97
ibid	30-6 (HIPed)	1.30	85.0	0.91
ibid	30-7 (HIPed)	2.20	73.0	0.93
ibid	30-8 (HIPed)	2.55	65.0	0.97
45°	45-2	0.10	134.0	0.98
ibid	45-3	1.90	73.0	0.94
ibid	45-4	2.70	85.0	0.99
ibid	45-5 (HIPed)	1.50	76.0	0.99
ibid	45-6 (HIPed)	2.40	90.0	0.98
ibid	45-8 (HIPed)	3.10	128.0	0.99
ibid	60-1	3.00	61.0	0.99
ibid	60-2	1.90	74.0	0.99
60°	60-3	0.10	98.3	0.92
ibid	60-6 (HIPed)	3.70	70.0	0.99
ibid	60-7 (HIPed)	3.70	116.0	0.97
ibid	60-8 (HIPed)	5.00	140.0	0.99
ibid	90-2	1.95	93.8	0.85
90°	90-3	5.90	49.7	0.98
ibid	90-4	5.01	51.0	0.99
ibid	90-5 (HIPed)	4.00	80.0	0.98
ibid	90-6 (HIPed)	3.80	123	0.99
ibid	90-8 (HIPed)	6.20	168.0	0.96

To further illustrate how the variability in these tests can be captured using Equation (2) Figure 1 also presents plots of the computed curves for specimens 30-3 and 90-8 as well as for specimen 45-8, which is HIPed and which represents a mid-range (in the context of the present study) crack growth curve. These computed curves were determined using Equation (2) together with the values of A and ΔK_{thr} given in Table 2.

To help quantify the effect of the HIPing process Table 3 presents the mean and standard deviations associated with specimens both with and without HIPing. As in [13,14] we see that in contrast to conventionally manufactured specimens the variability associated with the cyclic fracture toughness (A) is quite large. The mean value of A for the annealed specimens of approximately 79 MPa $\sqrt{m}$ is similar to the mean value of approximately 83 MPa $\sqrt{m}$ associated with the forty different AM Ti-6Al-4V specimens analysed in [8,13]. The standard deviation associated with the SLM specimens is approximately 23 MPa $\sqrt{m}$. This value is lower than the value of 49 MPa $\sqrt{m}$ obtained for the specimens analysed in [8,13]. This is to be expected since the later value covers specimens fabricated using a variety of different AM processes, viz. SLM, Direct Metal Deposition (DMLS), Laser Engineered Net Surface (LENS), etc., and includes specimens left as built, after annealing, and/or after HIPing. When specimens that were either left in the as fabricated state or HIPed are removed from the data being considered then the mean value increases slightly to approximately 89 MPa $\sqrt{m}$ with a standard deviation of approximately 57 MPa $\sqrt{m}$.

Table 3. Mean and standard deviations of ΔK_{thr} and A associated with SLM Ti-6Al-4V specimens with and without HIPing.

	Mean Value	Standard Deviation
Annealed at 735 °C		
A (MPa × $\sqrt{m}$)	78.6	23.0
ΔK_{thr} (MPa × $\sqrt{m}$)	3.2	2.4
Annealed at 735 °C and then HIPed for 2 hrs at 920 °C 1000 bar		
A (MPa × $\sqrt{m}$)	96.6	31.6
ΔK_{thr} (MPa × $\sqrt{m}$)	3.8	1.9

Table 3 reveals that the mean value of A for the SLMS specimens that were both annealed and HIPed is approximately 96.6 MPa $\sqrt{m}$. At first glance this would suggest that HIPing is advantageous. However, the standard deviation associated with these two sets of SLM specimens is quite large, and hence caution is urged with respect to this observation.

Table 3 also reveals that the mean values of the fatigue threshold term ΔK_{thr} associated with the annealed (only) and the annealed and HIPed specimens are 3.2 and 3.8 MPa $\sqrt{m}$, respectively. The corresponding standard deviations are 2.4 and 1.9 MPa $\sqrt{m}$. These values are comparable with the mean values of 3.3 and 3.5 MPa $\sqrt{m}$ associated with all of the forty different AM Ti-6Al-4V specimens analysed in [8,13], and with the value obtained when specimens that either were left in the as fabricated state or HIPed are removed from the data being considered. However, the mean values are misleading in that in several of these tests the value of ΔK_{thr} was significantly lower.

Crack Growth in ASTM Compact Tension SLM Ti-6Al-4V Specimens as Function of Crack Orientation Relative to the Build Direction

Let us next consider the $R = 0.1$ crack growth histories presented in [20] for crack growth in a 10 mm thick ASTM compact tension (CT) specimen with cracks at 0°, 45°, and 90° to the build direction in SLM Ti-6Al-4V. The measured and computed crack growth histories are shown in Figure 3 where we see excellent agreement. The values of ΔK_{thr} and A used in Figure 3 are given in Table 4.

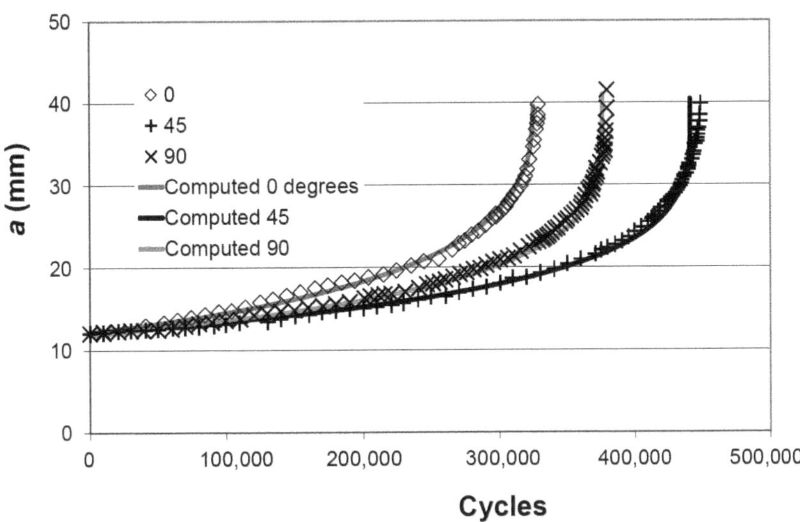

Figure 3. Measured and computed crack growth histories for the SLM Ti-6Al-4V tests reported in [20].

Table 4. Values used in Figure 3.

Build Direction	ΔK_{thr} (MPa × $\sqrt{m}$)	A (MPa $\sqrt{m}$)
0°	2.0	71
45°	3.8	52
90°	3.15	48.5

4. Implications for the Durability Analysis of AM Parts

USAF Structures Bulletin EZ-19-01 [3] explains that durability analysis is essential to the certification of AM parts. Lincoln and Melliere [21], as part of the USAF F-15 program, and [11,14] have shown that a durability analysis necessitates the use of the associated small crack da/dN versus ΔK curve (a similar statement is contained in Appendix X3 of the ASTM fatigue test standard E647-15 [19]). Structures Bulletin EZ-19-01 [3] also requires the use of a minimum equivalent initial damage size (EIDS) of 0.254 mm (0.01 inch). (This value is taken from the Joint Services Structural Guidelines 2006 [3].) Whereas the paper by Virkler and Hillberry [7] is acknowledged to be first to illustrate the variability associated with long cracks in metals, the paper by Kundu et al. [16], which presented the crack growth histories associated with twenty three small surface breaking cracks with length scales of the order of 0.254 mm in AA7050-T7451 aluminium alloy specimens, was (to the best of the authors knowledge) the first to examine the variability in the growth of small surface breaking cracks with sizes comparable to that of the EIDS required in [3]. This study revealed that the variability in the crack growth histories was accurately captured allowing for variability in ΔK_{thr}. The resultant variability in the da/dN versus ΔK curves associated with these twenty-three (small) surface breaking cracks is shown in Figure 4.

Of course, the variability seen in Figure 4 is associated with a limited data set, and as such it may not necessarily capture the extent of the true variability in the material properties. To account for such limited data sets Niu [22] and Rouchon [23] suggest adopting a statistical approach whereby the 'A basis' and 'B basis' properties are determined. An 'A basis' mechanical property value equals the mean value minus three standard deviations and is the value above which at least 99% of the population of values is expected to fall with a confidence of 95% [22]. A 'B basis' mechanical property value equals the mean value minus two standard deviations and is the value above which at least 95% of the population of values is expected to fall with a confidence of 95% [22]. The values of ΔK_{thr} given in [16] for these twenty-three cracks are given in Table 5. The mean value of ΔK_{thr} is approximately 0.80 MPa$\sqrt{m}$ and the standard deviation (σ) is approximately 0.24 MPa$\sqrt{m}$. This yields a Mean- 3σ of approximately 0.1 MPa $\sqrt{m}$. This curve is also shown in Figure 4. Interestingly the Mean- 3σ curve shown in Figure 4 for these size EIDS is close to that given in [24–26] for the growth of "small" cracks from small near micron size surface discontinuities in 7050-T7451. It is also the same as the values determined in [17] for the growth of small surface breaking cracks in LPBF Ti-6Al-4V.

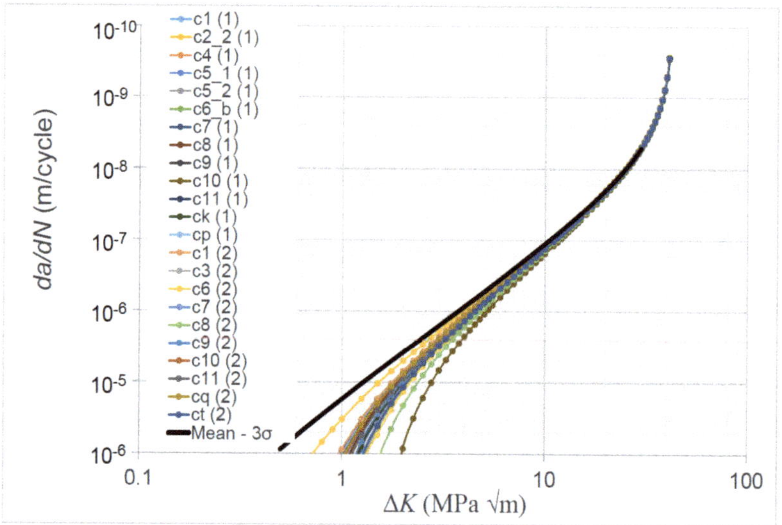

Figure 4. Variability in the crack growth curves for the twenty-three small surface breaking cracks in 7050-T7451 specimens [16]. The labelling convention used for these twenty-three cracks is taken from [16].

Taking the results of this study into the variability of near EIDS size surface breaking cracks into consideration, and noting that the fastest growing long crack in the SLM Ti-6Al-4V tests given in [9] can be approximated by using Equations (1) and (2) together with a low value for the fatigue threshold term ΔK_{thr}, it is hypothesised that this phenomena, i.e., that the worst case curve associated with surface breaking cracks with dimensions as per the minimum allowable EIDS given in [3] for the durability analysis of an AM part would also resemble the corresponding small crack curve, may also hold for AM parts. However, testing is required to evaluate this hypothesis.

Table 5. Values of the term ΔK_{thr} given in [16].

Crack Descriptor	ΔK_{thr} (MPa $\sqrt{m}$)
c1	0.7
c2_2	0.35
c4	0.75
c5_1	0.8
c5_2	0.6
c6_b	0.75
c7	0.85
c8	0.72
c9	0.65
c10	1.6

Table 5. *Cont.*

Crack Descriptor	ΔK_{thr} (MPa $\sqrt{\text{m}}$)
c11	0.61
ck	0.83
cp	0.63
c1	0.6
c3	0.95
c6	1
c7	0.9
c8	1.2
c9	0.8
c10	0.68
c11	0.75
cq	0.66

5. Conclusions

USAF Structures Bulletin EZ-19-01 states that the most difficult challenge for AM structural is to establish an "accurate prediction of structural performance" specific to DADT. It also notes the importance of being able to account for the variability in the crack growth rates associated with AM parts. To meet this challenge the present has examined the variability in the crack growth rates associated with two studies into crack growth in SLM Ti-6Al-4V. One of the studies analysed, involved thirty single edge notch tension specimens with five build orientations and two different post heat treatment methods. The other study involved ASTM standard CT specimens with three different build directions. The results of this analysis highlight the ability of the Hartman–Schijve crack growth equation to capture the variability and the anisotropic behaviour of crack growth in SLM Ti-6Al-4V. This seminal finding represents the first time that any fracture mechanics-based study has been shown to be able to capture the underlying response in such a large cross section of tests on AM specimens built using a single AM facility. This development is central to meeting the certification requirements delineated in MIL-STD-1530D and EZ-19-01.

It is also hypothesised that the worst-case curve associated with surface breaking cracks with dimensions as per the minimum allowable EIDS given in EZ-19-01 for the durability analysis of an AM part should resemble the corresponding small crack curve. However, additional testing is required to evaluate this hypothesis. This finding, once further validated, will have significant implications for the economic life/durability certification of AM parts.

Author Contributions: Conceptualisation and initial analysis—R.J.; Tests on Ti-6Al-4V with different build directions and analysis of the associated data—C.R.; Methodology for determining the constants in the Hartman–Schijve equations—J.G.M.; Analysis of crack growth data to obtain the optimized constants—A.P.I.; Analysis of the test data associated with the ASTM Compact Tension SLM Ti-6Al-4V specimens—D.P.; Overview and evaluation of the final report and its relationship to aircraft certification—N.P. All authors have read and agreed to the published version of the manuscript.

Funding: John Michopoulos and Athanasios Iliopoulos acknowledge support for this work by the Office of Naval Research (ONR) through the Naval Research Laboratory's core funding.

Institutional Review Board Statement: Not applicable.

Informed Consent Statement: Not applicable.

Data Availability Statement: Data sharing is not applicable to this article.

Conflicts of Interest: The authors declare no conflict of interest.

References

1. Gorelik, M. Regulatory Considerations for AM Qualification and Status of FAA Roadmap. In Proceedings of the Additive Manufacture for Reactor Materials & Components: Public Meeting, Bethesda, MD, USA, 28–29 November 2017. Available online: https://www.nrc.gov/docs/ML1733/ML17338A886.pdf (accessed on 2 February 2021).
2. Gorelik, M. Lessons Learned for Structural Alloys and Implications for Metal AM F&DT Considerations. In Proceedings of the Indian Structural Integrity Society, 3rd Structural Integrity Conference and Exhibition (SICE), IIT, Mumbai, India, 11 December 2020.
3. Structures Bulletin EZ-SB-19-01, Durability and Damage Tolerance Certification for Additive Manufacturing of Aircraft Structural Metallic Parts, Wright Patterson Air Force Base, OH, USA. 10 June 2019. Available online: https://daytonaero.com/usaf-structures-bulletins-library/ (accessed on 2 February 2021).
4. Hrabe, N.; Barbosa, N.; Daniewicz, S.; Shamsaei, N. Findings from the NIST/ASTM Workshop on Mechanical Behavior of Additive Manufacturing Components, ASTM-NIST Workshop on Mechanical Behavior of Additive Manufactured Components, NIST Advanced Manufacturing Series 100-4. 2016. Available online: https://doi.org/10.6028/NIST.AMS.100-4 (accessed on 2 February 2021).
5. MIL-STD-1530D, Department of Defense Standard Practice Aircraft Structural Integrity Program (ASIP). 13 October 2016. Available online: http://everyspec.com/MIL-STD/MIL-STD.../download.php?spec=MIL-STD-1530D (accessed on 2 February 2021).
6. Department of Defense Joint Service Specification Guide, Aircraft Structures, JSSG-2006. October 1998. Available online: http://everyspec.com/USAF/USAF-General/JSSG-2006_10206/ (accessed on 10 July 2020).
7. Virkler, D.A.; Hillberry, B.M.; Goel, P.K. *The Statistical Nature of Fatigue Crack Propagation*; Technical Report AFFDL-TR-78-43; Air Force Wright Aeronautical Laboratory: Wright Patterson Ai Force Base, OH, USA, 1978. Available online: https://apps.dtic.mil/dtic/tr/fulltext/u2/a056912.pdf (accessed on 2 February 2021).
8. Iliopoulos, A.P.; Jones, R.; Michopoulos, J.G.; Phan, N.; Rans, C. Further Studies into Crack Growth in Additively Manufactured Materials. *Materials* **2020**, *13*, 2223. [CrossRef] [PubMed]
9. Rans, C.; Michielssen, J.; Walker, M.; Wang, W.; Hoen-Velterop, L. Beyond the orthogonal: On the influence of build orientation on fatigue crack growth in SLM Ti-6Al-4V. *Int. J. Fatigue* **2018**, *116*, 344–354. [CrossRef]
10. Molent, L.; Jones, R. The influence of cyclic stress intensity threshold on fatigue life scatter. *Int. J. Fatigue* **2016**, *82*, 748–756. [CrossRef]
11. Jones, R. Fatigue crack growth and damage tolerance. *Fatigue Fract. Eng. Mater. Struct.* **2014**, *37*, 463–483. [CrossRef]
12. Schwalbe, K.H. On the Beauty of Analytical Models for Fatigue Crack Propagation and Fracture-A Personal Historical Review. *J. ASTM Int.* **2010**, *7*, 3–73. [CrossRef]
13. Jones, R.; Michopoulos, J.G.; Iliopoulos, A.P.; Singh Raman, R.K.; Phan, N.; Nguyen, T. Representing Crack Growth in Additively Manufactured TI-6AL-4V. *Int. J. Fatigue* **2018**, *111*, 610–622. [CrossRef]
14. Jones, R.; Raman, R.S.; Iliopoulos, A.P.; Michopoulos, J.G.; Phan, N.; Peng, D. Additively manufactured Ti-6Al-4V replacement parts for military aircraft. *Int. J. Fatigue* **2019**, *124*, 227–235. [CrossRef]
15. Iliopoulos, A.P.; Jones, R.; Michopoulos, J.G.; Phan, N.; Singh Raman, R.F.K. Crack growth in a range of additively manufactured aerospace structural materials, Special Issue, Civil and Military Airworthiness: Recent Developments and Challenges. *Aerospace* **2019**, *5*, 118. [CrossRef]
16. Kundu, S.; Jones, R.; Peng, D.; Matthews, N.; Alankar, A.; Singh Raman, R.K.; Huang, P. Review of Requirements for the Durability and Damage Tolerance Certification of Additively Manufactured Aircraft Structural Parts and AM Repairs. *Materials* **2020**, *13*, 1341. [CrossRef] [PubMed]
17. Jones, R.; Molaei, R.; Fatemi, A.; Peng, D.; Phan, N. A note on computing the growth of small cracks in AM Ti-6Al-4V. *Procedia Struct. Integr.* **2020**, *111*, 364–369. [CrossRef]
18. Sanaei, N.; Fatemi, A. Defect-based fatigue life prediction of L-PBF additive manufactured metals. *Eng. Fract. Mech.* **2021**, *244*. [CrossRef]
19. ASTM. *Measurement of Fatigue Crack Growth Rates*; ASTM: West Conshohocken, PA, USA, July 2016.
20. Sun, W.; Huang, W.; Zhang, W.; Qian, X. Effects of build direction on tensile and fatigue performance of selective laser melting Ti6Al4V titanium alloy. *Int. J. Fatigue* **2020**, *130*. [CrossRef]
21. Lincoln, J.; Melliere, R.A. Economic Life Determination for a Military Aircraft. *AIAA J. Aircr.* **1999**, *36*, 737–742. [CrossRef]
22. Niu, M.C.Y. *Composite Airframe Structures: Practical Design Information and Data*; Conmilit Press: Hong Kong, China, 1992.
23. Rouchon, J. Fatigue and Damage Tolerance Evaluation of Structures: The Composite Materials Response, National Aerospace Laboratory NLR, NLR-TP-2009-221, Rotterdam, The Netherlands. Available online: https://reports.nlr.nl/handle/10921/224 (accessed on 2 February 2021).
24. Main, B.; Evans, R.; Walker, K.; Yu, X.; Molent, L. Lessons from a Fatigue Prediction Challenge for an Aircraft Wing Shear Tie Post. *Int. J. Fatigue* **2019**, *123*, 53–65. [CrossRef]
25. Jones, R.; Molent, L.; Barter, S. Calculating crack growth from small discontinuities in 7050-T7451 under combat aircraft spectra. *Int. J. Fatigue* **2013**, *55*, 178–182. [CrossRef]
26. Tan, J.L.; Chen, B.K. Prediction of fatigue life in aluminium alloy (AA7050-T7451) structures in the presence of multiple artificial short cracks. *Theor. Appl. Fract. Mech.* **2015**, *78*, 1–7. [CrossRef]

MDPI

Article

Integrating Geometric Data into Topology Optimization via Neural Style Transfer

Praveen S. Vulimiri [1], Hao Deng [1], Florian Dugast [1], Xiaoli Zhang [2] and Albert C. To [1,*]

[1] Department of Mechanical Engineering & Materials Science, University of Pittsburgh, Pittsburgh, PA 15260, USA; praveen.vulimiri@pitt.edu (P.S.V.); had50@pitt.edu (H.D.); fld5@pitt.edu (F.D.)
[2] Department of Mechanical Engineering, Colorado School of Mines, Golden, CO 80401, USA; xlzhang@mines.edu
* Correspondence: albertto@pitt.edu

Abstract: This research proposes a novel topology optimization method using neural style transfer to simultaneously optimize both structural performance for a given loading condition and geometric similarity for a reference design. For the neural style transfer, the convolutional layers of a pre-trained neural network extract and quantify characteristic features from the reference and input designs for optimization. The optimization analysis is evaluated as a single weighted objective function with the ability for the user to control the influence of the neural style transfer with the structural performance. As seen in architecture and consumer-facing products, the visual appeal of a design contributes to its overall value along with mechanical performance metrics. Using this method, a designer allows the tool to find the ideal compromise of these metrics. Three case studies are included to demonstrate the capabilities of this method with various loading conditions and reference designs. The structural performances of the novel designs are within 10% of the baseline without geometric reference, and the designs incorporate features in the given reference such as member size or meshed features. The performance of the proposed optimizer is compared against other optimizers without the geometric similarity constraint.

Keywords: topology optimization; neural network; neural style transfer; additive manufacturing

Citation: Vulimiri, P.S.; Deng, H.; Dugast, F.; Zhang, X; To, A.C. Integrating Geometric Data into Topology Optimization via Neural Style Transfer. *Materials* **2021**, *14*, 4551. https://doi.org/10.3390/ma14164551

Academic Editor: Tuhin Mukherjee

Received: 27 May 2021
Accepted: 3 August 2021
Published: 13 August 2021

Publisher's Note: MDPI stays neutral with regard to jurisdictional claims in published maps and institutional affiliations.

1. Introduction

With recent advances in additive manufacturing, it is now more feasible to fabricate complex designs generated by topology optimization. Topology optimization is a mathematical analysis of a design space, optimizing the material distribution to improve performance for a given metric (i.e., compliance, stress, etc.). Beginning with the work by Bendsoe and Kikuchi [1], the field has grown to include new approaches such as solid isotropic material with penalization (SIMP), level set (LS), and bi-directional evolutionary structural optimization (BESO) [2,3] and various design problems such as static, dynamic, thermo-elastic, and manufacturability [4,5]. These advances all improve the functional use of the optimized design. In fields where aesthetics add to the use of a design, such as architecture or art, the visual features of the design also contribute to the value. Such features have yet to be considered in detail for topology optimization.

The task of modifying the topology optimized design for geometric style then becomes the designer's goal. Depending on the application, the visual appeal of a design also contributes to the overall performance, such as the air intake vents for a vehicle [6]. As a designer iterates between possible solutions, the final design may greatly deviate from the optimized result and decrease performance to satisfy the desired aesthetics. Rather than have a designer post-process the topology optimized result for form, it would be suitable for the analysis to simultaneously optimize for both performance and geometric features.

For example, texture synthesis integrated with topology optimization has been researched previously to perform a structural performance and geometric style optimization simultaneously [7,8]. These works sample regions from a given example texture and the optimized design to compare the appearance energies. The optimizer then minimizes the appearance energy subject to compliance and volume constraints. As these works are based on direct region similarity, there are still issues with applying the desired design concepts to the new design. The input must have patterns of similar size to the search region and have stochastic features. Otherwise, the final design would have many disconnected members not contributing to the performance. A full review of stylized design and fabrication can be found in Bickel et al. [9].

Although Wu et al. does not use by-example texture synthesis methods, the work constrains the amount of material in local regions of the design to achieve trabecular lattice structures [10]. A local mass constraint in these regions rather than using an example input creates these structures [11,12]. The work presents an excellent example of biomimicry for topology optimization. However, it is limited to only this lattice structure.

As an alternative to single design outputs, generative design is a recent design process to produce multiple outputs rather than a single result to satisfy a given condition. Genetic algorithms have been used to produce many different designs [13]. The genetic algorithm tests many samples within the design space and iterates on the highest performing samples, adjusting the variables according to their performance until an optimal sample is found. Autodesk has invested much in generative design research with multi-objective genetic algorithms [14]. However, producing these many designs requires large computational resources to achieve a variety of designs from which a designer should choose the best candidate.

Generative adversarial networks (GAN) have been shown to produce a wide variety of designs [15]. A GAN consists of two neural networks, a generator, and a discriminator, competing with each other. The generator attempts to design new structures similar to a database for a discriminator to discern which are from the database and which are from the generator. Following training, the generator produces structures indistinguishable from those in the database. One such work uses generative design for topology optimization [16]. The examples from the work show a wide variety of designs; however, the training database required hundreds of designs suited for the chosen design problem. This would be infeasible for a generic design tool, as thousands of designs would be needed to cover all design problems. Other examples of machine learning for topology optimization have focused on improving the computational time necessary to complete the analysis through training convolutional neural networks (CNN) [17–20]. These also require large amounts of training data to solve a specific design problem.

In this work, a pre-trained image classifier CNN provides differentiable extraction of geometric features in a design and reference. The geometric features and performance of a design are optimized simultaneously using a weighted objective loss function including neural style transfer [21] of a user-defined reference and topology optimization constraints including compliance, mass, and standard deviation. The neural style transfer uses the convolutional layer activations of a previously trained CNN to quantify the style of input. From the CNN's previous training, the convolutional layers are accurate filters to extract the characteristics of the input. The calculation is performed efficiently on a global scale to quantify the geometric style. Through tuning of the objective weights, the optimized design can have a balance of the optimal structural performance and the desired visual geometric features as determined by the user. Numerical examples are included to demonstrate and validate these methods. Using the current approach, three-dimensional optimization is not implemented, but the methods are similar and will be considered for future work.

The work accomplishes the following:

(a) The geometric result from a topology optimization analysis can be influenced efficiently using a reference design rather than directly copying the reference structures.
(b) The work is an early example using a deep learning model to define objectives and/or constraints for topology optimization, expanding available design objectives and/or constraints.
(c) The weighted objective formulation presents a simple method to add additional constraints to the problem for use with new optimizers developed for machine learning and increased performance.

The paper is organized as follows. In Section 2, the optimization framework with the neural style transfer algorithm, topology optimization formulation, and post-processing filter are described. In Section 3, three cases are presented demonstrating the method with different design problems and reference inputs. In Section 4, the performance of the method and the effects of the neural style transfer parameters and post-processing filter are discussed. Section 5 provides conclusions and future improvements to the method.

2. Materials and Methods

This section describes the weighted objective approach for topology optimization developed for this work. Each objective and constraint, including the structural objective and neural style transfer loss, are summed to develop a single loss function to be optimized. Figure 1 illustrates how each iteration of the analysis runs, starting with the design variable, ϕ, and concluding with the total loss sent to the optimizer. The unconstrained design variable is converted with the sigmoid function to be between 0 and 1. The result is used to calculate the loss, $\mathcal{L}$, for the given design problem with the objective function value from finite element analysis and comparing the geometric style with the reference image(s) using neural style transfer through mean squared error (MSE). Both results are multiplied by a weight, w, and summed to form the total loss used in the optimizer. After the optimization is complete, the result is post-processed using a physics-based filter to remove extraneous artifacts left from the optimization and smooth the result for manufacturing.

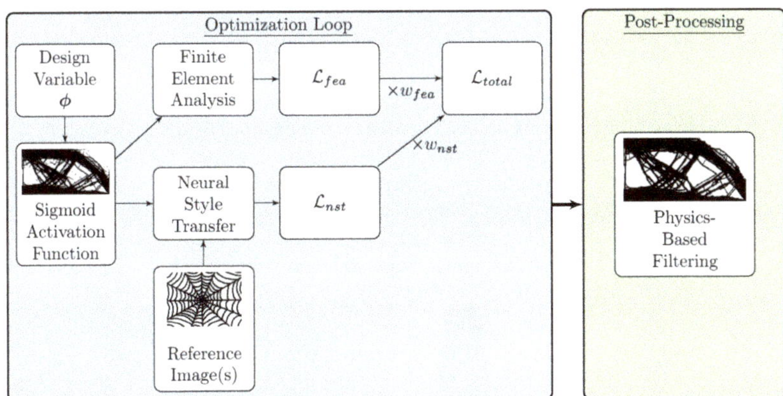

Figure 1. Op-Level Flowchart for each iteration.

Traditionally, the topology optimization analysis is performed as a single-objective, multi-constraint problem such as:

$$\begin{aligned}
\min \quad & f(x) \\
\text{s.t.} \quad & g_i(x) = c_i, \quad \text{for} \quad i = 1, \ldots, n, \\
& h_j(x) \geq d_j, \quad \text{for} \quad j = 1, \ldots, m
\end{aligned} \tag{1}$$

where $f(x)$ is the objective function to be optimized (e.g., compliance, mass, etc.), $g_i(x)$ are the equality constraints for n equations, $h_j(x)$ are the inequality constraints for m equations, and x is the design variable constrained between 0 and 1. Although this formulation is suitable for current structural optimizers such as Optimality Criterion or Method of Moving Asymptotes (MMA) [22], optimizers for machine learning problems such as Adam [23] are formulated as a single loss function. Adam was developed to be computationally and memory-efficient for a large number of parameters and use on graphics processing units (GPU). Performing neural network operations on GPUs greatly improves the speed per iteration and reduces the overhead of the neural style transfer calculation [24]. With these considerations and the prevalence of the use of the Adam optimizer, it was chosen for this work.

The loss function is therefore defined as a weighted objective function, where each constraint and objective equation from Equation (1) is considered as a loss term, $\mathcal{L}_i$, with a corresponding weight, w_i. The summation of each loss term and weight then forms the function to be optimized, see Equation (2).

$$\mathcal{L}_{total} = \sum_i w_i \mathcal{L}_i \tag{2}$$

With the machine learning optimizers, the design variables are not constrained between 0 and 1 as seen with the current structural optimizers presented. An activation function is used to convert the design variable, ϕ, to the elemental densities, x. For this work, the sigmoid function is used as it is continuously differentiable, see Equation (3).

$$\mathrm{Sigmoid}(\phi) = x = \frac{1}{1 + \exp(\phi)} \tag{3}$$

The neural style transfer method is based upon the work by Gatys et al. [21]. The convolutional filters of a pre-trained convolutional neural network are used as local feature extractors for an input. Within the work, the VGG-19 neural network [25] is trained to classify images to one of the over 80,000 sets found in the ImageNet database [26]. The network is modified for use with neural style transfer, using average pooling layers rather than max pooling. With such a variety of images, the convolutional layers are better suited to recognize features on a large variety of inputs. Early convolutional layers in the network (i.e., conv1_1) capture close local features of the input. Deeper layers, through stacked convolutional layers and pooling (i.e., conv5_1), correspondingly extract features from a larger region of the input. The different region scales from the extracted layers smoothly integrate the geometric features to the new design. Figure 2 shows how the neural network can extract the features of an input. This network architecture, also pretrained for classification of the ImageNet database, is also used for this work. The reader is referred to Simonyan et al. [25] and Gatys et al. [21] for the VGG-19 network architecture and modifications for neural style transfer.

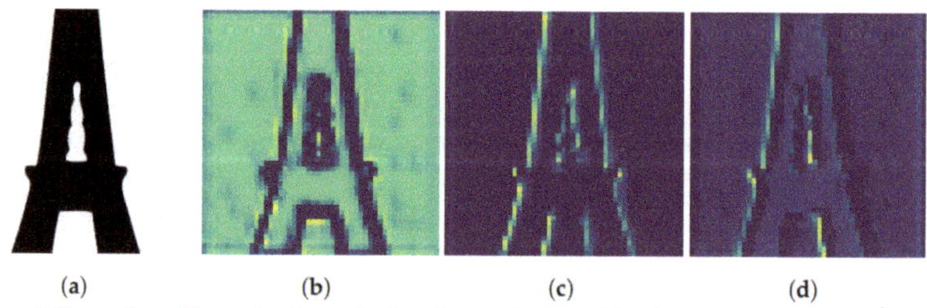

(a)   (b)   (c)   (d)

Figure 2. Examples of the activations of selected convolutional filter layers within the VGG-19 neural network for an (**a**) example input. The filters shown exhibit greater activation for (**b**) neighborhood of similar pixels, (**c**) the left outer border of the structure, and (**d**) the right inner border of the structure.

Using solely the convolutional layer for comparison between the input to be optimized and the reference, the optimizer would modify the input to be a copy of the reference to best reduce the loss. To determine the style representation, the activation across the entire selected filter must be used for comparison. To achieve this, the Gram matrix, G_{ij}^l, is calculated as the inner product of the vectorized feature map, F_{ij}^l, for layer l, where F_{ij}^l is the activation of the ith filter at position j of layer l, see Equation (4).

$$G_{ij}^l = \sum_k F_{ik}^l F_{jk}^l \tag{4}$$

To calculate the loss function for the style representation for layer l, the mean squared error loss between the Gram matrices of the input and style images is used, where N_l and M_l are the lengths of the Gram matrix of the input to be optimized, G_{ij}^l, and the Gram matrix of the reference input, A_{ij}^l, for layer l, respectively. This, as well as the derivative for the design sensitivities, are shown in Equations (5) and (6). The convolutional layers conv1_1, conv2_1, conv3_1, conv4_1, and conv5_1 of the VGG-19 network are used to represent the style transfer loss. Liu et al. [27] provides the mathematical representations of the convolutional layers used in the VGG-19 network.

$$\mathcal{L}_l = \frac{1}{4N_l^2 M_l^2} \sum_{i,j} (G_{ij}^l - A_{ij}^l)^2 \tag{5}$$

$$\frac{\mathrm{d}\mathcal{L}_l}{\mathrm{d}F_{ij}^l} = \frac{1}{4N_l^2 M_l^2} ((F^l)^T (G^l - A^l))_{ji} \tag{6}$$

The objective function of the topology optimization analysis for this work is based on the 88 line MATLAB script by Andreassen et al. [28]. The script details an efficient two-dimensional topology optimization problem using a Cartesian mesh. The structural objective presented in the paper is to minimize compliance, or maximize stiffness, for the design and load conditions. This formulation is self-adjoint which simplifies the sensitivity analysis. The equations for compliance and the sensitivities are as follows:

$$\mathcal{L} = \boldsymbol{U}^T \boldsymbol{K} \boldsymbol{U} = \sum_{e=1}^{N} (E_{min} + (E_0 - E_{min})x_e^p) \boldsymbol{u}_e^T \boldsymbol{k}_0 \boldsymbol{u}_e \tag{7}$$

$$\boldsymbol{K} \boldsymbol{U} = \boldsymbol{F} \tag{8}$$

$$\frac{\mathrm{d}\mathcal{L}}{\mathrm{d}x_e} = -px_e^{p-1}(E_0 - E_{min})\boldsymbol{u}_e^T \boldsymbol{k}_0 \boldsymbol{u}_e \tag{9}$$

where x is the input variable vector containing element densities x_e; K, U, and F are the global stiffness matrix, displacement vector, and force vector, respectively; u_e and k_0 are the element displacement vector and element stiffness matrix, respectively; p is a penalty term for the element densities; E_0 and E_{min} are the maximum and minimum allowable Young's moduli for solid and void material, respectively; and N is the number of elements used for the domain. To avoid checkerboard patterns, a convolutional filter is applied to the sensitivities of the compliance calculation [28]. The convolution is defined as follows:

$$\frac{\widehat{\mathrm{d}\mathcal{L}}}{\mathrm{d}x_e} = \frac{1}{\max(\gamma, x_e) \sum_{i \in N_e} H_{ei}} \sum_{i \in N_e} H_{ei} x_i \frac{\mathrm{d}\mathcal{L}}{\mathrm{d}x_i} \tag{10}$$

$$H_{ei} = \max(0, r_{min} - \Delta(e, i)) \tag{11}$$

where N_e is the set of elements with a center-to-center distance between the current element for the sensitivity, x_e, and an additional element, x_i, less than the user defined radius, r_{min}, γ is a small value equivalent to the void density to avoid division by zero, and H_{ei} is the weight factor for the additional element, x_i.

Two additional constraints were considered with the total loss function to achieve the desired results for the analysis: volume fraction and standard deviation. For the volume fraction, it is defined as the mean absolute error between the current volume of the design, $V(x)$, and the desired volume of the design, V_0, see Equation (12). Other formulations including the mean squared error were considered but the mean absolute error achieved results closer to the desired volume fraction. The standard deviation term is used to encourage the design to achieve a true 0/1 distribution and is defined as the standard deviation of the element densities. If inputs with intermediate densities are used, this ensures the final design achieves a 0/1 distribution rather than incorporating the intermediate densities from the reference input. This term would be subtracted from the total loss rather than summed, see Equation (13), where x is the input variable vector containing element densities x_e, μ is the mean of the input vector, and N is the number of elements in the input vector.

$$\mathcal{L} = |V(x) - V_0| \tag{12}$$

$$\mathcal{L} = -\sqrt{\frac{\Sigma_i (x_i - \mu)^2}{N}} \tag{13}$$

Although topology optimization alone encourages smooth designs to satisfy design constraints, the additional neural style transfer objective with a large weight or poorly suited reference design can introduce objects disconnected or minimally connected to the main design. Martinez et al. and Hu et al. have also presented this issue in their works [7,8]. To overcome the issue, each used an additional constraint within the optimization to discourage the formation of these objects. Martinez et al. suggests adding self-weight to the design problem. However, it is determined the design may not converge properly without relaxation of other constraints [7]. Hu et al. propose two adaptive regulations for the texture appearance weight. The weight would be calculated for each neighborhood of texture, reducing the appearance weight in void regions and avoiding disconnected or minimally connected objects [8]. However, the method described would not be appropriate for this work. The geometric features for this work are calculated on a global scale, not local. Integrating the method would add to the computational cost of each optimization iteration.

To avoid disconnected or minimally connected objects in the final design without great additional computational cost, a post-processing filter is introduced. Image processing techniques such as erosion and dilation were found to remove important load-carrying members or close features introduced from the reference. Similar to Groen and Sigmund [29], a physics-based filter is introduced to remove these disconnected objects without affecting the load-carrying members.

Through experimentation, it was found the equivalent von Mises stress at each element, σ_e^{vM}, in the disconnected objects of the final design was minimal compared to the fully connected objects of the design. The formulation is derived from the stress constrained topology optimization method by Holmberg et al. [30]. Using the result from last iteration of the weighted objective optimization, the equivalent stress for each element was calculated as follows:

$$\sigma_e(x_e) = \mathbf{EB}u_e \tag{14}$$

$$\sigma_e(x_e) = \begin{pmatrix} \sigma_{xx} & \sigma_{yy} & \tau_{xy} \end{pmatrix}^T \tag{15}$$

$$\sigma_e^{vM}(x_e) = (\sigma_{xx}^2 + \sigma_{yy}^2 - \sigma_{xx}\sigma_{yy} + 3\tau_{xy}^2)^{\frac{1}{2}} \tag{16}$$

where $\mathbf{E}$ is the constitutive matrix, $\mathbf{B}$ is the strain-displacement matrix, u_e is the element displacement vector for element x_e found from Equation (8), $\sigma_e(x_e)$ is the two-dimensional stress tensor with components for a Cartesian coordinate system, and σ_e^{vM} is the equivalent von Mises stress.

From the analysis, the elements with an equivalent stress value below a threshold would be set as void material. For the numerical examples presented in this work, it was found a threshold of 10% of the elements with the least stress produced favorable results. Examples demonstrating the effectiveness of the filter are presented in Section 4.

3. Numerical Examples

In this section, multiple two-dimensional examples are presented to show the capabilities of the proposed work. The examples were created using a Python script built around the PyTorch machine learning library implementing the method. Table 1 details the parameters and design domain used for each of the examples. The design problems presented include the MBB-beam and the cantilever beam, where black and white represent solid and void material, respectively. Both problems use the same values, as the values produce quality results for the examples and simplify the problems for the reader to reproduce the presented results. Figure 3 shows the reference inputs used for the examples.

Table 1. Variable Initialization.

Variable	Value
Elements in x-direction	400
Elements in y-direction	200
Filter Radius for Sensitivity Analysis	1.5 elements
Mass Penalty for Finite Element Analysis	3.0
Young's Modulus	10^{-6}–1.0
Poisson's Ratio	0.3
Force	1.0
Structural Compliance Weight	1
Neural Style Transfer Weight	10^3–10^5
Volume Fraction Weight	0.1
Standard Deviation Weight	0.1
Number of Iterations	500
Step Size for Adam Optimizer	0.08

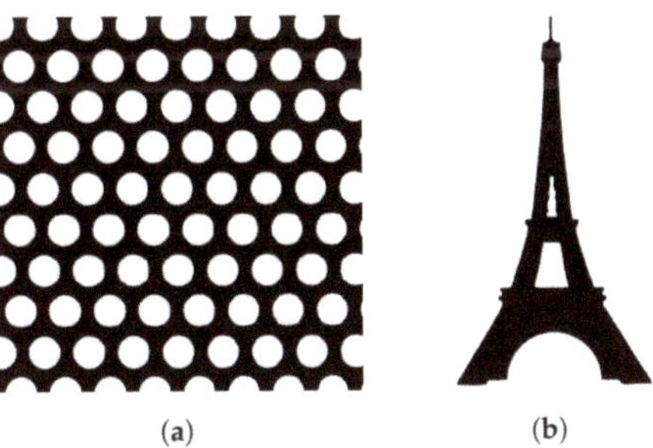

(a)   (b)

Figure 3. Reference design inputs for examples include (**a**) circular mesh and (**b**) Eiffel tower.

3.1. MBB Beam

The MBB beam is a common design problem among topology optimization research as a benchmark for new methods [28]. The design problem is illustrated in Figure 4. The left edge of the beam has zero horizontal displacement, and the bottom right node has zero vertical displacement. The force is applied to the top-left node. Using the parameters described in Table 1, the baseline design and designs influenced by the style input are shown in Figure 5.

Figure 4. Conditions for MBB Beam.

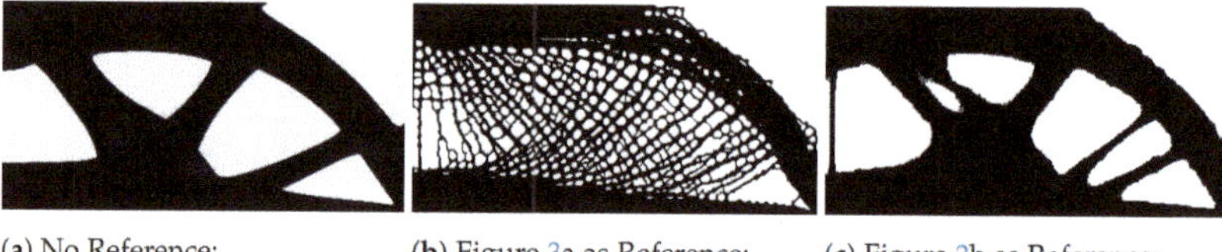

(**a**) No Reference:
Compliance: 80.384

(**b**) Figure 3a as Reference:
Compliance: 85.510

(**c**) Figure 3b as Reference:
Compliance: 81.062

Figure 5. Examples of the MBB beam examples with the compliance values for each design.

Figure 5a shows the optimized design for the MBB design problem without any reference input. The optimized design is characterized by three large supporting members inside the design envelope to support the force.

Observing the reference for the result shown in Figure 5b, the reference input is composed of a repeating circular mesh structure. Through the neural style transfer objective, the corresponding mesh is applied to the inner structure of the design, replacing the three supporting members found in Figure 5a. Although the size of mesh beams more closely reflect the reference image, the directions of the beams closely follow the standard result members to satisfy the compliance. This compromise results in irregular holes, rather than circular, in the final design. This may not fully achieve the desired geometric features, but the beams within the mesh better align with the ideal direction to support the load with the influence of the many holes from the reference input. The outer envelope of the design better matches the standard result, not incorporating the mesh from the reference input. As the outer envelope has larger members, it is determined the region greatly contributes to the structural performance. The optimizer converged to a solid region for improved structural performance, rather than including the holes from the reference design. When increasing the weight of the neural style transfer loss, a greater portion of the design incorporates the mesh, ultimately encompassing the full design space. Although this would satisfy the ideal geometric style, the structural performance is greatly diminished.

Figure 5c uses a reference input composed of a tower. The input is symmetrical but does not have many repeating elements as found for Figure 5b. The beams found in the reference are slender, with some material removed as it converges near the top of the tower. In the optimized design, the outer envelope is similar to the standard result. However, five supporting members are used inside the design envelope, rather than three found in the standard result. The voids are rounder to match the smooth curves of the reference and incorporate a hole in the leftmost member which is also found in the reference.

Reviewing all three designs reveals common elements among them, notably the outer envelope and the directions of the inner members. Even with the different reference inputs, these elements were considered crucial to maintaining the structural performance of the final design. Figure 5b heavily applies the mesh to the inner members to satisfy the neural style transfer objective. The reference input for Figure 5c is comparably simpler and the optimized result is nearly identical to the standard result except for the additional inner members. Although there are design differences between all three results, the structural performances of both stylized results are still maintained within 10% of the baseline design.

3.2. Cantilever Beam

The cantilever beam design is inspired by the work by Wu et al. [10] for infill optimization. The design problem is illustrated in Figure 6. The left edge of the beam is fixed in all directions. A downward vertical force is applied to the middle of the right edge of the beam. Using the parameters described in Table 1, the baseline design and designs influenced by the style input for this design problem are shown in Figure 7.

Figure 6. Conditions for Cantilever Beam.

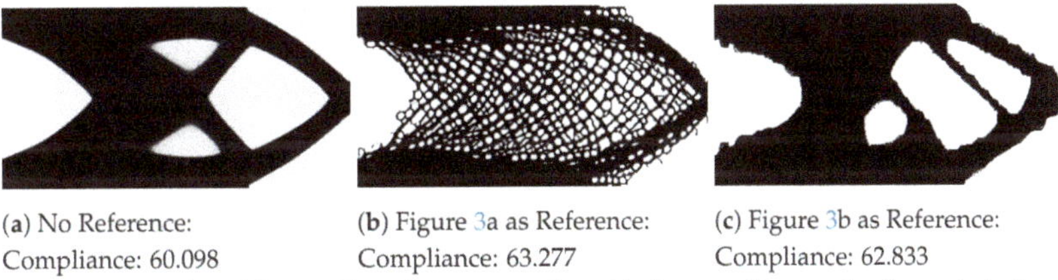

(**a**) No Reference: Compliance: 60.098

(**b**) Figure 3a as Reference: Compliance: 63.277

(**c**) Figure 3b as Reference: Compliance: 62.833

Figure 7. Examples of the cantilever beam examples with the compliance value for each design.

Figure 7a shows the optimized design with no reference input. Although the problem is not symmetrical, the optimized design is. Two small members inside the outer envelope provide additional rigidity to improve the structural performance. For this example, the reference inputs used for the MBB beam designs are used here.

Figure 7b uses the circular mesh pattern as the reference input. As seen in Figure 5b, the outer envelope of the design is very similar to the standard result. In this design, the mesh is incorporated at the edges of the envelope to satisfy the neural style transfer objective. To further reduce the neural style transfer loss, much of the interior structure is composed of circular mesh elements. The beams follow the directions of the two small members in the standard result and are tightly packed to resemble the reference input.

Figure 7c follows the simple design of the standard result but incorporates more features from the reference compared with Figure 5c. The interior members are correspondingly thinner and are present in only one direction to support the asymmetric load. The void areas also have round edges compared with the sharper corners in Figure 7a.

3.3. Using Multiple Reference Designs

A benefit of this method is multiple reference designs can be utilized for the neural style transfer and balance the geometric features of multiple sources. The additional reference is added as another objective for the multi-objective formulation. Figure 8 shows the results for two inputs.

Performing the optimization did not impact the performance of the optimizer. The average amount of time for each iteration was 1.8765 s. As presented in Table 2, this is comparable between two common optimizers used for TO solving similar design problems without the neural style transfer: Optimality Criterion and MMA [22]. The reason for this is that the gram matrices for each reference are stored and not recalculated for each iteration. The mean squared error calculation is a relatively fast operation and does not impact the optimization speed.

The result in Figure 8 does show resemblance to both provided references. The smooth edges of the tower are present at the solid-void boundaries in the optimized design. Smaller members are used as well as found in Figure 7c. To satisfy the circular hole pattern, many small holes were added to the members as found in Figure 7b but are not as prevalent as they would deviate from the tower reference input in Figure 7c.

Figure 8. Cantilever beam design using both reference inputs from Figure 3; Compliance: 61.147.

4. Discussion

4.1. Performance

Figure 9 shows the convergence history for the result shown in Figure 7c for each function: compliance, mass fraction, standard deviation, and style loss for each layer. All presented results follow similar histories. At the start of training, the optimizer greatly improves the compliance value and style losses and exceeds the desired volume fraction. After some steps, more material is removed and the volume fraction correspondingly decreases until it falls below the desired volume fraction, with improving compliance and style loss values. Towards the later stages, the improvements to the compliance and style loss diminish. For each oscillation of the volume fraction, the compliance and style loss values correspondingly oscillate. The values continue to improve at a much slower rate until the given number of steps is reached. Although the losses have not converged after the training procedures, the overall structure changed minimally at the end of the optimization, only varying in particular regions. These regions were processed using the physics-based filter to complete the optimization. With different parameters from Table 1, more or less iterations may be required to achieve a final design.

The results presented in this work are characterized as deterministic, and therefore do not require a statistical analysis. The CNN used for neural style transfer is pre-trained and is not updated between results. The initialization parameters, shown in Table 1, are also identical for repeated results, including the initial density of the design space. As the initial state for a set of parameters is always identical, the gradients for descent are also equal and the optimizer follows identical convergence paths for repeated analyses.

Table 2 shows the comparison of the Optimality Criterion optimizer and the MMA optimizer [22] using the MATLAB implementation of the 88 line topology optimization script by Andreassen et al. [28] with the Python implementation of this work using the Adam optimizer [23] with and without the neural style transfer constraint. This was performed with an 8-core Intel Xeon 3.7GHz processor, 128 GB RAM, and an NVIDIA Quadro RTX 6000 GPU.

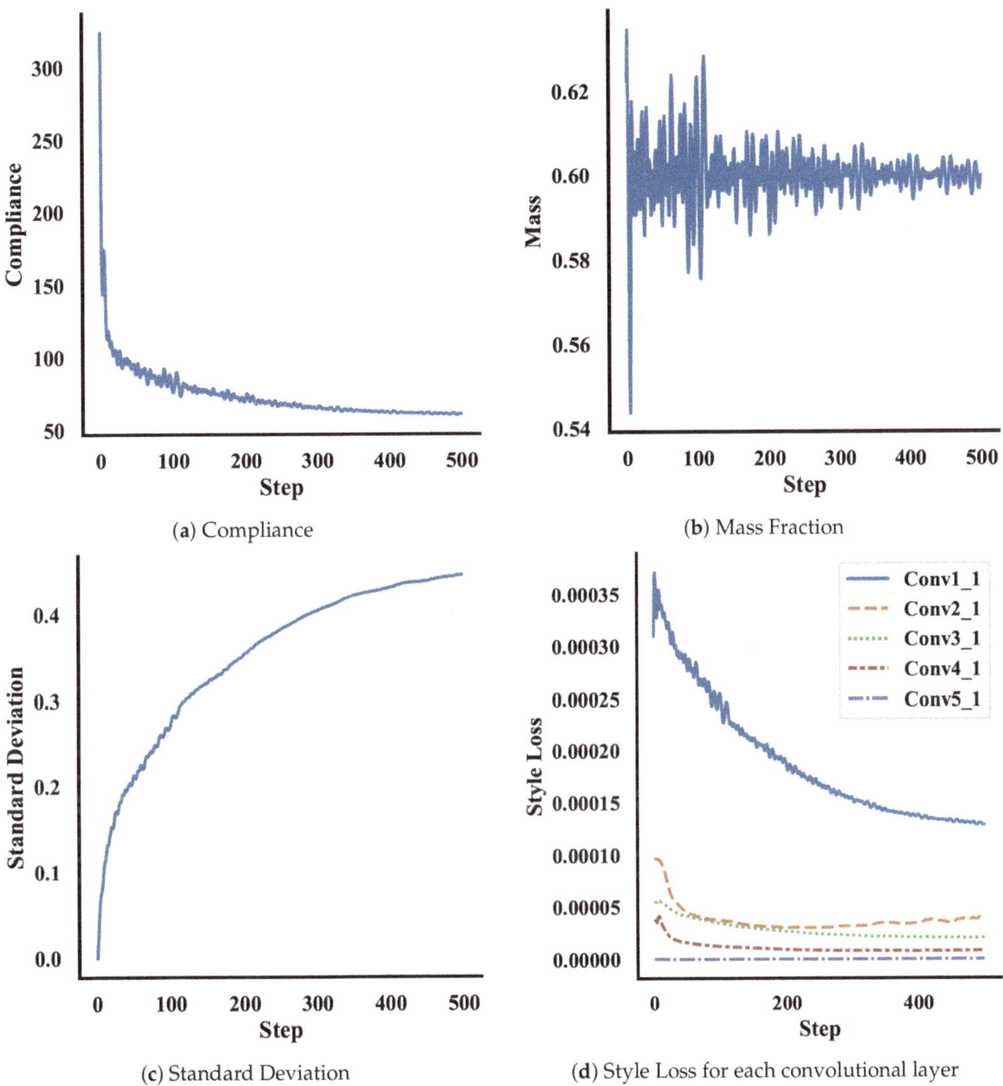

(**a**) Compliance

(**b**) Mass Fraction

(**c**) Standard Deviation

(**d**) Style Loss for each convolutional layer

Figure 9. Optimization convergence plots for result in Figure 7c.

Table 2. Average Time per Iteration of Optimizers.

Optimizer	Time (s)
Optimality Criterion (Top88 Formulation) [28]	1.1208
Method of Moving Asymptotes without Neural Style Transfer [22]	2.8790
Adam without Neural Style Transfer	1.0360
Adam with Neural Style Transfer	1.8965

Using the Adam optimizer with the given equipment is on par with the linear Optimality Criterion optimizer and much faster than the MMA algorithm. Although it is faster, more iterations are necessary to ensure a good result. It would be beneficial to use techniques for machine learning to speed up the accuracy of the network for this. One such example includes transfer learning from a coarse result. The design could be optimized for a low-resolution domain. The design would then be scaled to the finer resolution to complete the optimization in fewer steps. Another example would be learning rate annealing. High learning rate values result in large improvements early in the optimization process. After many steps, the high learning rate starts to overshoot and the accuracy fails to improve further. Learning rate annealing would reduce the learning rate for this situation. The smaller learning rate would help perform smaller steps to better achieve the minimum. A small learning rate could be used at the beginning of the optimization, but many more steps would be required compared with the annealed method.

4.2. Post-Processing

Without the neural style transfer objective, the optimized design is smooth and continuous with no artifacts left from the optimization. The neural style transfer objective, however, acknowledges the full design domain to optimize the geometric style. If a large area of the design does not contribute to the overall structural performance, the optimizer could satisfy the neural style transfer objective by adding material with the geometric features of the reference.

Figure 10a shows the result from Figure 5b immediately after optimization. The outer envelope contains many small members minimally connected to the design but still following the circular reference design. In the void material near the bottom right support, a disconnected member is also present. Traditional image processing techniques would not work for this design, as a filter that would remove the artifacts would also impact the desired circular mesh, introducing more minimally connected objects.

It was then found the minimally connected objects in both areas experience an equivalent von Mises stress, similar to the void material areas surrounding them. Figure 10b shows the equivalent von Mises stress for the result in Figure 10a. Although the disconnected members are visible in the structural result, the members exhibit very low stress and are indistinguishable from the void stress values. Using the threshold method described in Section 2, the elements with stress values below the threshold, including the disconnected members, are set to void material and removed from the final result.

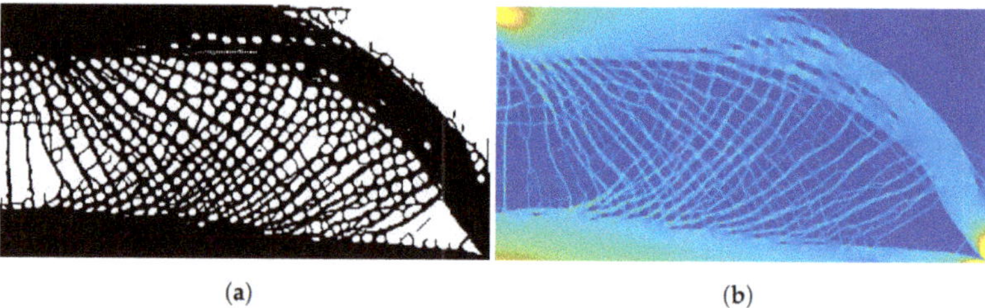

(a)   (b)

Figure 10. Effects of the stress-based post-processing filter (**a**) before and (**b**) the equivalent von Mises stress result. The post-processed result is found in Figure 5b.

4.3. Connectivity

As seen in Figure 10, some structures are disconnected from the main structure after the optimization analysis. The post-processing filter can remove these objects effectively. However, future work should be done to limit these artifacts during the optimization.

Such improvement could come from the addition of a stress constraint to the weighted objective function. The effectiveness of the post-processing filter shows the objective would reduce the number of the minimally connected objects. As discussed in Section 2, the calculation of the stress would impact the performance. The efficiency of the calculation would have to be considered during the implementation.

Through repeated training procedures, the weights of the individual style layers would be adjusted rather than using a single weight applied to all layers to improve the result. As shown in Figures 11 and 12, each layer contributes a different aspect of the reference input to the design, and adjusting the layer weights would change the final result. The images in the figures were not post-processed with the stress-based filter to show the result after optimization. Using conv1_1, it is understood a mesh would satisfy the reference input. Using conv2_1 and conv3_1, the thickness of the members is found. Using conv4_1 and conv5_1, the ideal angles of the members emerge in the design. Through the use of the system, the results and parameters of satisfactory designs would be saved to a database. Through searching or training another machine learning network of the database, parameter weights for each layer would be suggested to achieve ideal results.

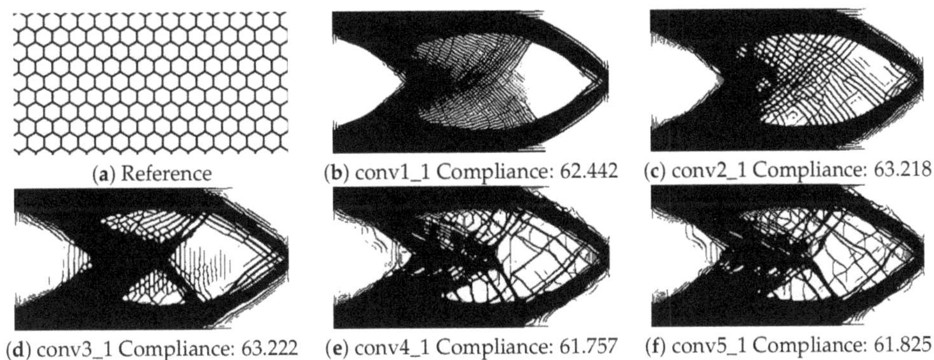

(**a**) Reference (**b**) conv1_1 Compliance: 62.442 (**c**) conv2_1 Compliance: 63.218

(**d**) conv3_1 Compliance: 63.222 (**e**) conv4_1 Compliance: 61.757 (**f**) conv5_1 Compliance: 61.825

Figure 11. Examples of the cantilever beam using different layers of the neural network.

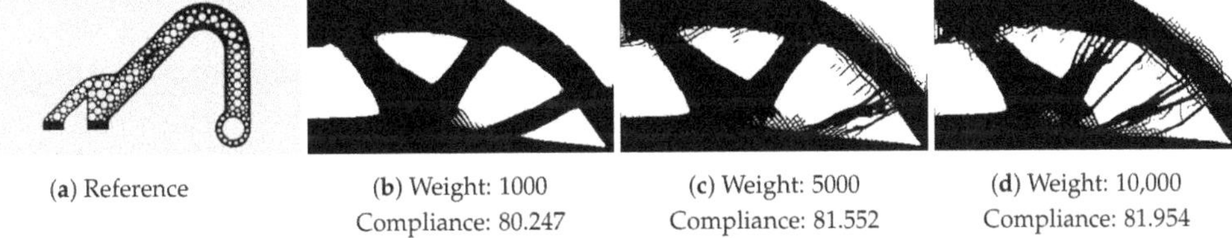

(**a**) Reference (**b**) Weight: 1000 (**c**) Weight: 5000 (**d**) Weight: 10,000
Compliance: 80.247 Compliance: 81.552 Compliance: 81.954

Figure 12. Examples of the MBB beam using different weights for the neural style transfer.

Additionally, this work uses the original neural style transfer formulation by Gatys et al. [21]. As described in Jing et al. [31], newer implementations are in development to improve the results. Jing summarizes the extensions to the original approach and various loss functions of the layer activations to improve the results. Adjusting the neural style transfer to one of the described methods would require additional investigation for further research.

5. Conclusions

In this work, a novel approach to generate topology optimized structures is proposed using a pre-trained neural network to quantify the desired geometric style of the optimized design. The conclusions drawn are as follows:

(a) The neural style transfer quantifies the geometric features of the reference and optimized designs efficiently using a Gram matrix calculation of the pre-trained convolutional filter activations for a neural network classifier. As such, the features of the input are replicated rather than directly copied in the optimized design, which expands the number of applicable inputs.

(b) The weighted objective formulation presents a simple method to add additional constraints to the problem and tune the influence for each constraint. The formulation also allows the utilization of new optimizers developed for machine learning and increases performance.

Although compliance is used for the objective, it is possible to use other structural objectives such as natural frequency or stress minimization using the same optimizer. Performing these analyses would further validate and improve the usefulness of the proposed method.

Additionally, recent neural style transfer methods have been developed which improve upon the original formulation used in this work; see Jing et al. [31] for a comprehensive review. Berger and Memisevic translate the feature maps to better capture spatial and symmetric arrangements [32]. A recent work by Gatys et al. allows for spatial control, limiting the style transfer to regions of the design and not the global structure [33]. Using these methods could improve the final results of the optimization process, eliminating the need for the stress-based post-processing filter. Similarly, the optimization need not be done as a linear summation of all objectives. Prioritized optimization is an example of an alternative method [34]. Using this method, multiple optimal solutions of one objective function are used to find equally optimal solutions for an additional objective function. Rather than using weights as found in the current linear summation, the method will iterate through solutions that limit the error for each objective function. These improvements will be left as future work for the authors.

Although topology optimization is most useful for three-dimensional CAD design, the neural style transfer implementation used is limited to two-dimensional input. Using three-dimensional voxelized inputs or other deep geometric learning methods, a three-dimensional CNN classifier for CAD files with a similar architecture compared with the network used in this work can be used for neural style transfer. The implementation described by Gatys et al. [21] can be replicated for three-dimensional CNNs. This method is under investigation by the authors.

Supplementary Materials: The following are available online at https://www.mdpi.com/article/10.3390/ma14164551/s1: Source Code.

Author Contributions: Conceptualization, P.S.V. and X.Z.; methodology, P.S.V. and F.D.; software, P.S.V.; validation, P.S.V.; formal analysis, P.S.V.; investigation, P.S.V.; resources, P.S.V.; data curation, P.S.V.; writing—original draft preparation, P.S.V.; writing—review and editing, P.S.V., H.D., F.D., X.Z., and A.C.T.; visualization, P.S.V.; supervision, A.C.T.; project administration, A.C.T.; funding acquisition, A.C.T. All authors have read and agreed to the published version of the manuscript.

Funding: This research was funded by the National Science Foundation (CMMI-1634261).

Institutional Review Board Statement: Not applicable.

Informed Consent Statement: Not applicable.

Data Availability Statement: The data presented in this study are available in the supplementary material.

Acknowledgments: The authors would like to acknowledge the support from the National Science Foundation (CMMI-1634261) and the University of Pittsburgh. The authors also thank the anonymous reviewers for their contributions to this work.

Conflicts of Interest: The authors declare no conflict of interest. The funders had no role in the design of the study; in the collection, analyses, or interpretation of data; in the writing of the manuscript, or in the decision to publish the results.

References

1. Bendsøe, M.P.; Kikuchi, N. Generating optimal topologies in structural design using a homogenization method. *Comput. Methods Appl. Mech. Eng.* **1988**, *71*, 197–224. [CrossRef]
2. Sigmund, O.; Maute, K. Topology optimization approaches. *Struct. Multidiscip. Optim.* **2013**, *48*, 1031–1055. [CrossRef]
3. Deng, H.; To, A. Topology optimization based on deep representation learning (DRL) for compliance and stress-constrained design. *Comput. Mech.* **2020**, *66*, 449–469. [CrossRef]
4. Meng, L.; Zhang, W.; Quan, D.; Shi, G.; Tang, L.; Hou, Y.; Breitkopf, P.; Zhu, J.; Gao, T. From Topology Optimization Design to Additive Manufacturing: Today's Success and Tomorrow's Roadmap. *Arch. Comput. Methods Eng.* **2020**, *27*, 805–830. [CrossRef]

5. Liu, J.; Gaynor, A.; Chen, S.; Kang, Z.; Suresh, K.; Takezawa, A.; Li, L.; Kato, J.; Tang, J.; Wang, C.; Cheng, L.; Liang, X.; To, A. Current and future trends in topology optimization for additive manufacturing. *Struct. Multidiscip. Optim.* **2018**, *57*, 2457–2483 . [CrossRef]

6. Sylcott, B.; Michalek, J.; Cagan, J. Towards understanding the role of interaction effects in visual conjoint analysis. In Proceedings of the ASME Design Engineering Technical Conference, Portland, ON, USA, 4–7 August 2013; Volume 3 A. [CrossRef]

7. Martínez, J.; Dumas, J.; Lefebvre, S.; Wei, L.Y. Structure and appearance optimization for controllable shape design. *ACM Trans. Graph.* **2015**, *34*, 1–11 . [CrossRef]

8. Hu, J.; Li, M.; Gao, S. Texture-guided generative structural designs under local control. *CAD Comput. Aided Des.* **2019**, *108*, 1–11. [CrossRef]

9. Bickel, B.; Cignoni, P.; Malomo, L.; Pietroni, N. State of the Art on Stylized Fabrication. *Comput. Graph. Forum* **2018**, *37*, 325–342. [CrossRef]

10. Wu, J.; Aage, N.; Westermann, R.; Sigmund, O. Infill Optimization for Additive Manufacturing-Approaching Bone-Like Porous Structures. *IEEE Trans. Vis. Comput. Graph.* **2018**, *24*, 1127–1140. [CrossRef]

11. Guest, J.K.; Prévost, J.H.; Belytschko, T. Achieving minimum length scale in topology optimization using nodal design variables and projection functions. *Int. J. Numer. Methods Eng.* **2004**, *61*, 238–254. [CrossRef]

12. Guest, J.K. Imposing maximum length scale in topology optimization. *Struct. Multidiscip. Optim.* **2009**, *37*, 463–473. [CrossRef]

13. Zimmermann, L.; Chen, T.; Shea, K. A 3D, performance-driven generative design framework: automating the link from a 3D spatial grammar interpreter to structural finite element analysis and stochastic optimization. *Artif. Intell. Eng. Des. Anal. Manuf.* **2018**, *32*, 189–199. [CrossRef]

14. Kazi, R.H.; Grossman, T.; Cheong, H.; Hashemi, A.; Fitzmaurice, G. DreamSketch: Early Stage 3D Design Explorations with Sketching and Generative Design. In Proceedings of the 30th Annual ACM Symposium on User Interface Software and Technology, UIST '17, Quebec City, QC, Canada, 22–25 October 2017 ; Association for Computing Machinery: New York, NY, USA, 2017; pp. 401–414. [CrossRef]

15. Creswell, A.; White, T.; Dumoulin, V.; Arulkumaran, K.; Sengupta, B.; Bharath, A. Generative Adversarial Networks: An Overview. *IEEE Signal Process. Mag.* **2018**, *35*, 53–65. [CrossRef]

16. Oh, S.; Jung, Y.; Kim, S.; Lee, I.; Kang, N. Deep Generative Design: Integration of Topology Optimization and Generative Models. *J. Mech. Des.* **2019**, *141*, 111405 . [CrossRef]

17. Cang, R.; Yao, H.; Ren, Y. One-shot generation of near-optimal topology through theory-driven machine learning. *Comput. Aided Des.* **2019**, *109*, 12–21. [CrossRef]

18. Gaymann, A.; Montomoli, F. Deep Neural Network and Monte Carlo Tree Search applied to Fluid-Structure Topology Optimization. *Sci. Rep.* **2019**, *9*, 1–16 . [CrossRef]

19. Yu, Y.; Hur, T.; Jung, J.; Jang, I.G. Deep learning for determining a near-optimal topological design without any iteration. *Struct. Multidiscip. Optim.* **2019**, *59*, 787–799. [CrossRef]

20. Banga, S.; Gehani, H.; Bhilare, S.; Patel, S.; Kara, L. 3D Topology Optimization using Convolutional Neural Networks. *arXiv* **2018**, arXiv:1808.07440 .

21. Gatys, L.A.; Ecker, A.S.; Bethge, M. Image Style Transfer Using Convolutional Neural Networks. In Proceedings of the 2016 IEEE Conference on Computer Vision and Pattern Recognition (CVPR), Las Vegas, NV, USA, 27–30 June 2016; pp. 2414–2423.

22. Svanberg, K. The method of moving asymptotes—A new method for structural optimization. *Int. J. Numer. Methods Eng.* **1987**, *24*, 359–373. [CrossRef]

23. Kingma, D.P.; Ba, J. Adam: A Method for Stochastic Optimization. In Proceedings of the 3rd International Conference on Learning Representations, ICLR, San Diego, CA, USA, 7–9 May 2015.

24. Shi, S.; Wang, Q.; Xu, P.; Chu, X. Benchmarking State-of-the-Art Deep Learning Software Tools. In Proceedings of the 2016 7th International Conference on Cloud Computing and Big Data (CCBD), Macau, China, 16–18 November 2016; pp. 99–104.

25. Simonyan, K.; Zisserman, A. Very Deep Convolutional Networks for Large-Scale Image Recognition. In Proceedings of the 3rd International Conference on Learning Representations, ICLR, San Diego, CA, USA, 7–9 May 2015.

26. Deng, J.; Dong, W.; Socher, R.; Li, L.; Fei-Fei, L. ImageNet: A large-scale hierarchical image database. In Proceedings of the 2009 IEEE Conference on Computer Vision and Pattern Recognition, Miami, FL, USA, 20–25 June 2009; pp. 248–255.

27. Liu, Q.; Zhang, N.; Yang, W.; Wang, S.; Cui, Z.; Chen, X.; Chen, L. A Review of Image Recognition with Deep Convolutional Neural Network. In *Intelligent Computing Theories and Application*; Huang, D.S., Bevilacqua, V., Premaratne, P., Gupta, P., Eds.; Springer International Publishing: Cham, Switzerland , 2017; pp. 69–80.

28. Andreassen, E.; Clausen, A.; Schevenels, M.; Lazarov, B.S.; Sigmund, O. Efficient topology optimization in MATLAB using 88 lines of code. *Struct. Multidiscip. Optim.* **2011**, *43*, 1–16. [CrossRef]

29. Groen, J.P.; Sigmund, O. Homogenization-based topology optimization for high-resolution manufacturable microstructures. *Int. J. Numer. Methods Eng.* **2018**, *113*, 1148–1163. [CrossRef]

30. Holmberg, E.; Torstenfelt, B.; Klarbring, A. Stress constrained topology optimization. *Struct. Multidiscip. Optim.* **2013**, *48*, 33–47. [CrossRef]

31. Jing, Y.; Yang, Y.; Feng, Z.; Ye, J.; Yu, Y.; Song, M. Neural Style Transfer: A Review. *IEEE Trans. Vis. Comput. Graph.* **2019**, *26*, 3365–3385. [CrossRef] [PubMed]

32. Berger, G.; Memisevic, R. Incorporating long-range consistency in CNN-based texture generation. In Proceedings of the 5th International Conference on Learning Representations, ICLR 2017-Conference Track Proceedings, International Conference on Learning Representations, ICLR, Toulon, France, 24–26 April 2017.
33. Gatys, L.; Ecker, A.; Bethge, M.; Hertzmann, A.; Shechtman, E. Controlling perceptual factors in neural style transfer. In Proceedings of the 30th IEEE Conference on Computer Vision and Pattern Recognition, CVPR 2017, Honolulu, Hawaii, USA, 21–26 July 2017; pp. 3730–3738. [CrossRef]
34. de Lasa, M.; Hertzmann, A. Prioritized optimization for task-space control. In Proceedings of the 2009 IEEE/RSJ International Conference on Intelligent Robots and Systems, St. Louis, USA, 11–15 October 2009; pp. 5755–5762. [CrossRef]

Review

Recent Trends and Innovation in Additive Manufacturing of Soft Functional Materials

Jaime Eduardo Regis [1,2,*], Anabel Renteria [1,2], Samuel Ernesto Hall [1,2], Md Sahid Hassan [1,2], Cory Marquez [1,2] and Yirong Lin [1,2]

1   Department of Mechanical Engineering, The University of Texas at El Paso, El Paso, TX 79968, USA; arenteriamarquez@miners.utep.edu (A.R.); sehallsanchez@miners.utep.edu (S.E.H.); mhassan2@miners.utep.edu (M.S.H.); cmarquez10@miners.utep.edu (C.M.); ylin3@utep.edu (Y.L.)
2   W.M. Keck Center for 3D Innovation, The University of Texas at El Paso, El Paso, TX 79968, USA
*   Correspondence: jeregis@miners.utep.edu

Abstract: The growing demand for wearable devices, soft robotics, and tissue engineering in recent years has led to an increased effort in the field of soft materials. With the advent of personalized devices, the one-shape-fits-all manufacturing methods may soon no longer be the standard for the rapidly increasing market of soft devices. Recent findings have pushed technology and materials in the area of additive manufacturing (AM) as an alternative fabrication method for soft functional devices, taking geometrical designs and functionality to greater heights. For this reason, this review aims to highlights recent development and advances in AM processable soft materials with self-healing, shape memory, electronic, chromic or any combination of these functional properties. Furthermore, the influence of AM on the mechanical and physical properties on the functionality of these materials is expanded upon. Additionally, advances in soft devices in the fields of soft robotics, biomaterials, sensors, energy harvesters, and optoelectronics are discussed. Lastly, current challenges in AM for soft functional materials and future trends are discussed.

Keywords: additive manufacturing; soft materials; smart materials; stretchable devices

Citation: Regis, J.E.; Renteria, A.; Hall, S.E.; Hassan, M.S.; Marquez, C.; Lin, Y. Recent Trends and Innovation in Additive Manufacturing of Soft Functional Materials. *Materials* **2021**, *14*, 4521. https://doi.org/10.3390/ma14164521

Academic Editor: Tuhin Mukherjee

Received: 30 June 2021
Accepted: 6 August 2021
Published: 12 August 2021

Publisher's Note: MDPI stays neutral with regard to jurisdictional claims in published maps and institutional affiliations.

check for **updates**

1. Introduction

Soft materials have developed as the key materials to address challenges in engineering fields where flexibility, large motions, and lightweight are desired. These types of materials can be easily deformed by thermal and mechanical stresses owing to their low Young's modulus at room temperature (<100 MPa) [1] and high elongation without breaking. Additionally, soft materials can be found in various states such as colloids, liquids, gels, and colloids and polymers. Many soft materials display inherent structures or properties that can be significantly altered through external stimuli in a controlled manner and can be regarded as functional materials. Examples of these intrinsic functional properties include shape memory, dielectric, self-healing, and color-changing properties. Some soft materials may display more than one of these properties or can respond to multiple stimuli and are considered to be multifunctional. Additionally, soft material can be given magnetic, piezoelectric, or piezoresistive properties by incorporating functional filler.

Additive manufacturing (AM) of soft materials has been gaining popularity in recent years due to the growing interest in wearable electronics, tailored biomedical implants, and soft robotics. In the rapidly increasing market for soft devices, the one-shape-fits-all approach may soon no longer be the standard. Because of the limitations and cost associated with creating new geometries through traditional manufacturing methods, AM has been considered the future for soft material processing. In AM, a computer designed structure is sliced into individual layers and then these layers are physically realized through a variety of methods such as deposition of build material through a nozzle, jetting of binder onto a build material substrate, photopolymerization of a material vat, or thermal coalescence of powder particles. AM techniques enable freedom of geometrical design, customization with no added cost, and fabrication of complex geometries.

Materials that are otherwise rigid can be engineered to achieve softness and stretchability through advanced geometries and then be considered engineering soft materials [2]. Advanced geometries such as lattice and auxetic structures have previously been used to tune the mechanical properties of polymer and polymer composites which has allowed for some rigid polymers to have larger deformations without rupture. Examples of these are seen in many biomedical applications where material compliance to the geometry of the human body is necessary. The geometric freedom granted through AM has led to progress in these advanced designs; for this reason, engineering soft materials will be considered in this review.

The many AM technologies that have been developed have aided in the development of functional materials through intelligent design of structures, the development of stimuli-responsive 3D geometries, and functionality gradients through multi-material use in each individual layer. Next, different AM technologies and their application in soft functional materials manufacturing will be discussed. Then, the recent advancements, future trends and governing mechanism for AM processed functional materials will be detailed. Lastly, advances and challenges in applying these functional materials will be discussed.

1.1. Material Extrusion

Material Extrusion (ME) is an AM method that selectively deposits material through a nozzle onto a movable substrate in a layer-by-layer fashion to produce a three-dimensional part. ME is the most commonly used methodology, especially for rapid prototyping due to low cost, ease of use, and market availability. The most two common techniques for ME are fused deposition modeling (FDM) and direct ink write (DIW) also known as robocasting, paste extrusion (PE) or bio-extrusion (when applied for biomedical purposes).

FDM and DIW differ in the materials they use, and the way solidification is achieved. FDM uses thermoplastic filaments and has a fast solidification process through the cooling of the printed material below its glass transition temperature [3]. DIW does not require the temperature to achieve solidification and instead works with feed materials that flow due to a shear-thinning effect and retain their shape after deposition thanks to a high storage modulus (G').

FDM has been used to successfully fabricate soft functional devices using thermoplastic polyurethanes (TPU), and shape memory polymers (SMP). In contrast, since DIW only has the requirement of appropriate rheological properties for its print materials, it has expanded to include a large soft material selection including polymers with a wide range of molecular weights, liquid crystal elastomers hydrogels, and SMPs [4].

ME techniques are also of interest since they can incorporate multi-head nozzles to realize devices with elaborate designs with the use of selectively deposited support material or can produce devices with structural and functional regions by carefully placing different materials together using the different nozzles [5]. Common applications for these AM technologies have been sensors, soft robotics, biomedical, and wearable devices [6,7].

1.2. Vat Photopolymerization

Vat photopolymerization (VP) is an AM method in which a vat of liquid photopolymer resin is selectively cured by a light source in a layer-by-layer manner to construct a three-dimensional object. VP includes various techniques that differ in the way that the light source, typically ultraviolet (UV) light, is projected onto the liquid photopolymer reservoir [8]. In processes such as stereolithography (SLA), micro-stereolithography (u-SL), and two-photon polymerization (TPP), light is projected from a single-point source that traces the part's shape until a full layer is built. Processes such as digital light processing (DLP) and continuous liquid interphase production (CLIP) operate with a specialized projector that cures an entire layer at once.

VP is one of the more promising AM methods for soft material printing due to the ability to fabricate delicate objects with high resolution, close tolerances, and smooth surface finishes. Additionally, the ability to incorporate filler material into the resin allows for tunability of physical and mechanical properties leading to better control over the functionality of the final part. Despite the numerous advantages, there is a lack of available soft materials compatible with the technology. Moreover, end-use parts manufactured through VP often require more extensive post-processing steps than other AM methods and have limited production size and strength, narrowing the use of the technology for many practical applications.

VP has been previously used to print soft materials through free radical polymerization (FRP) by attaching acrylate and methacrylate functional groups to elastomer monomer, allowing for the formation of covalent crosslinks upon UV exposure. In recent years, however, research on soft material printing through VP has been moving towards click chemistries, especially thiol-ene reactions, due to the mild reaction conditions, insensitivity to oxygen or water, rapid polymerization rate, high efficiency, and low cytotoxicity. Thiol-ene click chemistries have become an efficient tool to covalently crosslink polysiloxanes into silicone elastomers, gels, and encapsulants, and have received great attention in 3D printing technologies since it exhibits high resolution and accuracy through instantaneous formations of crosslinked networks only at the local and temporal exposure to UV-radiation [9].

1.3. Material Jetting

Material jetting (MJ) is an AM method in which droplets of liquid photopolymer are selectively deposited and cured successively layer by layer. MJ offers fewer manufacturing difficulties than other AM methods such as vat polymerization, which ensures similar resolution between prints and a higher rate of production. MJ also provides a more efficient method of deposition, line-wise deposition, compared to the other AM methods previously discussed in which deposition is point-wise. Thus, the technology can achieve high accuracy and smooth surface finishes often without the need for post-processing.

MJ consists of several techniques, including drop-on-demand (DOD), PolyJet printing, and nanoparticle jetting (NPJ). DOD printers operate by accurately depositing photopolymer resins on a substrate and subsequently curing through UV-radiation layer by layer until the full structure is created. In PolyJet printing, an ultra-thin layer of photopolymer resin is sprayed on the build platform and cured through UV light. Gel-like support materials that can easily be removed by hand or dissolved are used in this technique to support complex geometries. Lastly, NPJ nanoparticles or support nanoparticles are incorporated in a resin that is sprayed onto a build platform in the form of tiny droplets. Solvents used for flow in the nanoparticle resin are then evaporated by the high temperatures inside of the machine, leaving behind structures made from nanoparticle materials.

MJ systems have become popular in recent years due to their capability for multi-material printing through DOD [10]. Although used for more advanced purposes, the MJ printing methods share many similarities to traditional 2D document printing systems making it possible for 3D printers to easily be used in an ordinary office rather than a laboratory environment. Advances in MJ technology have led to the fabrication of technology such as heat-responsive active composite structures [11] and functionally graded actuators for soft robots [12].

Although MJ is a promising AM method for soft materials, the technology, just like other AM methods, suffers from limited availability of materials that are printable. Additionally, wax-like materials, which are widely used due to their compatibility with MJ tend to be rather fragile, limiting the application for the structure produced through this method.

1.4. Other Additive Manufacturing Methods

Table 1 summarizes the AM methods previously discussed. The remaining AM methods (binder jetting, directed energy deposition, and sheet lamination) will not be discussed in further detail since these methods are typically reserved for processing metals, ceramics, or hard polymers and are not used for the development of soft structures (inherent and engineered) as previously described.

Table 1. Summary of advantages and disadvantages of additive manufacturing methods and techniques used for the fabrication of soft functional materials.

Printing Technology	Specific Methodology	Deposition	Feature Size	Materials	Features	Drawbacks
Material Extrusion	Fuse Deposition Modeling	Line	≈200 µm [13]	Thermoplastics	• Low maintenance • Low cost • Simplicity • Potential for multi-material printing	• Voids • Limited to complex geometric prints
	Direct Ink Write	Line	≈120 µm [14]	Thermoplastics Thermosets Elastomers Hydrogels Nanoparticles	• Large availability of materials • Rapid prototyping • Potential for multi-material printing	• Warping • Cracks • Post-processing
Vat Photopolymerization	Stereolithography	Light single point	≈50 µm [15]	Thermoset Elastomers Acrylate resins Nanoparticles	• High resolution • Close tolerances • Smooth surface finishes • Complex geometric prints	• Low material availability • Long printing times • Extensive post-processing
	Micro-Stereolithography	Light single point	≈10 µm [16]			
	Two-Photon Polymerization	Light single point	≈0.3 µm [17]			
	Continuous Liquid Interface Production	Light entire layer	≈0.4 µm [18]	Thermoplastic Acrylate resins Nanoparticles	• High resolution • Wavelength multiplexing	• Low material availability • Post-processing
	Digital Light Processing	Light entire layer	≈200 µm [19]			
Material Jetting	Drop on Demand	Drop	≈32 µm [20]	Polymers Thermoplastic Acrylate resins Elastomers Nanoparticles	• High accuracy • Little to no post-processing • Capability to fabricate functionally graded materials • Potential for multi-material printing	• Inconsistent material droplet spread
	Nanoparticle Jetting	Drop	≈10 µm [21]			

2. Materials and Methods

2.1. Shape Memory Polymers

Shape memory polymers (SPs) are stimulus-responsive materials that can store different geometries in memory as a "temporary shape" and then return to their permanent shape by applying an external stimulus. SMPs have a high entropy state at their permanent shape, however, when stress is applied and reaches plastic deformation at their transition temperature (T_{trans}), this allows them to reach a low entropy state. When cooling to temperatures below their T_{trans} without removing the stress, the entropy is frozen at a low state, which allows a temporary shape to be fixed by trapping the kinetic energy [22,23]. Lastly, when a stimulus is applied at T_{trans}, the chain mobility of the material is reactivated and returned to its high entropy state [22,23]. SMPs must have a soft and a hard component, forming an interlinked polymer chain to activate functionalization of the material with a reversible shape memory effect (SME). The soft component allows elastic deformations or the shape morphing at T_{trans}, while the hard component, usually a crosslinker, determines the permanent shape of the material. The performance of the SME for shape-memory materials is typically characterized by measuring their shape fixity ratio (R_s), which measures the material's capability to be deformed into a temporary shape, and their shape recovery ratio (R_r). Table 2 further summarizes AM processed SMPs, the AM method used for their fabrication, glass transition temperature, elastic modulus, elongation at break, durability, shape fixity and shape recovery.

Table 2. Summary of recent studies in the fabrication of SMPs using AM technologies and their impact on mechanical properties including shape fixity and shape recovery.

Materials	Technique	Glass Transition Temperature (°C)	Elastic Modulus (MPa)	Elongation at Break (%)	Durability (Cycles)	Shape Fixity (%)	Shape Recovery (%)	Elastic Modulus (MPa)	Elongation at Break (%)	Ref.
tBA, DEGDA with nanosilica fillers	DLP	56.23	-	85.2	10	100	90-97	-	85.2	[24]
filaflex embedded with polycaprolactone	FDM	70	48	700	10	76	97	48	700	[25]
N-butyl Acrylate	UV-assisted DIW	95.2	610	25.4 (Ultimate strain)	3	97.1	98.5	610	25.4 (Ultimate strain)	[26]
Polycyclooctene with boron nitrate and MWCNT	FDM	70	3.85 (Storage modulus)	-	-	98.9	99.2	3.85 (Storage modulus)	-	[27]
Polylactic acid (PLA)/Fe$_3$O$_4$ composites	FDM	66.6	1600 (Storage modulus)	-	-	96.8	96.3	1600 (Storage modulus)	-	[28]
poly(dimethyl acrylamide-costearyl acrylate and/or lauryl acrylate) (PDMAAm-co-SA)	SLA	-	-	-	3	99.8	87.6	-	-	[29]
2-Methacryloyloxy 4-formylbenzoate	DLP	57	57 ± 4.0	39.30 ± 1.0	3	97.5 ± 0.30	91.4 ± 0.20	57 ± 4.0	39.30 ± 1.0	[30]
Poly(ethylene terephthalate) (PET)	FDM	85–100	-	45	7	100	90–98	-	45	[31]
poly(ethylene glycol) dimethacrylate (PEGDMA), isobornyl acrylate and 2-ethylhexyl acrylate	DLP	125	-	-	10	92.6	95.3	-	-	[32]

SMPs have recently gained significant attention due to diverse advantages such as their light weight, flexibility of programming mechanisms, high shape deformability, biocompatibility, and biodegradability for actuator applications. SMPs are also attractive materials to fabricate diverse types of sensors for soft robotics and aerospace applications and for minimally invasive surgery devices for biomedical purposes. 3D printing offers an effortless way to fabricate more elaborate designs, AM has the flexibility to control some of the factors that have been found to affect SME's performance including geometry, print path direction, and thickness of the sample [33]. VP has been of interest for processing SMPs due to its in-situ polymerization process that allows the fabrication of elaborate geometries for very specific applications. SLA and DLP have been reported in the fabrication of origami structures, biomimetic, and soft robotic devices. Choong et al. evaluated the use of nanosilica dispersed in tBA-co-DEGDA photocurable resin using DLP to enhance nucleation and accelerate the polymerization rate, which significantly reduced fabrication time with Rs of 100% and Rr of 87% [24]. For ME, FDM and UV-assisted DIW have been reported in the fabrication of SMPs with biomedical purposes and soft robotics application [25]. Villacres et al. used the FDM technique to print a semi-crystalline TPU where they evaluated the effect of printing orientation and infill on the SME. It was found that a print angle orientation of 60° and 100% infill resulted in an increment of failure strain and strength where the infill content had a higher influence on mechanical properties [34]. Chen et al. fabricated tough epoxy and N-butyl Acrylate SMPs composites by UV-assisted DIW with Rs of 97.1% and Rr of 98.5% [26]. Lastly, Jeon et al. fabricated multicolored photo responsive SMP structures through Polyjet printing that showed different geometries when triggered by assorted color lights of different wavelengths [35].

There are two main classifications by stimuli response known as thermal-responsive and chemo-responsive. Thermal-responsive SMPs are triggered by applying heat to the material and raising their temperature up to their Ttrans. However, using a direct heating method could restrict their applications, which has led to the use of functional fillers to fabricate SMPs composites that trigger SME by alternative methods, such as electricity, magnetism, light, and ultrasound. Liu et al. fabricated multi-responsive SMPs using Polycyclooctene with boron nitride and multi-wall carbon nanotubes (MWCNTs) by the FDM technique. By using 20 wt.% MWCNTs in the composite, the SME could be triggered by heat (under water at 70 °C), light (100 mW·cm^2), and electricity (5 V) with outstanding properties, Rs of 98.9% and Rr of 99.2% [27] Zhang et al. fabricated PLA-Fe$_3$O$_4$ composites by FDM using magnetism (27.5 kHz) as an alternative stimulus for SME. It was found that a higher content of Fe$_3$O$_4$ led to a higher Rr, where PLA-Fe$_3$O$_4$-20% mass fraction gave the best results with Rs of 96.8% and Rr of 96.3% [28].

In chemo-responsive SMPs, the SME is triggered by altering the ionic strength to promote plasticizing and lower Ttrans below room temperature [36]. The most common method consists of submerging the SMP in a medium, such as an organic solvent or water, that triggers the plasticizing. Recovery time can be decreased by reducing the dimensions of the polymers to micro-fibers [37]. Solvent-responsive SMPs commonly report the use of organic solvents such as ethanol, dimethyl sulfoxide methanol, and N-N dimethylformamide (DMF) [3, 12]. Some water responsive SMPs includes hydrogels (Polyvinyl alcohol, polyethylene glycol) [38] and TPUs [37]. Shiblee et al. fabricated water-responsive shape memory gels by SLA process using poly (dimethyl acrylamide-costearyl acrylate and/or lauryl acrylate) (PDMAAm-co-SA) by incorporating hydrophilic and hydrophobic monomers in the formulation. This shape-memory gel showed an Rs of 99.8% and Rr of 87.6% after the first cycle and Rr of 99.8% after the second cycle, the authors attributed the change of Rr to the training phenomenon [29].

Thermal-responsive SMPs are mainly activated by hot programming, which consists of heating the material to its T$_{trans}$. The main advantage of hot programming is a high R$_s$, a minimal springback and it usually requires a small amount of applied stress to produce a plastic deformation [39]. Li et al. fabricated Bisphenol-A glycerolate diacrylate (BPAGA) SMPs by DLP technique using hot programming method at glass transition temperature obtaining R_r of about 97% and R_s of 100% [40]. On the other hand, chemo responsive SMPs are activated by either hot or cold programming. Cold programming is possible below T_{trans} and usually occurs at room temperature. However, cold programming is usually more challenging since some thermosets are brittle at their rubbery point leading to possible fractures [39]. Keshavarzan et al. evaluated hot programming and cold programming methods for BCC and rhombic structures using 3DM-LED.W, which is a commercial SMP resin for DLP. It was found that cold programming is beneficial for higher energy absorption while hot programming obtained a higher shape fixity ratio, it was also noticed that rhombic structures have a better energy absorption and recovery due to higher strength and stiffness [32].

SMPs show different behaviors according to the number of geometries that can be stored in memory, which depends on the network elasticity of the material [36]. Lastly, multi-SMPs are materials that can learn more than three geometries additionally to their permanent shape. Peng et al. synthesized triple SMPs by using poly(ethylene glycol) dimethacrylate (PEGDMA), isobornyl acrylate and 2-ethylhexyl acrylate through DLP. The SMPs were able to store two different geometries in memory with an Rs of 92.6% and Rr of 95.3% without a significant degradation after 10 cycles, proving the effectiveness of SMPs [41].

SMPs with chemical crosslinking are usually thermosets and have stronger bonds than physically crosslinked polymers and higher shape recovery. However, SMPs with chemical crosslinks cannot be reprocessed unless they have dynamic bonds. Some SMPs with chemical crosslinking take advantage of dynamic chemistries such as transesterification, transcarbamoylation, Diels–Alder bonds, disulfide bonds, diselenide bonds, and imine bonds [38]. Thermadapts are a type of SMPs with dynamic covalent bonds that have recently gained attention due to their capability to change the temporary shape after curing. Some of the dynamic covalent bonds used to fabricate SMPs are hindered urea bonds and triazolinedione. Miao et al. developed thermadapt SMPs (2-Methacryloyloxy and 4-formylbenzoate) with dynamic imine covalent bonds using DLP that allowed changing the temporary shape after printing for different actuation purposes that can be useful for soft robotics applications [30]. Davidson et al. used LCEs to develop thermadapt SMPs by radical-mediated dynamic covalent bonds using the hot DIW technique. When exposed to UV light during actuation, the exchangeable bonds that allow the change of the permanent shape of LCE are activated [42]. Some SMPs that reported the use of physical crosslinking include hydrogen bonds, ionic bonds, π-stacking, charge transfer interactions [38]. Chen et al. synthetized PET copolyester using π-stacking synergistic crosslinking to induce enhance shape memory properties by the FDM technique. The optimal copolyester was P(ET-co-PN) 20 with an Rs of 100% and Rr of 98%, it was also found to have some levels of self-healing due to π-stacking crosslinking and flame retardant properties due to the nature of PET [31].

SMPs have a diverse range of applications due to their unique mechanism, where 3D printing contributes to the evolution of elaborate designs. Many efforts to control the responsiveness by alternative methods besides direct heating have been made. An interesting research direction could be the development of SMPs with dual responsive mechanisms for different purposes that expand their fields of application. Furthermore, multi-material printing may allow the fabrication of SMPs that can store multiple geometries in memory. The evaluation of 3D printing structures to fabricate reprocessable SMPs with dynamic covalent bonds is another interesting research direction that can redefine SMPs' functionality.

2.2. Self-Healing Materials

Self-healing polymers are a branch of functional materials designed to take advantage of intricate physical or chemical processes to reform broken bonds caused by mechanical damage. The ability of self-healing polymers to respond to damage that may be difficult to detect, helps prevent the propagation of cracks or ruptures that result from the polymer's exposure to fatigue, abrasion, and other deteriorating forces during regular operation. Recent advances in AM have led to an increase in the development of self-healing materials that overcome the design limitations of traditional casting methods, resulting in self-healable structures with increased complexity and tunable properties. For this reason, the application of self-healing polymers has extended beyond protective coatings to wearable devices, implantable biomedical devices, health monitors, and electronic skins.

The healing efficiency of AM processed self-healing polymers is often measured through the restoration percentage of physical properties such as fracture strain and corresponding tensile or compressive stress. Additionally, the recovery time can also be an indicator of performance and in the case of non-autonomic processes, the activation energy required to trigger the self-healing process, which can be calculated through the Arrhenius equation. Table 3 further summarizes AM processed self-healing polymers, the AM method used for their fabrication, their functional chemistry, recovery performance, recovery conditions, and applications.

Table 3. Summary of recent studies on AM processable self-healing polymers highlighting mechanical robustness and healing efficiency.

Materials	Tensile Strength	Max Strain	Technique	Application	Self-Healing	Stimulus	Healing Time	Efficiency	Ref.
Semi-interpenetrating polymer network elastomer	5 MPa	600%	UV-assisted DIW	Biomedical Devices	Embedded semicrystalline thermoplastic	Heat at 80 °C	20 min	<30%	[43]
Ferrogel	-	288%	DIW Bioprinting	Drug Delivery and Tissue Engineering	Reversible Imine Bond Formation	No Stimulus	10 min	~95%	[44]
Dynamic Covalent Polymer Networks	3.3 MPa	140%	FDM	-	Diels–Alder Reaction	Heat at 80 °C Deionized Water at RT	12 h	96% ~70%	[45]
Photoelastomer Ink	16 kPa	130%	SLA	Soft Actuators, Structural Composites, Architected Electronics	Disulfide Exchange	Heat at 60 °C	2 h	100%	[46]
Fluid Elastic Actuators	13–129 kPa	45–400%	SLA	Soft Robotics	Unreacted Prepolymer Resin	Sunlight ~15,000 cd m^2	30 s	-	[47]
Physically Crosslinked Hydrogels	95 kPa	1300%	SLA	Flexible Devices, Soft Robotics, Tissue Engineering	Hydrophobic Association	Contact	6 h	~100%	[48]
Silicone Elastomer	~225 kPa	~330%	SLA	Endurable Wearables, Flexible Electronics	Ionic Bonding	Heat at 100 °C	12 h	>90%	[49]
Host–Guest Supramolecular Hydrogel	0.3–0.5 MPa	70%	DIW	Biomedical	Host–Guest Interactions	Mechanical Stress	1 h	Up to 80%	[50]
PEDOT:PSS with Polymeric Surfactant	3 MPa	35%	DIW	Energy Harvesting	Surfactant	Electrical Current 10–3 A	1 s	85% electrical output	[51]
Polyurethane Elastomer	3.39 ± 0.09 MPa	400.38%	DLP	-	Disulfide Exchange	Heat at 80 °C	12 h	95% First healing	[52]

In the past two years, there has been a surge in the AM of crosslinked gels with reversible imine Schaff bonding. For example, Kim et al. demonstrated the design and preparation of biocompatible, polysaccharides-based, self-healing hydrogels [53]. These hydrogels were processed through extrusion-based bioprinting to form stable geometries such as donuts, disks, and filamentous structures. The hydrogels displayed autonomic healing ability in ambient conditions (room temperature in air). The healing was based on reversible crosslinks comprising of imine bonds and hydrazine bonds, that were capable of completely restoring functionality within 10 min. The same research group took the concept one step further by incorporating iron oxide nanoparticles. This ferrogel was capable of fully recovering autonomously after gel breakage in a lapse of 10 min in three different conditions (in air at room temperature, in a buffer solution, and under a magnetic field). Additionally, this ferrogel also demonstrated shape memory capabilities that were triggered by the presence of a magnetic field [44]. In another example, Lei et al. synthesized a gelatin-based self-healing hydrogel from dialdehyde carboxymethyl cellulose and amino-modified gelatin. The hydrogel showed good fatigue resistance by recovering its original strength during 10 cyclic compressive loading and unloading tests and by having a healing efficiency of up to 90% after being heated for 1 h at 37 °C. Additionally, the hydrogel possessed ideal hemocompatibility and cytocompatibility, making it a prospective candidate for injectable tissue engineering scaffolds [54].

While self-healing gels are mainly processed through material extrusion AM techniques, most self-healing elastomers are processed through UV-based methods such as VP and MJ. Due to the nature of the technology and the rigidity of elastomers compared to gels, AM processed self-healing elastomers can obtain more complex designs, higher resolutions, better surface quality, and overall faster printing speeds, especially through the use of click chemistries such as thiol-ene photolymerization.

Pertaining to VP, Liu et al. demonstrated the fabrication of hydrolysis-resistant silicone elastomers through photopolymerization conversion of vinyl in thiol-ene photoreactions in a stereolithography process [49]. This self-healing silicone elastomer demonstrated a healing efficiency higher than 90% when healed at 100 °C for 12 h. The silicone elastomers could be healed multiple times through reversible ionic crosslinks without losing significant strength. Furthermore, the silicone elastomers were shown to be reprocessable, retaining 85% of their original strength when pulverized and re-casted. Similarly, Yu et al. developed a photocurable PDMS-based elastomer through the same photopolymerization and AM technique but with dynamic covalent crosslinks [46]. Through disulfide exchange, this photoelastomer was able to completely regain 100% of its original strength in significantly less time (2 h) and under milder conditions (60 °C) than the silicone elastomer Liu et al. prepared, however, with less translucency.

In regard to UV-light-assisted methods, Kuang et al. developed a novel semi-interpenetrating polymer (semi-IPN) composite for UV-light-assisted DIW that displayed self-healing properties [43]. Photocurable resin composed of urethane diacrylate, and n-butyl acrylate was incorporated to assist in the shape retention of the printed beads of the material when post-cured after extrusion of every layer. This allowed the group to fabricate complex structures with high stretchability like an Archimedean spiral capable of stretching over 300% strain with negligible in-plane anisotropy. The healing in the semi-IPN elastomer is based mainly on the diffusion of a semicrystalline polymer, polycaprolactone, and partially from hydrogen bonding between urethanes. As such, the elastomer demonstrated the ability to heal micro-cracks (3 mm long and 30 μm wide) after heat treatment at 80 °C for 20 min leaving only slight scarring. The elastomer was also able to heal larger cracks such as a notched gap, however, the healing efficiency was relatively low (<30%) in comparison to the self-healing materials previously discussed. Additionally, the elastomer possessed shape memory capabilities associated to the crystalline component of the elastomer. While self-healing elastomers still have a long way to go to be ready for consumer-based applications, in terms of mechanical robustness, significant progress has been made over the past few years to achieve faster healing under mild conditions (low temperatures/pressure) and better printability.

Despite self-healing materials being some of the most varied, AM processable self-healing polymers are still relatively limited due to the novelty of AM technology. One of the shortcomings of self-healing polymers is their fragility, especially for gels. Through free-standing AM techniques such as DIW and MJ, the hydrogels are susceptible to collapsing under their own weight. Although the use of support particulate gel beds is not uncommon, the manufacturing process of the particle can significantly affect the quality of the gel. For this reason, Senios et al. developed a fluid-gel bed that provides support to 3D extruded structures and prevents them from collapsing under their own weight prior to being crosslinked [55]. This process, known as suspended layer additive manufacturing (SLAM), was shown to be able to overcome limitations associated with printing low viscosity inks such as spread when depositing and sagging in multilayered structures, thus allowing for the fabrication of hydrogels with even more intricate designs than previously achievable. The development of AM assisting techniques such as SLAM, will help broaden the material catalog and allow for their use in a wider range of applications and push for commercial goods such as wearable electronics.

2.3. Electronic Polymers

Electronic soft polymers are those that exhibit changes to their electronic properties, such as polarization, capacitance, or resistance when exposed to mechanical, thermal, light, or pH stimulus. These characteristics have made electroactive polymers useful for actuating [56], sensing [57], and energy harvesting [58] applications, and to develop as a very important research area.

There exist various families of electronic polymers defined by the governing physical mechanism of their functionality. Examples of electronic polymers include dielectric elastomers, piezoresistive polymers, and piezoelectric polymers. These polymer families have been explored for AM to produce conformable, smart structures with programmed sensing and actuating behaviors.

Dielectric elastomers (DEs) are electroded elastomers that respond with large actuation to applied electric fields due to their high compliances. The Coulombic forces that result from the applied electric field cause a reduction in the thickness and anisotropic expansion of the electrode area of the elastomer. DEs are quite useful as actuators for biomedical [59] and soft robotics [56] because of their large strains, low noise, and quick response times. However, DEs have inherent disadvantages such as requiring large electric fields upwards of 100 kV/mm [60] and having isotropic non-directed deformations.

AM processes have been used in recent years to obtain three-dimensional dielectric elastomer actuators (DEAs). AM provides advantages to DEs such as the ability to build different elements such as the DE films, the electrodes, or the rigid frames together using a wide variety of deposition methods. ME and MJ technologies have been mainly used for the manufacturing of DEs [61], Progress in MJ of DEAs has been driven by the development of rubbers suitable for jetting into films with controlled distribution and thickness. For example, AM patterned films were obtained through MJ using commercial silicone rubber by diluting them in a solvent to obtain suitable jetting properties prior to printing [61]. Careful design of the printing inks resulted in prints with comparable properties to traditionally casted films. Another approach to building DEAs using MJ that has been explored is the aerosol jetting of graphene oxide electrodes onto DE films [62]. The capability to pattern electrodes onto dielectric elastomer films allowed for electrode patterns that were unaffected by substrate stretching and it was proposed that stacked DEAs could be produced by alternating layers of silicone and graphene oxide jetting.

Multi-material manufacturing approaches enabled by AM have been used to fabricate fully functional DE actuators. In one example, a unimorph cantilever was built by selective deposition of an active barium titanate/silicon DE layer, a passive stiff silicon layer, and ionogel electrodes [63]. All different elements were built using a single AM process and the resulting cantilever bent upon the application of an electric field. In a similar multi-material approach, 3D actuators were manufactured by printing rigid thermoplastic frames on a prestretched acrylic DE substrate using FDM [64]. Stretched DEs were used as a substrate so that once the print was finished and the substrate was released, the contraction of the substrate would transform into bending due to the mismatch of modulus across the thickness. The resulting curled-up structures followed predictions using minimization of energy approach and showcased how three-dimensional complex actuators could be built using intelligent design of printed patterns onto a DE substrate. Similarly, honeycombs of TPU were patterned using FDM into both sides of acrylic DE substrates to obtain anisotropic unidirectional actuators [65]. Higher degrees of anisotropy were obtained by varying the rib angle of the honeycombs and increasing the pre-stretch ratios of the DEs during print. Under optimal conditions, an axial strain of 15.8% with only -0.97% transverse strain was achieved by loading with 7.5 kV.

Overall, the necessity to pre-stretch DEs to obtain useful actuation behaviors limits the incorporation of these materials into all AM processes. DEs will see further development only in multi-material approaches where complete actuators can be built in one AM process. Still, smart design for AM will continue to develop and result in actuators with efficient electromechanical energy transfer.

Piezoresistive materials respond with a change in electrical resistance when strained. In the case of piezoresistive polymer composites, a network of electrically conductive fillers embedded in a thermoplastic or thermoset elastomer matrix is disturbed by strain causing a variation in the electrical conductivity of the composites [66]. This variation in electrical resistance as a function of strain if large, can be used to accurately measure strain. For piezoresistive polymer composites, sensitivity is maximum when the concentration of the filler approaches what is known as the percolation threshold. At this concentration, it is possible for agglomerations to form and impact the sensitivity. Thus, manufacturing methods for piezoresistive composites must ensure that agglomerations do not occur. Among the suitable manufacturing processes for piezoresistive composites, AM has emerged as one of the most important prospects because of the freedom in design, and because of the control of filler alignment possible in processes such as ME.

Piezoresistive soft polymer composites have been manufactured through ME using different matrix and filler systems [67–71]. For example, sensors of TPU with CNT fillers were manufactured using FDM [68]. The use of AM enabled new or enhanced properties in some cases. For example, biaxial strain sensors made of TPU with CNT fillers were manufactured using FDM [68]. The different patterns of CNT electrode deposition allowed for different designs, each with its own sensitivities to axial and transverse deformations, with largely unaffected mechanical properties (all materials exhibited $\approx 50\%$ axial strain at 4 MPa loading). In another work, a hierarchically porous lattice of TPU was printed and bonded to a stretchable matrix of the same TPU and used as a conformable sensor [70]. The macroscale porosity was controlled by the spacing of the struts, while intermediate and small-scale porosities were achieved through sacrificial fillers burned out after printing. The hierarchical porosity achieved anisotropy in response allowing the sensor to be bonded to curved substrates without affecting its pressure sensitivity. This was not previously possible for conventionally casted sensors. Other approaches used in AM of piezoresistive polymer composites to enhance their performance have consisted of using particle–matrix interface modifiers [67] and embedding CNTs to a printed elastomer lattice using partial melting [69].

Piezoresistive soft polymer composites have been manufactured through ME using different matrix and filler systems [67–71]. The use of AM enabled new or enhanced properties in some cases. For example, biaxial strain sensors made of TPU with CNT fillers were manufactured using FDM [68]. The different patterns of CNT electrode deposition allowed for different designs, each with its own sensitivities to axial and transverse deformations, with largely unaffected mechanical properties (all materials exhibited $\approx 50\%$ axial strain at 4 MPa loading). In another work, a hierarchically porous lattice of TPU was printed and bonded to a stretchable matrix of the same TPU and used as a conformable sensor [70]. The macroscale porosity was controlled by the spacing of the struts, while intermediate and small-scale porosities were achieved through sacrificial fillers burned out after printing. The hierarchical porosity achieved anisotropy in response allowing the sensor to be bonded to curved substrates without affecting its pressure sensitivity. This was not previously possible for conventionally casted sensors. Other approaches used in

AM of piezoresistive polymer composites to enhance their performance have consisted of using particle–matrix interface modifiers [67] and embedding CNTs to a printed elastomer lattice using partial melting [69]. Ionic gels are another class of piezoresistive materials that have been developed for AM. A group of researchers developed a shear-thinning ionic gel that could be patterned into 3D structures and studied how a reentrant honeycomb structure enabled 310% larger elongations and sensitivity as compared to a traditional film [72]. Another group further developed ionic gels for printing using eutectic solvents as the media for better stability post-printing [73]. Once again, the freedom of design from AM was used to construct auxetic structures that offered enhanced strain sensitivity with a max GF of 3.30 and a strain of 300%.

Piezoresistive polymers have been widely developed for AM using ME. Further control of the porosity, and design of metamaterial structures will enable enhanced sensitivity and a broader range of operation. However, other AM processes with higher resolutions such as VP will be needed to develop smaller piezoresistive polymer sensors for use in MEMS.

Piezoelectric materials possess a permanent polarization that when disturbed through mechanical loading produces a voltage across the material. These materials are electromechanically coupled so mechanical loads produce voltages and applied voltages cause strains. Because of this characteristic, piezoelectric materials can function as both electrically driven actuators [74] and as mechanical sensors [75].

Flexible piezoelectric materials have been developed with the help of AM by printing polymers as well as nanocomposites with piezoelectric ceramic fillers. Optimization of concentration, along with intelligent structure design, has allowed materials to exhibit larger coupling coefficients as well as increased elongations. Strategies for AM of soft piezoelectrics have focused on intelligent structure design to overcome the inherent stiffness of common piezoelectric materials such as PVDF and ceramics. For example, Li et al. used electrical field-assisted FDM to produce piezoelectric sheets with designed deformations made with nanocomposites of sodium niobate ceramics and PVDF [76]. Chiral patterns were built into the sheets to allow large deformations. Thus, once the sheets were rolled onto artery-like structures they could expand to sense radial pressures such as those found in blood flow inside the human body. Similarly, Yao et al. developed flexible and wearable piezoelectric sensors using lattice patterns through DLP [77]. Highly sensitive but soft piezoelectric lattices were possible thanks to surface functionalization of the piezoelectric ceramic fillers, which enhanced mechanical energy transfer at lower solid loadings. Three-dimensional honeycomb structures were printed using the high-resolution photopolymerization printing method and the performance of the sensors surpassed piezoelectric polymers in sensitivity and compliance. Another approach towards building compliant piezoelectric structures through AM consisted of infiltration of a ceramic lattice with PDMS elastomer [78]. This strategy allowed for complex structures to be built using only a ceramic-filled resin using SLA while still being able to obtain complaint structures afterwards.

Soft piezoelectric materials have been developed for AM despite the limited selection of polymer systems that exhibit piezoelectric polymers and the high stiffness of bulk nanocomposites of piezoelectric ceramics and elastomers. Continued development of metamaterial structures and identification of unique piezoelectric behaviors will continue to drive AM of soft piezoelectric structures composed of polymers and ceramic fillers.

Table 4 below summarizes the different efforts to print electronic polymers for sensing and actuation applications, highlights the specific elements fabricated using AM, their softness, and their individual performance.

Table 4. Summary of various soft electronic polymers recently manufactured through AM for use as sensors and actuators and their performances achieved.

Material	Application	Role of Printed Material	Young's Modulus or Elasticity	Max Strain	Printing Technique	Performance	Ref.
Silicone elastomer	Dielectric actuator	Actuating layer	$\approx$700 kPa	600%	DOD	A maximum areal strain of 6.1% at an electric field of 84.0 V/μm	[61]
Reduced graphene oxide-elastomer nanocomposites	Dielectric actuator	Flexible electrode layer	-	104%	Aerosol Jet Printing	Electrodes with a maximum stretchability of 100% could be bonded to dielectric layers without losing conductivity	[62]
Thermoplastic polyurethane/carbon nanotubes/silver nanoparticles composites	Dielectric actuator	Actuating material	3.44 MPa in print direction	Up to 800% in print direction	FDM	Dielectric constant of 6.32 and a radial extension of 4.69% at an applied 4.67 kV	[79]
Barium titanate filled silicone elastomer	Dielectric actuator	Actuating layer	39.82 kPa	>100%	DIW	Maximum tip displacement of 6 times its thickness at 5.44 kV Blocking force of 17.27 mN and deflection of 0.85 mm under a 0.12 g mass with a 5 kV applied	[63]
Thermoplastic elastomer	Dielectric actuator	Elastic frame	-	-	FDM	A tilt angle of 128° and a blocked force of 24 mN were measured when driven by 6 kV	[64]
Thermoplastic polyurethane	Dielectric actuator	Elastic frame	-	-	FDM	Honeycomb structures could achieve a longitudinal strain of 15.8% and transverse strain of -0.97% at a driving voltage of 7.5 kV	[65]
Thermoplastic polyurethane and multiwalled carbon nanotubes	Piezoresistive sensors	Sensor and electrodes	-	>100%	FDM	Anisotropic electrical resistance responses to strain with gauge factors between 1.5 and 3	[68]
Thermoplastic polyurethane and carbon nanotubes	Piezoresistive sensor	Sensor	$\approx$1 MPa at a strain of 30%	-	FDM	Consistent 50% change in developed current during 1500 cycles of 5% strain	[67]
Thermoplastic elastomer	Piezoresistive sensor	Sensor body	-	800%	FDM	Pressure sensitivity as high as 136.8 kPa^{-1} at pressures <200 Pa and GF of 6.85 at stretching	[69]
Thermoplastic polyurethane, carbon black, and silver composites	Piezoresistive sensor	Substrate, sensor, and electrode layers	-	600% for TPU, 120% for electrode layer	DIW	Low sensitivity to in-plane stretching of (R/R$_0$ < 7%) and pressure sensitivity of 5.54 kPa^{-1} at low pressures (<10 kPa)	[70]

Table 4. *Cont.*

Material	Application	Role of Printed Material	Young's Modulus or Elasticity	Max Strain	Printing Technique	Performance	Ref.
PDMS and multi-walled CNT	Piezoresistive sensor	Conductive pattern	-	>70%	DIW	GF of 13.01 with linear responses up to 70% tensile strain	[71]
Ionogels	Piezoresistive sensor	Sensor	Tensile strength of 0.81 MPa at 242% strain	242%	DIW	≈40% change in electrical current under a 29% elongation	[72]
Cellulose nanocrystals and deep eutectic solvent ionogel	Piezoresistive sensor	Sensor	0.20 MPa	At least 790%	DIW	Up to a 3.3 GF in the strain range up to 300% strain	[73]
Barium titanate	Piezoelectric sensor	Sensing lattice	-	-	Ceramic slurry DLP	Compressibility up to at least 10% and direct relationship between recorded voltage and applied strain	[78]
Polyvinylidene fluoride and sodium potassium niobate	Piezoelectric sensor	Sensor	<1.0 MPa for designed structure	40%	FDM	Pressure sensitivity of 2.295 mV kPa^{-1} and ability to self-power	[76]
Lead Zirconate Titanate	Piezoelectric sensor	Sensor	Compliance up to 3 × 10^{-8} Pa	-	Ceramic slurry DLP	Variable piezoelectric charge constant up to 110 pC/N and sensitivity to pressure and extension with high conformability to surfaces	[77]

2.4. Chromic Materials

Chromic materials have the ability to change in appearance in their refractive index (e.g., color, fluorescence, brightness, transparency) when applying different stimuli such as temperature, mechanical stress, electricity, pH concentration, among others. Chromic materials are of interest due to their reversible optical mechanism that can be incorporated into wearable devices for sensing, and for soft robotics. Table 5 further summarizes AM processed SMPs, the AM method used for their fabrication, applied stimuli, color change, absorbance wavelength and reversibility.

The use of AM for soft chromic materials provides an effortless method to explore different geometrical designs and obtain tunable, mechanically activated chromic (also known as mechanochromic) responses with reversible optical properties. For example, Rohde et al. used the DIW technique to explore different geometries for mechanochromic composite elastomers, using PDMS microbeads as a matrix with spiropyran aggregates as a functional filler [80]. Activation of the chromic mechanism was possible by applying either a low mechanical strain under uniaxial tension or compression. The soft elastomer composite showed reversible mechanochromic properties displaying a purple color in the area of applied mechanical force and returning to white after releasing such force. Chen et al. fabricated highly stretchable photonic crystal hydrogels with reversible mechanochromic properties by DIW technique [81]. The physically crosslinked poly(butylacrylate) (PBA) composites provided a high elongation at break of 2800% and reversible color change from blue to grey under tension and compression.

Table 5. Summary of chromic materials and their properties that have been recently processed through AM techniques.

Materials	Matrix	Response Type	Technique	Applied Stimuli	Color-Change	Absorbance Wave-length (nm)	Reversibility	Ref.
Spiropyran	Polydimethylsiloxane	Mechanochromic	DIW		Off-white to Purple		Yes	[80]
Poly(butyl acrylate)	Polyacrylamide	Mechanochromic and Hydrochromic	DIW	Compression: 5.7 kPa	Entire Wave-length Spectrum Colors	500–900	Yes	[81]
Polyethyleneimine-co-poly(acrylic acid)	Polyethylene glycol diacrylate	Hydrochromic	DIW		Blue-Green, and Red	400–650	Yes	[81]
Polyurethane acrylate	Isobornyl acrylate	Thermochromic	Projection Micro Stereolithography	74.2–81.7 °C	Black, Red, Blue, Yellow, White	400–800	Yes	[82]
Poly(N-isopropylacrylamide)	Silica-alumina based gel	Thermochromic and Electrochromic	DIW	>60 °C and 0.6–1.8 Amp (2~6 V)	Transparent to Opaque State	1400–1900	Yes	[83]
Poly(3,4 ethylenedioxythiophene)-poly(styrene sulfonate) (PEDOT:PSS)/silver (Ag)	Grid/polyethylene terephthalate	Electrochromic	IJ	−0.6–0 V	Transparent, Black	633	Yes	[84]

Additionally, AM has contributed to the fabrication of complex inks with moisture-activated chromic properties (often referred to as hydrochromism) that allow the obtainment of multiple color changes. For example, Yao et al. developed a 3D printable hydrogel ink for the DIW technique to develop soft actuators with shape memory and appearance tuning properties [85]. By using polyethyleneimine-co-poly (acrylic acid) (PEI-co-PAA), hydrochromism was produced, showing tunable luminescence from blue to green, which can be controlled by water absorption of the actuators in the range of 20% to 90% relative humidity. Moreover, by the incorporation of fluorophore-lanthanide an additional red color was also tunable by water absorption in the sample. The hydrochromic soft actuators also showed a reversible change in opacity from opaque to transparent produced by phase separation caused by dehydration.

Thermally activated chromic (often referred to as thermochromic) are the most reported chromic materials due to the simplicity of their chromism mechanism tuning. One example is Chen et al. who developed a 3D printable resin with polyurethane acrylate (PUAs) oligomer and isobornyl acrylate (IBOA) monomer with shape memory properties for the μSL process [82]. By the addition of thermochromic microcapsules, it was possible to fabricate self-actuating devices with reversible change colors from red to white with tunable glass transition temperature from 74.2 to 81.7 °C.

Electrically activated chromic materials (also known as electrochromic) have been of interest for soft functional devices as an alternative method to used temperature to trigger a color change. For example, Zhou et al. used DIW technique to fabricate a device using poly(N-isopropylacrylamide) (PNIPAm) as functional particles dispersed in an Si/Al sol-gel, producing a hybrid hydrogel (PSAHH) with reversible appearance properties from opaque to translucent that could be triggered by heat or electricity [83]. These reversible thermochromic and electrochromic properties could be triggered by heating the samples above 60 °C or by increasing current from 0.6 to 1.8 Amp (2~6 V). The change in the sample's appearance was due to the temperature-induced dehydration of

PNIPAm particles, which acted as light scattering fillers. Cai et al. used ink-jet printing to fabricate electrochromic WO3-PEDOT:PSS composites printed on flexible substrates with electrochromic properties [84]. The flexible device showed a fast electrochromic response even under bending conditions in the range of −0.6 to 0 V transitioning from transparent to black and with good electrochemical stability up to 10,000 cycles.

Chromic materials are an emerging research area that has recently found its way into AM techniques for the fabrication of soft functional devices. Due to the high potential of chromic to fabricate wearable devices, it is expected to see future research trends taking advantage of AM to develop multi-responsive devices with reversible chromic properties.

2.5. Multifunctional Soft Materials

Multifunctional materials are those that present two or more functionalities due to their inherent properties or when combined with other functional materials as composites. The functionalities that make up multifunctional materials can be a combination of shape memory, self-healing, actuation, sensing, optical, biological, elastic, etc. In engineered multifunctional material systems, properties are carefully selected to achieve the desired multifunctionalities based on the field of applications. For example, multifunctional biomaterials must first present a therapeutic functionality and then may present added functionality including sensing of body temperature and pressure, or actuation [86]. Other engineered multifunctional materials can provide structural support in demanding environments while providing additional functionality to address very strict requirements [87]. The applications of such materials include energy [88], medicine [87], nanoelectronics, aerospace, defense, semiconductor, and other industries.

Multifunctional materials reduce system complexity by having one material perform functions that would be otherwise performed by multiple different materials. This is beneficial in applications such as soft robotics where weight reduction and simplicity are some of the key characteristics and the use of the least number of materials ensures the best performance possible [89]. The integration of multifunctional materials into structures requires material compatibility and adhesion between different components. AM allows a seamless transition from structural to functional sections through material gradients [90]. Thus, the combined development of materials with multiple functionalities together with the development of AM techniques that easily transition between materials allows for the simplest, most size effective structures. Table 6 below summarizes the different multifunctional materials and composites that have recently been developed using a variety of AM processes.

Table 6. Different multifunctional soft materials recently manufactured using AM techniques and their applications.

Materials	Modulus	Max Strain (%)	Technique	Application	Functionalities	Ref.
Polyacrylamide	17 KPa	574	DIW	Biocompatible Soft Robotics	Magnetic response	[91]
Polyacrylamide with Carbomer	40 KPa	260	DIW	Biocompatible Soft Robotics	Magnetic response	[91]
PLA-PEA	125 MPa	2.5	DIW	Actuation and Sensing	Shape memory effect and piezoelectric effect	[92]
Polypyrrole (PPy)	498 kPa	1500	DIW	Sensor	Self-healing	[93]
Polydimethylsiloxane (PDMS)	160 kPa	210	DIW	Sensor	Superhydrophobicity	[94]

One of the families of advanced soft materials with the greatest potential for multifunctionality enabled through AM is hydrogels. These materials possess molecular networks swollen in water that, when subject to stimuli such as temperature or strain, may undergo gelling. Gelling constitutes a physical change where the stiffness of the hydrogel, as well as other relevant properties such as the swelling behavior, are altered. Thus, these materials possess sensing and shape change capabilities. Additionally, due to their high inherent compliance, they offer the potential to form functional composites with metallic or carbon fillers without sacrificing their softness. Functional hydrogel composites with electrically [93] or magnetically [31,71,72] responsive fillers have been enabled through AM. For example, hydrogels with self-healing molecular networks have been combined with conductive carbon fillers and used to fabricate complex sensing and healing structures through DIW [93]. Similarly, self-healing hydrogels with magnetic iron particle fillers were printed using DIW [44]. The structures could heal damage over time through reversible imine-bond formations, and the printed structures also exhibited macroscopic shape change in the presence of a magnetic field. Additionally, hydrogels were printed simply through DIW with the use of carbomer as a rheological modifier and their multifunctional properties were showcased [91]. The printed hydrogels showed time-dependent actuation in a hot (50 °C) water environment due to a phase transition and deswelling of one of the printed layers at the water bath temperature. The showcased hydrogels were also mixed with magnetic particle fillers and printed to form a biologically inspired octopus structure. This structure could locomote in the presence of a moving magnetic field and demonstrated the potential to obtain soft robotic nature mimicking structures through DIW [91]. Though magnetic hydrogel composites are capable of motion when a magnetic field is applied, their magnetic response is still considered weak. Thus, Tang et al. instead obtained 3D printed actuating structures that worked through the magnetothermal effect by introducing an alterning magnetic field causing heat and degradation of the hydrogel networks [95]. The printed parts were able to both encase and kill cancer cells after an oscillating magnetic field was applied due to the actuation of the heated hydrogel-filled elastomeric arms.

Composites with SMPs enabled through multi-material AM have shown potential to develop into multifunctional actuators. Bodkhe and Ermanni designed a piezoelectric SMP that changed its shape with temperature and could simultaneously measure the extent of its deformation through the development of a proportional voltage signal [92]. The multifunctional composite was enabled through the deposition of rigid and soft sections with multi-material DIW. The possibilities to tune actuation temperatures from 100 °C down to body temperatures were explored, and a robust sensor capable of withstanding temperatures ranging from 23 °C to 100 °C was presented, and to over 5000 operation cycles. This shape-memory composite had a shape recovery rate of ~98% and the sensor had a linear voltage response in the force range of 0.1–1 N [92]. Ge et al. combined hydrogels and SMPs through a multi-material DLP print method. UV curable hydrogels were developed by the creation of a water-soluble photoinitiator [96]. To obtain heterogeneous structures of elastomer and hydrogel, a moving stage with "puddles" of the different precursor solutions was placed under the UV light at different times per layer and air-jetted off in-between material exchanges. The multi-material structures were able to perform multiple functions owing to the combination of advanced materials. For example, stents were built that could be cold programmed to a small diameter size and inserted into blood vessels and later would expand through the body temperature induced shape recovery process.

AM has the potential to provide additional functionality through careful geometrical design. As an example, multifunctional silicone structures with hierarchical porosities were built using a combination of micrometer sized sacrificial pore forming fillers and macroscopic infill gaps through DIW [94]. The 3D printed structures that had millimeter sized gaps on their infill attained super hydrophobicity and super olephilicity due to their capacity to entrap air at the pores inside and around the printed struts. Moreover, the silicone inks used to print the porous structures were able to be mixed with CNTs, showcasing their capabilities as resistive sensors. The additional functionality of these structures as sensors was demonstrated by wetting them in a CNT bath, subjecting them to cyclic compressions, and measuring the linear electrical resistance response. A linear response was observed up to 10% strain although the highly porous structures could easily be compressed cyclically without loss of elasticity.

3D printed multifunctional materials and structures will continue to develop through the development of AM compatible chemistries, as well as the incorporation of multi-material printing to the different AM technologies. Careful design of new materials will aim to achieve multifunctionality without compromising AM compatibility. The incorporation of heterogeneous materials into AM feed has been highly developed, and thus, the blueprints for a future of yet unrealized multifunctional composites are in place. Multifunctional materials may begin to outpace single function materials due to the benefits they provide including system simplicity and reduction of mass. However, careful design must ensure that coexisting functionalities are not compromised by the presence of one another. AM provides pathways to seamlessly transition from structural to functional elements, and the high resolution of various printing technologies will enable careful control of material deposition to ensure functionalities are not compromised. Together, the fields of AM and advanced materials will continue to develop hand in hand to realize a future with efficient, responsive, and intelligent structures for various industries.

3. Applications

3.1. Biomaterials

Bioprinting has seen a surge in recent years helping progress the efforts in tissue and organ engineering through the incorporation of live cells into 3D printable inks that are engineered to interact with biological systems. Recent advancements in biological inks either promote the growth and interaction of cells and tissues or by becoming part of the cellular structure. In a recent example, Zhange et al. developed bioinks with the same biological activity as original cartilage extracellular matrices, which could be used to construct scaffolds for cartilage tissue engineering [97]. In another example, Butler et al. developed a bioink from a blend of chitosan and starch [98]. This bioink showed great printability and biocompatibility with great promise for neural tissue engineering applications.

Traditional methods for fabricating biomaterials such as casting, suffer from significant drawbacks such as suboptimal engraftment, poor cell survivability (~1%) and uncontrolled differentiation which limit their use in in vivo applications such as stem-cell therapy [99]. The ability to design and control the internal structure of biomaterials through AM has recently been seen as a method to overcome some of these barriers. Microarchitectural control in biomaterials, enabled through AM, can lead to better control of drug delivery and antibacterial properties of the material resulting in fewer infections and a lower likelihood of rejection by the immune system [100]. For example, Tytgat et al. reported on the potential of norbornene-functionalized gelatin and thiolated gelatin processed through the ME technique with a cell viability above 90% up to 14 days [101].

Bioprinting has primarily been used to produce higher quality models of tissue and organ to study complex diseases and anomalies such as cancer in vitro. For example, Zhu et al. bioprinted prevascularized tissues with complex 3D microarchitectures without the need of sacrificial material [102]. The use of bioprinting allowed the group to study different architectural features of the vascular network to produce a pre-vascularized tissue with high cell viability in vitro and in vivo. In another example, Mollica et al. developed a 3D biomimetic in vitro model through ME technique, using extracellular matrix from decellularized rat and human breast tissues [103]. This model was capable of sustaining large structural growths such as bioprinted organoid and tumaroid formations, or formations resembling organs and tumors. Another group from Georgia Institute of Technology and Emory University demonstrated the potential of producing anatomically correct, live, and functional organs through patient-derived models of the human heart at different stages of development through DLP of gelatin methacrylate bioinks (Figure 1) [104]. The development of advanced models can help in the discovery of diseases and treatments as well as significantly improve the reliability of those laboratory discoveries. In addition, the use of accurate and functional tissue and organ models could lead to reduced reliance on animal models.

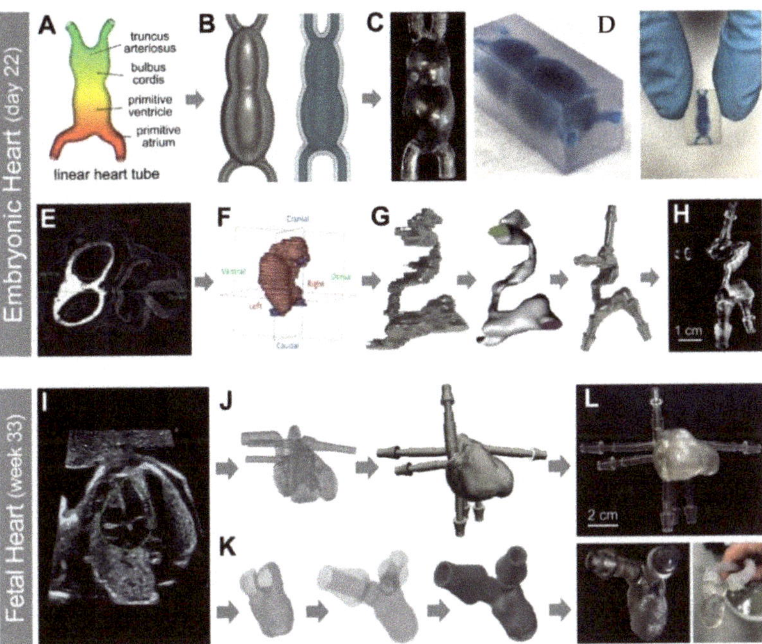

Figure 1. "Experimental workflow used to create patient-derived 3D printed heart models at varying developmental stages. A 3D model of an embryonic human heart at Carnegie stage 10 (21–23 days old, e-HT) was made either through (**A–D**) CAD modeling of an idealized structure, or (**E–H**) using the actual human data obtained in collaboration with the 3D Atlas of Human Embryology. The anatomical heart tube model contained the (**E**) primitive cardiovascular system which was (**F–H**) trimmed to only include the linear heart structure. (**I–L**) A patient-specific fetal left ventricle (f-LV) was acquired from fetal echocardiography of the human heart (week 33). The (**I**) full heart model was (**J–L**) trimmed at the mitral and aortic valves to only include the inner surface of the LV. For both models, the geometries were optimized, hollowed, and smoothed using the Meshmixer. Flow extensions were appended using AutoDesk Fusion 360 at the trimmed inlets and outlets to simulate adjacent vasculature. Different scales of (**C,H**) e-HT and (**L**) f-LV constructs were 3D printed by a Form 2 printer using the clear resin."—[A. D. Cetnar et al., "Patient-Specific 3D Bioprinted Models of Developing Human Heart," Adv. Healthc. Mater., vol. n/a, no. n/a, p. 2001169. Reproduced with permission.].

One of the challenges in fabricating scaffolds for soft tissue and organs through AM is the lack of available biomaterials with the necessary mechanical and physical properties to be compatible with AM technology [105]. Most common biomaterials such as collagen, gelatin, fibrin, chitosan, hyaluronic acid, and silk suffer from poor mechanical strength which results in melting, dissolution and warping of the printed structure [106]. Recent approaches to better the rheological properties of bioinks have been seen through the incorporation of polymer microspheres [107]. Additionally, advances in the AM technology such as the development of Suspended Layer Additive Manufacturing (SLAM) for ME-based bioprinting have helped to achieve a higher complexity in the printed structures as shown by Figure 2 [55].

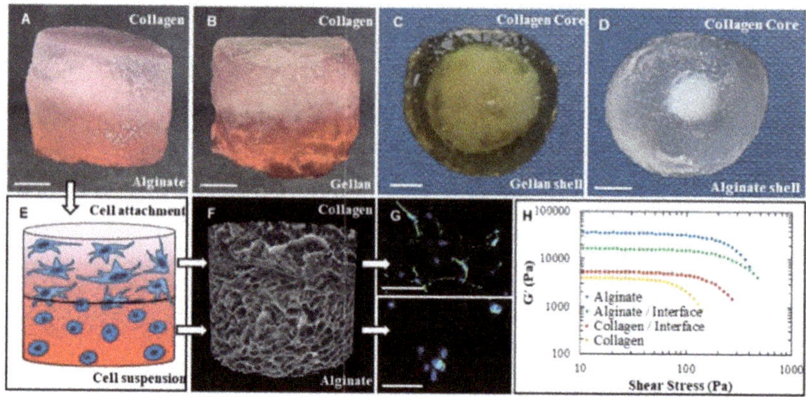

Figure 2. "3D printing multilayer gradient scaffolds by SLAM. (**A**,**B**) Images of bilayer scaffolds using combinations of (**A**) collagen–alginate and (**B**) collagen–gellan. (**C**) Large collagen–core gellan–shell scaffold and (**D**) small collagen–core alginate–shell scaffold (scale bars = 5 mm). (**E**) Schematic of diagram showing control of cell behavior with attachment motif bearing complexes in the upper collagen gel and no attachment motifs for cell suspension within an alginate gel. (**F**) Micro-CT showing gradient porosity within a lyophilized collagen–alginate scaffold. (**G**) Confocal micrographs of Hoechst/actin cell staining of HDFs attached in the collagen layer (upper) and suspended in the alginate (lower) regions of a dual layer scaffold (scale bars = 100 μm). (**H**) Stress versus G′ showing variations in gel strength and elasticity across a collagen–alginate scaffold including the interfacial regions illustrating a gradient in mechanical strength across the printed part." [J. J. Senior, M. E. Cooke, L. M. Grover, and A. M. Smith, "Fabrication of Complex Hydrogel Structures Using Suspended Layer Additive Manufacturing (SLAM)," Adv. Funct. Mater., vol. 29, no. 49, p. 1904845, 2019. Reproduced with permission.].

3.2. Sensors

Precise sensing is critical for various applications including biomedical, structural health monitoring, aerospace, and chemical. Various conditions of interest to be sensed include strain, pressure, temperature, and the presence of a gas or liquid. Soft materials that develop electric responses to these stimuli are of interest because their high stretchability makes them suitable for attachment onto various types of surfaces, including human bodies as wearable sensors and electronic skins [108]. Conventionally manufactured soft sensors have been limited by their geometries and material arrangements. These types of sensors cannot fully explore the capabilities of their constituent sensing materials. For example, it has been shown that enhanced stretchability and sensitivity could be achieved with the use of designed cellular structures [109]. These kinds of structures are difficult to achieve using conventional manufacturing methods without the use of specialized tools.

AM techniques have been used as methods to develop both, casing and sensing elements, to build soft sensors with expanded material and geometric selection [110–112]. Gao et al. used a ME technique to create a substrate for a pressure sensor that had slots built into it where conductive elastomer composites were later cast [110]. This sensor was able to be attached to different objects with curved surfaces and could distinguish the different gripping forces exerted to lift objects. A similar approach was used by Zhou et al. who printed a piezoresistive strain sensor using SLA where the upper layer of the printed structure had channels that were filled with strain-sensitive Galinstan after the printing process [111]. The asymmetry of the printed structures enabled by the different patterns on each cured layer allowed for the determination of the type of bending motion experienced.

Complete sensors have been completely built using AM taking advantage of the multi-material capabilities of processes such as ME, where multiple nozzles are filled with a unique material that can also be used in a single print [113]. The process of building a complete strain sensor using AM consisted of first depositing elastomer layers to act as insulators, then depositing electrode beads using a carbon nanotube-filled elastomer, and finally depositing a pressure-sensitive layer made from an ionic liquid-filled elastomer. The multi-material sensor performed reliably and the ability to deposit electrode traces in specific regions of the prior backing layers allowed for localization of pressure along the body of the sensor.

Various soft strain sensors have been recently built using a variety of AM techniques and materials [68,71,114–120]. Some of these developed sensors have shown enhanced out-of-plane sensitivity compared to casted sensors when 3D patterning using ME [117]. The use of print nozzles with a 100 µm diameter allowed for miniaturized flexible sensors that could sense very low pressures [117]. Different print patterns built using ME have also been used to develop strain sensors with augmented anisotropic strain sensitivities [68]. The capability to pattern complex structures out of sensing materials using AM was utilized by Lai and Yu who built wearable sensors with auxetic cellular geometries out of conductive hydrogels as shown in Figure 3 [73]. These sensors had vastly increased stretching limits when compared to solid geometries and could maintain contact with human skin even after muscle contractions and extensions. Yu et al. also demonstrated the advantage of building three-dimensional cellular geometries for strain sensors [121]. Lattices of thermoplastic polyurethane were built using FDM and then coated with carbon nanotubes to add strain-sensing functionality. The presence of carbon only on the surfaces as well as the inherent softness of the lattices resulted in very high sensitivity and stretchability. In a similar manner, Yin et al. used SLA to manufacture hydrogel electrodes with mesh geometries for capacitive strain sensors [122]. Their electrode designs enhanced the sensitivity of the sensors and they found that three-dimensional meshes exhibited higher strain sensitivities than two-dimensional meshes because of their higher compliance.

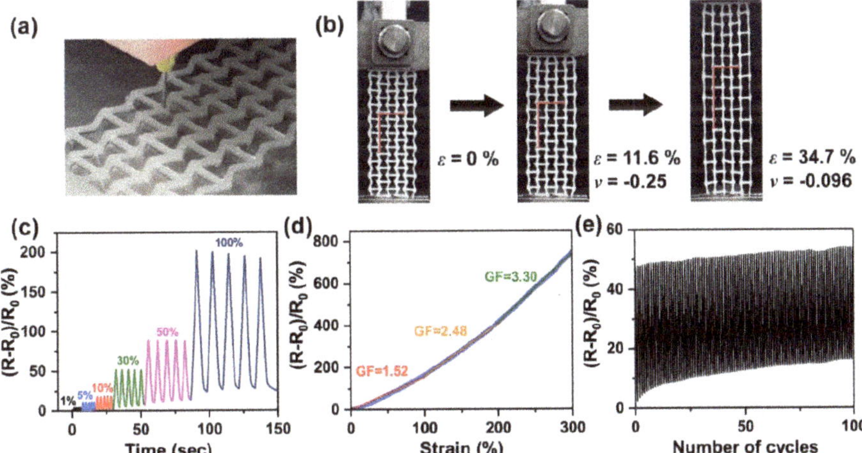

Figure 3. "Performance of 3D printed strain sensors with resistance change. (**a**) Photograph of the printing process. (**b**) Dimensional change in the auxetic sensor during stretching. (**c**) Relative resistance change in the sensor at different strains. (**d**) Variation in relative resistance vs. strain. The slope of linear fitting corresponds to the gauge factors in the range of strain from 0 to 100%, 100 to 200%, and 200 to 300%, respectively. (**e**) Relative resistance change in the sensor at 30% strain for 100 cycles." [Chun-Wei LaiChun-Wei Lai, "3D Printable Strain Sensors from Deep Eutectic Solvents and Cellulose Nanocrystals I ACS Applied Materials and Interfaces." https://pubs.acs.org/doi/10.1021/acsami.0c11152 (accessed 22 October 2020). Reproduced with permission.].

AM has enabled the development of soft sensors with enhanced and added functionalities. The capability to pattern unique cellular structures through AM was used to build a temperature sensor that was insensitive to strain, as shown by Figure 4 [123]. The use of cellular patterns built using ME allowed for the sensor to stretch without experiencing large amounts of elastic strain on the cellular trusses and thus reduced the sensitivity of its electrical properties to deformation. The built sensors had the ability to sense temperature in the same manner as undeformed sensors despite being bent and twisted to conform to curved surfaces. AM has also aided in the development of hydrogel-based liquid sensors capable of distinguishing the position and volume of the target liquid. Electrically conductive hydrogels were printed using a high-resolution multi-material SLA process [124]. By depositing sections of hydrogel with and without conductive carbon fillers, the researchers were able to build a mesh that could detect the position and volume of liquid in contact with the sensor. The different sensitivities of the neat and carbon-filled hydrogels allowed for bidirectional sensing of liquid coming from leakage in different sections of a system.

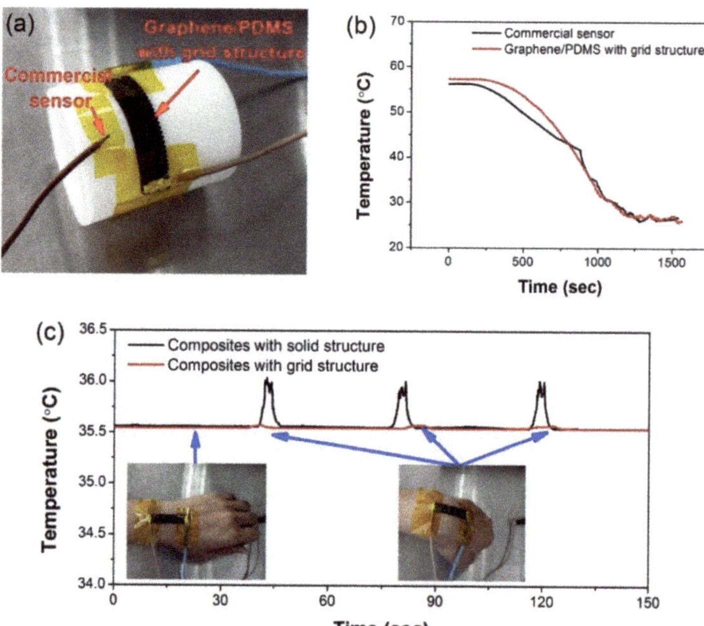

Figure 4. "Monitoring of the cooling process on a curved surface: (**a**) experimental setup; (**b**) comparison of the temperature performance between graphene/PDMS composites with a grid structure and commercial temperature sensor; and (**c**) simultaneous monitoring of wrist skin temperature and joint bending." [Z. Wang et al., "3D-Printed Graphene/Polydimethylsiloxane Composites for Stretchable and Strain-Insensitive Temperature Sensors," *ACS Appl. Mater. Interfaces*, vol. 11, no. 1, pp. 1344–1352, Jan. 2019.].

Soft sensors fabricated through AM have shown great potential as they allow for unique patterns with enhanced sensitivity to stimuli and unique multidirectional sensing. As new materials with better performance are developed for sensors, they will require manufacturing methods with great flexibility in the material selection that do not require complete retooling for the new materials. AM will continue to co-develop with emerging materials to result in highly sensitive and resistant sensors suitable for new operational environments.

3.3. Energy Harvesting

Energy harvesters are devices that can generate electrical energy from alternative sources of energy, such as thermal and mechanical. Energy harvesting can take the form of triboelectric, thermoelectric, and piezoelectric devices, the last of which are among the most common [125]. With the push for sustainable green energy in recent years, the use of soft engineered materials for energy harvesting has gained significant interest due to their capability to conform to different surfaces, making them attractive in many modern fields such as automotive, aerospace, and biomedical applications. The use of AM technology has contributed to improved designs of energy harvesting devices by expanding the geometric selection to include lattice structures with better mechanical properties such as higher compression and conversion of energy, and easier incorporation towards their application [126]. With great shape conformity and elasticity, AM-enabled energy harvesters may find great use in applications such as wearable electronics, textiles, and skins [127,128].

Various types of energy harvesters have been explored through AM using soft matrix materials that can respond to different stimuli such as motion, temperature, or stress. One type of energy harvester, the triboelectric device, generates electrical energy from materials that are electrically charged after they are separated from a different material. For example, Zou et al. fabricated a triboelectric device through DIW and were able to harvest the motion of muscles to power a sensor while underwater. This printed triboelectric device demonstrated good flexibility and stretchability for energy conversion while showing outstanding tensile fatigue resistance, as seen in Figure 5 [129]. In another example, Kaijuan et al. printed a triboelectric nanogenerator (TENG) energy harvester through DIW that could harvest mechanical energy from motion and power a sensor that was embedded in a shoe insole [130]. This sensor could be built into multilevel structures that could harvest energy due to the triboelectric effect of three sequential deformation stages as shown in Figure 6. The use of multi-material AM allowed for a single process programming of a TENG. Kaijuan et al. successfully provided a new approach to wearable devices by combining 3D printing with TENG devices for self-sustainable and comfortable embedded sensors. Another type of energy harvesting device is the thermoelectric device which can harvest energy from electric potentials generated from temperature differences across a distance Lazaros et al. printed an organic thermoelectric generator (TEG) with TPU and multi-walled carbon nanotubes through FDM for potential large-scale energy harvesting applications with stretchability [131]. One of the current challenges of thermoelectric energy harvesters is their limited use in industrial applications. However, the development of wearable devices where complex 3D architectures and customizability are required have opened new avenues for these harvesters.

One of the most common types of energy harvesters is piezoelectric devices which are capable of transforming applied mechanical stress into an electric potential. As an example of the use of AM to better the properties of these devices, Xiaoting et al. used a combined DOD and ME printer to fabricate a multilayer piezoelectric energy harvester that could harvest higher amounts of power than conventional flat harvesters [132]. Annealing of the piezoelectric polymers after each layer was deposited allowed for a higher amount of ferroelectric phase development and an overall higher piezoelectric performance. The multilayer harvesters were wrapped around a "rugby ball" shaped part made of soft PDMS polymer that was printed using ME at a separate print process. Overall, the AM harvesters performed 2.2 times better in open circuit conditions compared to flat single layered harvesters and could develop an output power 100 times greater than flat, single-layered harvesters. In another example, Desheng et al. were also able to fabricate embedded piezoelectric sensors through μSL and obtained high piezoelectric responsiveness and compliance via exploiting the effects of nanoparticle–matrix functionalization of the electromechanical performance of the piezoelectric nanocomposite [77]. This allowed them to achieve target flexibilities while keeping high piezoelectric responses via rational designs of inclusion morphologies and monomer stiffness of the constituent materials.

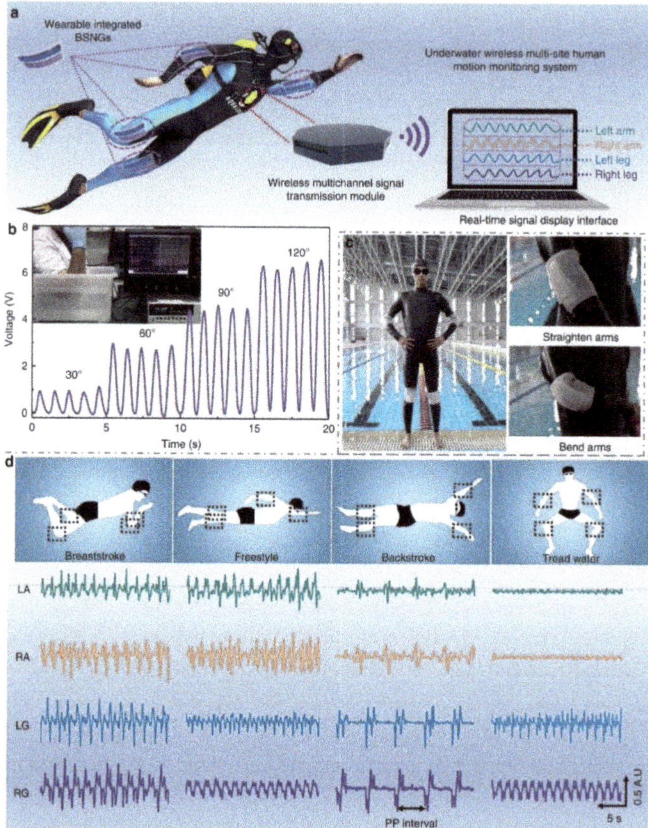

Figure 5. "Underwater wireless multi-site human motion monitoring system. (**a**) Illustration of underwater wireless multi-site human motion monitoring system based on bionic stretchable nanogenerator (BSNG). (**b**) Signal outputs of BSNG fixed on the elbow at different curvature motion. (**c**) Photographs of integrated wearable BSNG worn on the arthrosis of humans. (**d**) Signal outputs recorded by underwater wireless multi-site human motion monitoring system when the volunteer swam in different strokes (LA, RA, LG, RG represent left arm, right arm, left leg, right leg, respectively; PP interval represents the time interval between two peaks)" [Y. Zou et al., "A bionic stretchable nanogenerator for underwater sensing and energy harvesting," *Nat. Commun.*, vol. 10, no. 1, p. 2695, Dec. 2019].

AM has greatly contributed to the development of energy harvesters by the incorporation of soft functional materials that can help to increase the range of application by allowing the adaptation to different surfaces and shapes. Furthermore, AM has contributed to being able to fabricate devices with a high degree of customization without design limitations, which has shown improved functionality for energy conversion.

3.4. Soft Robotics

Soft robots and/or actuators are highly compliant devices with multiple degrees of freedom that enable object manipulation with minimal damage. Due to their interesting material properties such as flexibility and compliance, soft robotics have gained a significant attention in the fabrication of diverse types of grippers for manufacturing [133], nature inspired actuation [134,135], sea exploration [136], surgical devices [137,138], and rehabilitation devices [133,139]. Actuation of soft robots occurs usually through the application of pneumatic or hydraulic pressure [133], temperature [134,135,140], UV light [135,141], magnetic fields [134], or electric fields [137,138]. The most used materials for soft robotics applications include SMPs, hydrogels, ionic polymer–metal composites (IPMC), and elastomers.

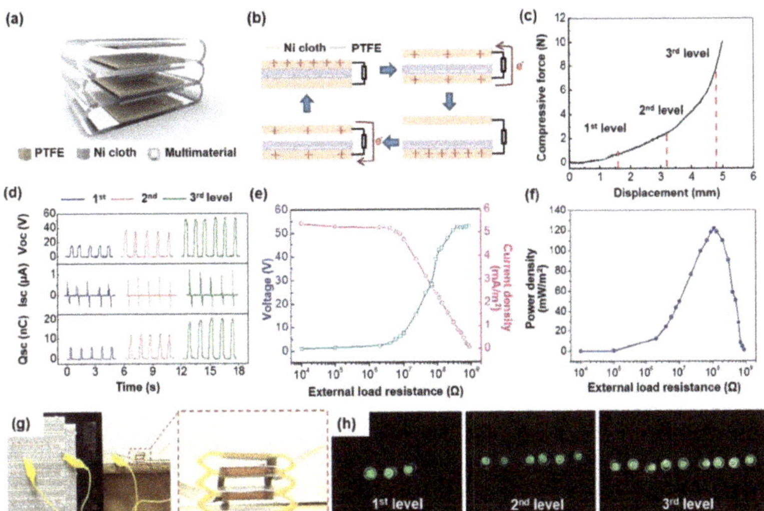

Figure 6. Design, working mechanism, and output performance of mTENG. (**a**) Schematic illustration of the 3D printing multi-material used as self-powered wearable electronics. (**b**) Schematic illustration of the electricity-generation process of the single-level contact separation. (**c**) Force–displacement curve of compressive test of the three-layer multi-material showing three stages of deformation. (**d**) The output performance of TENG in the different levels of contact-separation mode, including Voc, Isc, and Qsc. (**e**) Relation between the instantaneous output voltage and current density with the external load resistance. (**f**) Relation between the instantaneous output power density with the external load resistance. (**g**) The m-TENG was used as an energy harvester to lighten LEDs. (**h**) The mTENG can lighten 3, 6, and 9 LEDs by compression with the first, second, and third levels of deformation. [K. Chen et al., "Dynamic Photomask-Assisted Direct Ink Writing Multimaterial for Multilevel Triboelectric Nanogenerator," *Adv. Funct. Mater.*, vol. 29, no. 33, p. 1903568, 2019, Copyright Wiley-VCH GmbH. Reproduced with permission.].

AM enables the tuning of mechanical properties of soft actuators by controlling the material orientation and chemical composition without design limitations and with outstanding performance. Some researchers have used AM to print molds and then traditionally cast a soft material with the purpose of being used as a soft actuator. However, this method has several limitations, such as limited structure design, poor mechanical properties, and extended manufacturing processes. AM used for direct printing of soft robotics devices is a cost-effective method that contributes to a reduction in the manufacturing lead time as well as facilitating the fabrication of custom designs. For example, Ang et al. fabricated different bellow-type of soft actuators (Figure 7) by FDM methods using NinjaFlex, which is a TPU-based filament, with the purpose of controlling bending and increasing length of actuation. At 175 kPa the bellow-type soft actuators could withstand a weight of 0.94 Kg and handle delicate objects such as food without damaging them [133]. Furthermore, Ang et al. demonstrated that besides being used as grippers for manufacturing, NinjaFlex soft actuators could also be used for locomotion, and wearable devices applications.

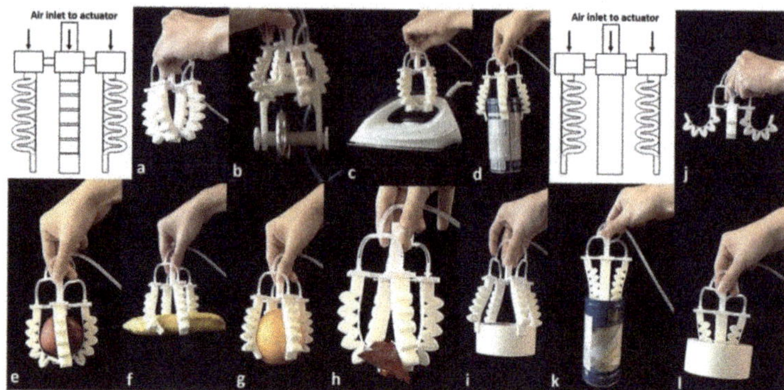

Figure 7. "Demonstration of (**a**) robotic gripper equipped with 3D-printed pneumatic actuators picking up (**b**) a 0.94 kg structure and (**c–j**) items of different shapes, sizes, and weight using 175 kPa of pressure. (**j**) A second robotic gripper with outward-facing 3D-printed pneumatic actuators that can grasp (**k**) inside of a cylindrical can and (**l**) roll of tape." [B. A. W. Keong and R. Y. C. Hua, "A Novel Fold-Based Design Approach toward Printable Soft Robotics Using Flexible 3D Printing Materials," *Adv. Mater. Technol.*, vol. 3, no. 2, p. 1700172, 2018, Copyright Wiley-VCH GmbH. Reproduced with permission.].

In addition, AM facilitates the engineering of mechanical properties to obtain fast actuation and control of bending curvature, which determines the efficiency of soft actuators. Electroactive hydrogels (EAH) for example, are materials of interest for soft actuators that exhibit a linear deformation dependence on an applied electric field. Han et al. used PµSL to control bending curvature and actuation of EAH by varying the thickness of several types of structures (grippers, transporters, and human-like) that enabled manipulation and locomotion due to their elaborate designs [141]. Saed et al. developed molecularly engineered LCEs to modify their thermomechanical properties and control their actuation temperature and strain. The soft actuators were fabricated by DIW with an extended actuation temperature range from 20 °C to 100 °C [140]. Roach et al. took advantage of the shear force generated on the ink during extrusion in DIW to align the liquid crystal monomers in the direction of printing and obtain large actuations. The alignment of monomers was locked as the material was deposited by shining a UV light and locking it through the formation of crosslinks. They demonstrated that printing with smaller nozzle diameters contributed to a higher actuation strain, additionally, the influence of printing speed on the alignment of LCEs was evaluated [142].

Soft polymer composites with functional nanofillers are another interesting approach for soft robots that either provides an alternative actuation mechanism or additional functionalities. Some researchers have used nanofillers to develop nature-inspired or mimetic soft actuators. For example, Kim et al. developed 3D origami soft sensing robots using FDM composed of nanocomposite filaments of cellulose nanofiber, PLA, and TPU, as shown in Figure 8. These devices could detect electromyography signals for health monitoring [143]. Tognato et al. created a method to tune the anisotropy of robotic actuation by adding magnetic responsive iron oxide nanoparticles (IOPs) to a PEG matrix suitable for AM. Due to the high biocompatibility of IOPs, this device could be used as a bioactuator for cell reorientation with a multi-stimuli responsive mechanism like starfish-inspired structures [134]. Finally, Kim et al. developed several conductive soft actuators by using MWCNT dispersed in epoxy aliphatic acrylate with tunable mechanical properties fabricated through DLP [144].

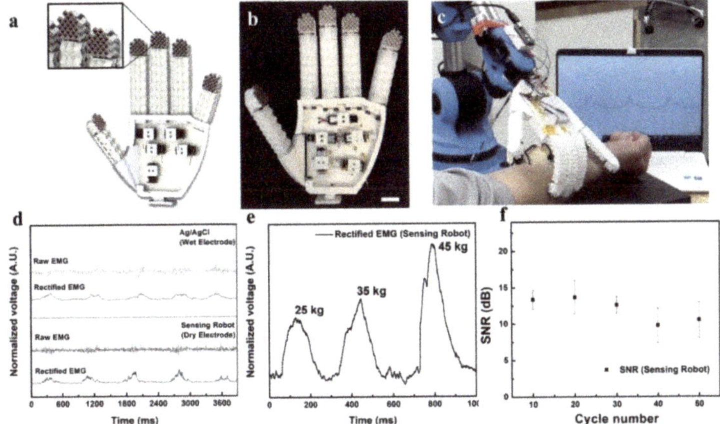

Figure 8. EMG assessment with origami-inspired robotic structures and materials. (**a**) Schematics and (**b**) actual photo image of electromyogram (EMG) sensing humanoid robot hand (scale bar: 2 cm). (**c**) Photo image of the sensing robot during the EMG measurement. (**d**) Typical raw and rectified EMG signals measured by wet Ag/AgCl electrodes (**top**) and the EMG sensing robot (**bottom**). (**e**) Rectified EMG signals at three different fist-holding forces of the participant. (**f**) SNR profiles of EMG sensing robot as a function of touching cycles to the human subject. [T.-H. Kim, J. Vanloo, and W. S. Kim, "3D Origami Sensing Robots for Cooperative Healthcare Monitoring," Adv. Mater. Technol., vol. n/a, no. n/a, p. 2000938, Copyright Wiley-VCH GmbH. Reproduced with permission].

AM has shown great capabilities to expand the functionality and applications of soft robotics for manufacturing and biomedical purposes compared with traditional manufacturing methods. Some future trends in soft robotics might expand to include multi-material for single 3D printing steps of soft actuators and electrodes, improving the adhesion of materials and mechanical properties. Another interesting future approach is the evaluation of actuation mechanisms and tunable mechanical properties for metamaterial designs, which are only possible by AM techniques. In addition, some future trends for biomedical purposes include the fabrication of soft actuators for surgical devices, which might allow the development of body temperature responsive devices.

3.5. Optoelectronics

Optoelectronic devices place a connection between optics and electronics by generating light from electrical energy or producing energy by capturing light through a semiconductor. Optoelectronics can be applied as photodiodes [145], solar cells [146], or light-emitting diodes [147].

AM has drawn more attention in the manufacturing of optoelectronic devices over traditional microfabrication technologies in recent years. This is due to AM's ability to extend the flexibility in the design and fabrication of 3D structured optoelectronics that results in high-performance integrated active electronic materials and devices. In addition, AM also allows for the integration of both, organic/inorganic/biological and conducting/semiconducting materials, as a single tool [148]. For example, Hu et al. [149] have shown black phosphorus as a functional ink platform for MJ of visible and near-infrared photoelectronic, including photodetectors. Additionally, AM's ability to incorporate functional fillers has demonstrated tunable optical properties. A group from the University of Warwick utilized FDM to fabricate several electromagnetic devices capable of controlling the propagated wave through a particularly graded refractive index [150]. Such devices like 3D printed gradient refractive index lens showed ways to manipulate and control an electromagnetic wave going through a boundary between two homogeneous media.

Soft optoelectronic devices printed by AM have a diverse field of application including omnidirectional light-sensing, and light-emitting which are typically used for detecting structural defects [151]. Optoelectronic structures fabricated through FDM of multi-material filaments have been shown by Loke et al. [151]. The specialized filaments were performed to contain photodetecting or light-emitting structures at their core. This allowed a good adhesion between the elements that made up the optoelectronic structures and the production of application-ready parts using a single-step process. The method for fabricating filaments was proposed for a wider range of materials and could prove to be the next evolution in the multi-material extrusion of functional devices. Another approach for building complete optoelectronic structures using AM consisted in depositing photodetectors onto flexible elastomeric substrates using a DIW method as shown in Figure 9 [148]. The photodetectors were realized through a multi-material deposition of the sensing and conductive elements, and the process was highly adaptable to various types of substrate materials both, rigid and soft, and geometries including flat and concave. The expanded flexibility of AM allowed for accurate photo sensors with new capabilities including onboard powering, as well as wearability.

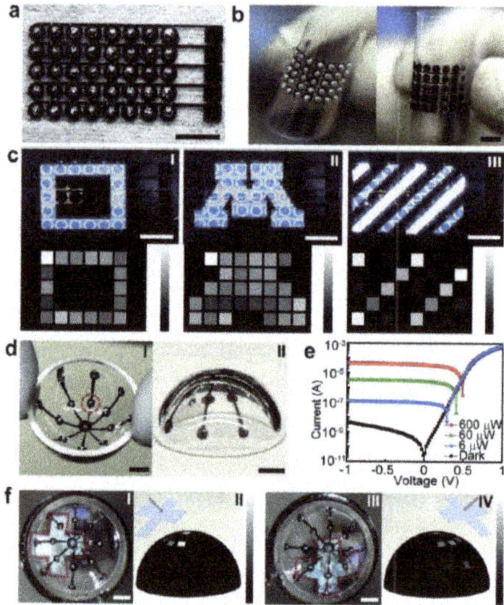

Figure 9. "3D-printed photodetector arrays printed on planar and spherical surfaces. (**a**) 5 × 8 photodetector array printed on PET. (**b**) Photographs of the bent photodetector array on PET films, viewed from both sides. (**c**) Characterization of the flexible photodetector array as an image sensor. Optical patterns projected onto the photodetector array and the reconstructed images: (I) square pattern; (II) letter "M;" and (III) white-light parallel strips with varying intensities. The range of the grayscale bars is from 0 (black) to 300 nA (white). (**d**) Photographs of the concentric photodetector array printed onto the inner surface of a hemispherical glass dome: (I) inside view and (II) outside view of the 3D-printed concentric photodetector array. (**e**) Current–voltage characteristics of the photodetector highlighted in (**d**). The excitation light source is a 405 nm laser of varying intensities. (**f**) Characterization of the spherical photodetector array as an image sensor. (I) the projected cross mark onto the devices (denoted by the dashed outline); (II) reconstructed cross mark pattern; (III and IV) cross mark pattern projected onto the spherical photodetector array rotated 90° and the reconstructed cross mark pattern. The range of the grayscale bar is from 0 (black) to 300 nA (white). Scale bars are 5 mm." [Park, S.H.; Su, R.; Jeong, J.; Guo, S.-Z.; Qiu, K.; Joung, D.; Meng, F.; McAlpine, M.C. 3D Printed Polymer Photodetectors. Adv. Mater. 2018, 30, 1803980, Copyright Wiley-VCH GmbH. Reproduced with permission].

Researchers and scientists have carried out significant work in the field of photonics and optoelectronic applications; however, there are still plenty of opportunities for expanding the properties, applications, and interconnectivity of optoelectronic devices. Future 3D printers' capability to control optical properties (like refractive index, reflectivity, transmittance, absorption, diffusion, etc.) will empower flourishing design space for sensing, display, and illumination.

4. Conclusions

A comprehensive review, including a brief description of AM techniques for soft functional materials, recent developments promoting their functional properties, and their use in various applications, is presented. AM technology has significantly contributed to the recent surge in the development of soft materials with functional properties such as the ability to self-heal, change color, program shape, serve as electronic devices, and multi-functionality. The versatility of AM technology has allowed for tunability of properties and greater freedom of design, boosting their potential in various applications, from tissue engineering to soft robotics. Additionally, the ability to integrate multi-materials in a single print through and flexibility in the design and fabrication of complex geometries such as lattice structures has helped to achieve high-performance integrated active electronic materials such as soft optoelectronic devices and energy harvesters.

The lower cost of manufacturing, rapid prototyping, fabrication of complex geometry and custom building has given AM technology a competitive edge over traditional methods in the manufacturing of functional materials. However, there are a few drawbacks that may require further research and development in the technology. The drawbacks include limited printable materials, limited use of multilaterals in AM of soft devices, and in some cases low printing speed. With enormous research efforts on the AM of soft functional materials, many challenges are being overcome in this field. For example, technology such as the SLAM technique is being developed to overcome the shortcomings of AM such as low material viscosity and allowing for materials to achieve more complex designs that would otherwise not be compatible with AM. Alternatively, manipulations of the molecular structures of materials or incorporation of functional fillers into existing materials have also been shown to improve the printability of soft functional materials and expand the list of printable materials in AM technology. Additionally, incorporating an expansion of design-based software with more informed inputs based on material genomics, multiscale modeling, topology optimization, and further use of multi-material AM technology will facilitate the integration and sophistication stage in AM soft structures. The increased amount of funding and massive research and development in 3D printing technology will bring a revolution in soft functional materials.

Author Contributions: Conceptualization, J.E.R., A.R., S.E.H.; investigation, J.E.R., A.R., S.E.H., M.S.H., C.M.; writing—original draft preparation, J.E.R., A.R., S.E.H., M.S.H., C.M.; writing—review and editing, J.E.R., A.R., S.E.H., M.S.H., C.M.; supervision, Y.L.; project administration, Y.L.; funding acquisition, Y.L. All authors have read and agreed to the published version of the manuscript.

Funding: This work was supported by the US Department of Energy's National Nuclear Administration under Grant No. DE-NA-0003865.

Institutional Review Board Statement: Not applicable.

Informed Consent Statement: Not applicable.

Data Availability Statement: Not applicable.

Acknowledgments: Jaime Regis would like to thank the National Science Foundation Bridge to the Doctorate Fellowship under Grant No. HRD-1810898 and the National Science Foundation Graduate Research Fellowship under Grant No. 1848741 for their support.

Conflicts of Interest: The authors declare no conflict of interest.

References

1. Rus, D.; Tolley, M.T. Design, Fabrication and Control of Soft Robots. *Nature* **2015**, *521*, 467–475. [CrossRef] [PubMed]
2. Zhou, L.-Y.; Fu, J.; He, Y. A Review of 3D Printing Technologies for Soft Polymer Materials. *Adv. Funct. Mater.* **2020**, *30*, 2000187. [CrossRef]
3. Truby, R.L.; Lewis, J.A. Printing Soft Matter in Three Dimensions. *Nature* **2016**, *540*, 371–378. [CrossRef]
4. Ma, S.; Zhang, Y.; Wang, M.; Liang, Y.; Ren, L.; Ren, L. Recent Progress in 4D Printing of Stimuli-Responsive Polymeric Materials. *Sci. China Technol. Sci.* **2020**, *63*, 532–544. [CrossRef]
5. Hamidi, A.; Tadesse, Y. Single Step 3D Printing of Bioinspired Structures via Metal Reinforced Thermoplastic and Highly Stretchable Elastomer. *Compos. Struct.* **2019**, *210*, 250–261. [CrossRef]
6. Wallin, T.J.; Pikul, J.; Shepherd, R.F. 3D Printing of Soft Robotic Systems. *Nat. Rev. Mater.* **2018**, *3*, 84–100. [CrossRef]
7. Cheng, Y.; Chan, K.H.; Wang, X.-Q.; Ding, T.; Li, T.; Lu, X.; Ho, G.W. Direct-Ink-Write 3D Printing of Hydrogels into Biomimetic Soft Robots. *ACS Nano* **2019**, *13*, 13176–13184. [CrossRef]
8. Liravi, F.; Toyserkani, E. Additive Manufacturing of Silicone Structures: A Review and Prospective. *Addit. Manuf.* **2018**, *24*, 232–242. [CrossRef]
9. Xiang, H.; Wang, X.; Ou, Z.; Lin, G.; Yin, J.; Liu, Z.; Zhang, L.; Liu, X. UV-Curable, 3D Printable and Biocompatible Silicone Elastomers. *Prog. Org. Coat.* **2019**, *137*, 105372. [CrossRef]
10. Saleh, E.; Zhang, F.; He, Y.; Vaithilingam, J.; Fernandez, J.L.; Wildman, R.; Ashcroft, I.; Hague, R.; Dickens, P.; Tuck, C. 3D Inkjet Printing of Electronics Using UV Conversion. *Adv. Mater. Technol.* **2017**, *2*, 1700134. [CrossRef]
11. Ding, Z.; Yuan, C.; Peng, X.; Wang, T.; Qi, H.J.; Dunn, M.L. Direct 4D Printing via Active Composite Materials. *Sci. Adv.* **2017**, *3*, e1602890. [CrossRef] [PubMed]
12. Hamidi, A.; Almubarak, Y.; Tadesse, Y. Multidirectional 3D-Printed Functionally Graded Modular Joint Actuated by TCPFL Muscles for Soft Robots. *Bio-Des. Manuf.* **2019**, *2*. [CrossRef]
13. Zhang, J.; Wang, X.Z.; Yu, W.W.; Deng, Y.H. Numerical Investigation of the Influence of Process Conditions on the Temperature Variation in Fused Deposition Modeling. *Mater. Des.* **2017**, *130*, 59–68. [CrossRef]
14. Pierin, G.; Grotta, C.; Colombo, P.; Mattevi, C. Direct Ink Writing of Micrometric SiOC Ceramic Structures Using a Preceramic Polymer. *J. Eur. Ceram. Soc.* **2016**, *36*, 1589–1594. [CrossRef]
15. Kuo, A.P.; Bhattacharjee, N.; Lee, Y.-S.; Castro, K.; Kim, Y.T.; Folch, A. High-Precision Stereolithography of Biomicrofluidic Devices. *Adv. Mater. Technol.* **2019**, *4*, 1800395. [CrossRef]
16. Mostafa, K.G.; Nobes, D.S.; Qureshi, A.J. Investigation of Light-Induced Surface Roughness in Projection Micro-Stereolithography Additive Manufacturing (PμSLA). *Procedia CIRP* **2020**, *92*, 187–193. [CrossRef]
17. Geng, Q.; Wang, D.; Chen, P.; Chen, S.-C. Ultrafast Multi-Focus 3-D Nano-Fabrication Based on Two-Photon Polymerization. *Nat. Commun.* **2019**, *10*, 2179. [CrossRef]
18. Janusziewicz, R.; Tumbleston, J.R.; Quintanilla, A.L.; Mecham, S.J.; DeSimone, J.M. Layerless Fabrication with Continuous Liquid Interface Production. *Proc. Natl. Acad. Sci. USA* **2016**, *113*, 11703–11708. [CrossRef]
19. Kadry, H.; Wadnap, S.; Xu, C.; Ahsan, F. Digital Light Processing (DLP) 3D-Printing Technology and Photoreactive Polymers in Fabrication of Modified-Release Tablets. *Eur. J. Pharm. Sci.* **2019**, *135*, 60–67. [CrossRef]
20. Bass, L.; Meisel, N.A.; Williams, C.B. Exploring Variability of Orientation and Aging Effects in Material Properties of Multi-Material Jetting Parts. *Rapid Prototyp. J.* **2016**, *22*, 826–834. [CrossRef]
21. Oh, Y.; Bharambe, V.; Mummareddy, B.; Martin, J.; McKnight, J.; Abraham, M.A.; Walker, J.M.; Rogers, K.; Conner, B.; Cortes, P.; et al. Microwave Dielectric Properties of Zirconia Fabricated Using NanoParticle Jetting™. *Addit. Manuf.* **2019**, *27*, 586–594. [CrossRef]
22. Sabahi, N.; Chen, W.; Wang, C.-H.; Kruzic, J.J.; Li, X. A Review on Additive Manufacturing of Shape-Memory Materials for Biomedical Applications. *JOM* **2020**, *72*, 1229–1253. [CrossRef]
23. Lu, H.B.; Huang, W.M.; Yao, Y.T. Review of Chemo-responsive Shape Change/Memory Polymers. *Pigment Resin Technol.* **2013**, *42*, 237–246. [CrossRef]
24. Choong, Y.Y.C.; Maleksaeedi, S.; Eng, H.; Yu, S.; Wei, J.; Su, P.-C. High Speed 4D Printing of Shape Memory Polymers with Nanosilica. *Appl. Mater. Today* **2020**, *18*, 100515. [CrossRef]
25. Estelle, K.; Blair, D.; Evans, K.; Gozen, B.A. Manufacturing of Smart Composites with Hyperelastic Property Gradients and Shape Memory Using Fused Deposition. *J. Manuf. Process.* **2017**, *28*, 500–507. [CrossRef]
26. Chen, K.; Kuang, X.; Li, V.; Kang, G.; Qi, H.J. Fabrication of Tough Epoxy with Shape Memory Effects by UV-Assisted Direct-Ink Write Printing. *Soft Matter* **2018**, *14*, 1879–1886. [CrossRef] [PubMed]
27. Liu, J.; Zhao, L.; Guo, Y.; Zhang, H.; Zhang, Z. Multi-Responsive Shape Memory Polymer Printed by Fused Deposition Modeling Technique. *Express Polym. Lett.* **2020**, *14*, 348–357. [CrossRef]
28. Zhang, F.; Wang, L.; Zheng, Z.; Liu, Y.; Leng, J. Magnetic Programming of 4D Printed Shape Memory Composite Structures. *Compos. Part Appl. Sci. Manuf.* **2019**, *125*, 105571. [CrossRef]
29. Islam Shiblee, M.N.; Ahmed, K.; Khosla, A.; Kawakami, M.; Furukawa, H. 3D Printing of Shape Memory Hydrogels with Tunable Mechanical Properties. *Soft Matter* **2018**, *14*, 7809–7817. [CrossRef]
30. Miao, J.-T.; Ge, M.; Peng, S.; Zhong, J.; Li, Y.; Weng, Z.; Wu, L.; Zheng, L. Dynamic Imine Bond-Based Shape Memory Polymers with Permanent Shape Reconfigurability for 4D Printing. *ACS Appl. Mater. Interfaces* **2019**, *11*, 40642–40651. [CrossRef]

31. Chen, L.; Zhao, H.-B.; Ni, Y.-P.; Fu, T.; Wu, W.-S.; Wang, X.-L.; Wang, Y.-Z. 3D Printable Robust Shape Memory PET Copolyesters with Fire Safety via π-Stacking and Synergistic Crosslinking. *J. Mater. Chem. A* **2019**, *7*, 17037–17045. [CrossRef]

32. Keshavarzan, M.; Kadkhodaei, M.; Forooghi, F. An Investigation into Compressive Responses of Shape Memory Polymeric Cellular Lattice Structures Fabricated by Vat Polymerization Additive Manufacturing. *Polym. Test.* **2020**, *91*, 106832. [CrossRef]

33. Choong, Y.Y.C.; Maleksaeedi, S.; Eng, H.; Wei, J.; Su, P.-C. 4D Printing of High Performance Shape Memory Polymer Using Stereolithography. *Mater. Des.* **2017**, *126*, 219–225. [CrossRef]

34. Villacres, J.; Nobes, D.; Ayranci, C. Additive Manufacturing of Shape Memory Polymers: Effects of Print Orientation and Infill Percentage on Mechanical Properties. *Rapid Prototyp. J.* **2018**, *24*, 744–751. [CrossRef]

35. Jeong, H.Y.; Woo, B.H.; Kim, N.; Jun, Y.C. Multicolor 4D Printing of Shape-Memory Polymers for Light-Induced Selective Heating and Remote Actuation. *Sci. Rep.* **2020**, *10*, 6258. [CrossRef] [PubMed]

36. Nam, S.; Pei, E. A Taxonomy of Shape-Changing Behavior for 4D Printed Parts Using Shape-Memory Polymers. *Prog. Addit. Manuf.* **2019**, *4*, 167–184. [CrossRef]

37. Wang, C.; Wang, M.; Ying, S.; Gu, J. Fast Chemo-Responsive Shape Memory of Stretchable Polymer Nanocomposite Aerogels Fabricated by One-Step Method. *Macromol. Mater. Eng.* **2020**, *305*, 1900602. [CrossRef]

38. Wang, K.; Jia, Y.-G.; Zhao, C.; Zhu, X.X. Multiple and Two-Way Reversible Shape Memory Polymers: Design Strategies and Applications. *Prog. Mater. Sci.* **2019**, *105*, 100572. [CrossRef]

39. Bodaghi, M.; Damanpack, A.R.; Liao, W.H. Triple Shape Memory Polymers by 4D Printing. *Smart Mater. Struct.* **2018**, *27*, 065010. [CrossRef]

40. Li, A.; Challapalli, A.; Li, G. 4D Printing of Recyclable Lightweight Architectures Using High Recovery Stress Shape Memory Polymer. *Sci. Rep.* **2019**, *9*, 7621. [CrossRef]

41. Peng, B.; Yang, Y.; Gu, K.; Amis, E.J.; Cavicchi, K.A. Digital Light Processing 3D Printing of Triple Shape Memory Polymer for Sequential Shape Shifting. *ACS Mater. Lett.* **2019**, *1*, 410–417. [CrossRef]

42. Davidson, E.C.; Kotikian, A.; Li, S.; Aizenberg, J.; Lewis, J.A. 3D Printable and Reconfigurable Liquid Crystal Elastomers with Light-Induced Shape Memory via Dynamic Bond Exchange. *Adv. Mater.* **2020**, *32*, 1905682. [CrossRef]

43. Kuang, X.; Chen, K.; Dunn, C.K.; Wu, J.; Li, V.C.F.; Qi, H.J. 3D Printing of Highly Stretchable, Shape-Memory, and Self-Healing Elastomer toward Novel 4D Printing. *ACS Appl. Mater. Interfaces* **2018**, *10*, 7381–7388. [CrossRef]

44. Ko, E.S.; Kim, C.; Choi, Y.; Lee, K.Y. 3D Printing of Self-Healing Ferrogel Prepared from Glycol Chitosan, Oxidized Hyaluronate, and Iron Oxide Nanoparticles. *Carbohydr. Polym.* **2020**, *245*, 116496. [CrossRef] [PubMed]

45. Zhou, Q.; Gardea, F.; Sang, Z.; Lee, S.; Pharr, M.; Sukhishvili, S.A. A Tailorable Family of Elastomeric-to-Rigid, 3D Printable, Interbonding Polymer Networks. *Adv. Funct. Mater.* **2020**, *30*, 2002374. [CrossRef]

46. Yu, K.; Xin, A.; Du, H.; Li, Y.; Wang, Q. Additive Manufacturing of Self-Healing Elastomers. *NPG Asia Mater.* **2019**, *11*, 1–11. [CrossRef]

47. Wallin, T.J.; Pikul, J.H.; Bodkhe, S.; Peele, B.N.; Murray, B.C.M.; Therriault, D.; McEnerney, B.W.; Dillon, R.P.; Giannelis, E.P.; Shepherd, R.F. Click Chemistry Stereolithography for Soft Robots That Self-Heal. *J. Mater. Chem. B* **2017**, *5*, 6249–6255. [CrossRef]

48. Chen, H.; Hao, B.; Ge, P.; Chen, S. Highly Stretchable, Self-Healing, and 3D Printing Prefabricatable Hydrophobic Association Hydrogels with the Assistance of Electrostatic Interaction. *Polym. Chem.* **2020**, *11*, 4741–4748. [CrossRef]

49. Liu, Z.; Hong, P.; Huang, Z.; Zhang, T.; Xu, R.; Chen, L.; Xiang, H.; Liu, X. Self-Healing, Reprocessing and 3D Printing of Transparent and Hydrolysis-Resistant Silicone Elastomers. *Chem. Eng. J.* **2020**, *387*, 124142. [CrossRef]

50. Wang, Z.; An, G.; Zhu, Y.; Liu, X.; Chen, Y.; Wu, H.; Wang, Y.; Shi, X.; Mao, C. 3D-Printable Self-Healing and Mechanically Reinforced Hydrogels with Host–Guest Non-Covalent Interactions Integrated into Covalently Linked Networks. *Mater. Horiz.* **2019**, *6*, 733–742. [CrossRef]

51. Kee, S.; Haque, M.A.; Corzo, D.; Alshareef, H.N.; Baran, D. Self-Healing and Stretchable 3D-Printed Organic Thermoelectrics. *Adv. Funct. Mater.* **2019**, *29*, 1905426. [CrossRef]

52. Li, X.; Yu, R.; He, Y.; Zhang, Y.; Yang, X.; Zhao, X.; Huang, W. Self-Healing Polyurethane Elastomers Based on a Disulfide Bond by Digital Light Processing 3D Printing. *ACS Macro Lett.* **2019**, *8*, 1511–1516. [CrossRef]

53. Kim, S.W.; Kim, D.Y.; Roh, H.H.; Kim, H.S.; Lee, J.W.; Lee, K.Y. Three-Dimensional Bioprinting of Cell-Laden Constructs Using Polysaccharide-Based Self-Healing Hydrogels. *Biomacromolecules* **2019**, *20*, 1860–1866. [CrossRef]

54. Lei, J.; Li, X.; Wang, S.; Yuan, L.; Ge, L.; Li, D.; Mu, C. Facile Fabrication of Biocompatible Gelatin-Based Self-Healing Hydrogels. *ACS Appl. Polym. Mater.* **2019**, *1*, 1350–1358. [CrossRef]

55. Senior, J.J.; Cooke, M.E.; Grover, L.M.; Smith, A.M. Fabrication of Complex Hydrogel Structures Using Suspended Layer Additive Manufacturing (SLAM). *Adv. Funct. Mater.* **2019**, *29*, 1904845. [CrossRef]

56. Schaffner, M.; Faber, J.A.; Pianegonda, L.; Rühs, P.A.; Coulter, F.; Studart, A.R. 3D Printing of Robotic Soft Actuators with Programmable Bioinspired Architectures. *Nat. Commun.* **2018**, *9*, 878. [CrossRef] [PubMed]

57. Chen, D.; Pei, Q. Electronic Muscles and Skins: A Review of Soft Sensors and Actuators. *Chem. Rev.* **2017**, *117*, 11239–11268. [CrossRef] [PubMed]

58. Liu, R.; Kuang, X.; Deng, J.; Wang, Y.-C.; Wang, A.C.; Ding, W.; Lai, Y.-C.; Chen, J.; Wang, P.; Lin, Z.; et al. Shape Memory Polymers for Body Motion Energy Harvesting and Self-Powered Mechanosensing. *Adv. Mater.* **2018**, *30*, 1705195. [CrossRef]

59. Christianson, C.; Goldberg, N.N.; Tolley, M.T. Elastomeric Diaphragm Pump Driven by Fluid Electrode Dielectric Elastomer Actuators (FEDEAs). In Proceedings of the Electroactive Polymer Actuators and Devices (EAPAD) XX, Denver, CO, USA, 5–8 March 2018; International Society for Optics and Photonics: Bellingham, WA, USA, 2018; Volume 10594, p. 105940O.

60. Zhao, Y.; Zha, J.-W.; Yin, L.-J.; Gao, Z.-S.; Wen, Y.-Q.; Dang, Z.-M. Remarkable Electrically Actuation Performance in Advanced Acrylic-Based Dielectric Elastomers without Pre-Strain at Very Low Driving Electric Field. *Polymer* **2018**, *137*, 269–275. [CrossRef]

61. McCoul, D.; Rosset, S.; Schlatter, S.; Shea, H. Inkjet 3D Printing of UV and Thermal Cure Silicone Elastomers for Dielectric Elastomer Actuators. *Smart Mater. Struct.* **2017**, *26*, 125022. [CrossRef]

62. Reitelshöfer, S.; Martin, S.; Nendel, F.; Schäfer, T.; Pham, D.; Franke, J. Accelerated Aerosol-Jet-Printing of Stretchable RGO-Electrodes for Stacked Dielectric Elastomers by Using a New Hybrid Atomizer. In Proceedings of the Electroactive Polymer Actuators and Devices (EAPAD) XXII, Online, 27 April–9 May 2020; International Society for Optics and Photonics: Bellingham, WA, USA, 2020; Volume 11375, p. 113751W.

63. Haghiashtiani, G.; Habtour, E.; Park, S.-H.; Gardea, F.; McAlpine, M.C. 3D Printed Electrically-Driven Soft Actuators. *Extreme Mech. Lett.* **2018**, *21*, 1–8. [CrossRef]

64. Zhou, F.; Zhang, M.; Cao, X.; Zhang, Z.; Chen, X.; Xiao, Y.; Liang, Y.; Wong, T.-W.; Li, T.; Xu, Z. Fabrication and Modeling of Dielectric Elastomer Soft Actuator with 3D Printed Thermoplastic Frame. *Sens. Actuators Phys.* **2019**, *292*, 112–120. [CrossRef]

65. Liu, K.; Chen, S.; Chen, F.; Zhu, X. A Unidirectional Soft Dielectric Elastomer Actuator Enabled by Built-In Honeycomb Metastructures. *Polymers* **2020**, *12*, 619. [CrossRef]

66. Costa, P.; Oliveira, J.; Horta-Romarís, L.; Abad, M.-J.; Moreira, J.A.; Zapiráin, I.; Aguado, M.; Galván, S.; Lanceros-Mendez, S. Piezoresistive Polymer Blends for Electromechanical Sensor Applications. *Compos. Sci. Technol.* **2018**, *168*, 353–362. [CrossRef]

67. Xiang, D.; Zhang, Z.; Han, Z.; Zhang, X.; Zhou, Z.; Zhang, J.; Luo, X.; Wang, P.; Zhao, C.; Li, Y. Effects of Non-Covalent Interactions on the Properties of 3D Printed Flexible Piezoresistive Strain Sensors of Conductive Polymer Composites. *Compos. Interfaces* **2020**, 1–15. [CrossRef]

68. Christ, J.F.; Aliheidari, N.; Pötschke, P.; Ameli, A. Bidirectional and Stretchable Piezoresistive Sensors Enabled by Multimaterial 3D Printing of Carbon Nanotube/Thermoplastic Polyurethane Nanocomposites. *Polymers* **2018**, *11*, 11. [CrossRef] [PubMed]

69. Yu, R.; Xia, T.; Wu, B.; Yuan, J.; Ma, L.; Cheng, G.J.; Liu, F. Highly Sensitive Flexible Piezoresistive Sensor with 3D Conductive Network. *ACS Appl. Mater. Interfaces* **2020**, *12*, 35291–35299. [CrossRef] [PubMed]

70. Wang, Z.; Guan, X.; Huang, H.; Wang, H.; Lin, W.; Peng, Z. Full 3D Printing of Stretchable Piezoresistive Sensor with Hierarchical Porosity and Multimodulus Architecture. *Adv. Funct. Mater.* **2019**, *29*, 1807569. [CrossRef]

71. Abshirini, M.; Charara, M.; Saha, M.C.; Altan, M.C.; Liu, Y. *Optimization of 3D Printed Elastomeric Nanocomposites for Flexible Strain Sensing Applications*; American Society of Mechanical Engineer: New York, NY, USA, 2020.

72. Wong, J.; Gong, A.T.; Defnet, P.A.; Meabe, L.; Beauchamp, B.; Sweet, R.M.; Sardon, H.; Cobb, C.L.; Nelson, A. 3D Printing Ionogel Auxetic Frameworks for Stretchable Sensors. *Adv. Mater. Technol.* **2019**, *4*, 1900452. [CrossRef]

73. 3D Printable Strain Sensors from Deep Eutectic Solvents and Cellulose Nanocrystals | ACS Applied Materials & Interfaces. Available online: https://pubs.acs.org/doi/10.1021/acsami.0c11152 (accessed on 22 October 2020).

74. Li, J.; Huang, H.; Morita, T. Stepping Piezoelectric Actuators with Large Working Stroke for Nano-Positioning Systems: A Review. *Sens. Actuators Phys.* **2019**, *292*, 39–51. [CrossRef]

75. Tuloup, C.; Harizi, W.; Aboura, Z.; Meyer, Y.; Khellil, K.; Lachat, R. On the Use of In-Situ Piezoelectric Sensors for the Manufacturing and Structural Health Monitoring of Polymer-Matrix Composites: A Literature Review. *Compos. Struct.* **2019**, *215*, 127–149. [CrossRef]

76. Li, J.; Long, Y.; Yang, F.; Wei, H.; Zhang, Z.; Wang, Y.; Wang, J.; Li, C.; Carlos, C.; Dong, Y.; et al. Multifunctional Artificial Artery from Direct 3D Printing with Built-In Ferroelectricity and Tissue-Matching Modulus for Real-Time Sensing and Occlusion Monitoring. *Adv. Funct. Mater.* **2020**, *30*, 2002868. [CrossRef]

77. Yao, D.; Cui, H.; Hensleigh, R.; Smith, P.; Alford, S.; Bernero, D.; Bush, S.; Mann, K.; Wu, H.F.; Chin-Nieh, M.; et al. Achieving the Upper Bound of Piezoelectric Response in Tunable, Wearable 3D Printed Nanocomposites. *Adv. Funct. Mater.* **2019**, *29*, 1903866. [CrossRef]

78. Song, X.; He, L.; Yang, W.; Wang, Z.; Chen, Z.; Guo, J.; Wang, H.; Chen, L. Additive Manufacturing of Bi-Continuous Piezocomposites With Triply Periodic Phase Interfaces for Combined Flexibility and Piezoelectricity. *J. Manuf. Sci. Eng.* **2019**, *141*, 111004. [CrossRef]

79. Gonzalez, D.; Garcia, J.; Newell, B. Electromechanical Characterization of a 3D Printed Dielectric Material for Dielectric Electroactive Polymer Actuators. *Sens. Actuators Phys.* **2019**, *297*, 111565. [CrossRef]

80. Rohde, R.C.; Basu, A.; Okello, L.B.; Barbee, M.H.; Zhang, Y.; Velev, O.D.; Nelson, A.; Craig, S.L. Mechanochromic Composite Elastomers for Additive Manufacturing and Low Strain Mechanophore Activation. *Polym. Chem.* **2019**, *10*, 5985–5991. [CrossRef]

81. Chen, J.; Xu, L.; Yang, M.; Chen, X.; Chen, X.; Hong, W. Highly Stretchable Photonic Crystal Hydrogels for a Sensitive Mechanochromic Sensor and Direct Ink Writing. *Chem. Mater.* **2019**, *31*, 8918–8926. [CrossRef]

82. Chen, L.; Zhang, Y.; Ye, H.; Duan, G.; Duan, H.; Ge, Q.; Wang, Z. Color-Changeable Four-Dimensional Printing Enabled with Ultraviolet-Curable and Thermochromic Shape Memory Polymers. *ACS Appl. Mater. Interfaces* **2021**, *13*, 18120–18127. [CrossRef]

83. Zhou, Y.; Layani, M.; Wang, S.; Hu, P.; Ke, Y.; Magdassi, S.; Long, Y. Fully Printed Flexible Smart Hybrid Hydrogels. *Adv. Funct. Mater.* **2018**, *28*, 1705365. [CrossRef]

84. Cai, G.; Cheng, X.; Layani, M.; Tan, A.W.M.; Li, S.; Eh, A.L.-S.; Gao, D.; Magdassi, S.; Lee, P.S. Direct Inkjet-Patterning of Energy Efficient Flexible Electrochromics. *Nano Energy* **2018**, *49*, 147–154. [CrossRef]

85. Yao, Y.; Yin, C.; Hong, S.; Chen, H.; Shi, Q.; Wang, J.; Lu, X.; Zhou, N. Lanthanide-Ion-Coordinated Supramolecular Hydrogel Inks for 3D Printed Full-Color Luminescence and Opacity-Tuning Soft Actuators. *Chem. Mater.* **2020**, *32*, 8868–8876. [CrossRef]

86. Lendlein, A.; Trask, R.S. Multifunctional Materials: Concepts, Function-Structure Relationships, Knowledge-Based Design, Translational Materials Research. *Multifunct. Mater.* **2018**, *1*, 010201. [CrossRef]

87. Salonitis, K.; Pandremenos, J.; Paralikas, J.; Chryssolouris, G. Multifunctional Materials: Engineering Applications and Processing Challenges. *Int. J. Adv. Manuf. Technol.* **2010**, *49*, 803–826. [CrossRef]

88. Read "Frontiers of Engineering: Reports on Leading-Edge Engineering from the 2004 NAE Symposium on Frontiers of Engineering". Available online: https://www.nap.edu/read/11220/chapter/1 (accessed on 15 November 2020).

89. Whitesides, G.M. Soft Robotics. *Angew. Chem. Int. Ed.* **2018**, *57*, 4258–4273. [CrossRef]

90. Additive Manufacturing of Functionally Graded Materials—A Review | Elsevier Enhanced Reader. Available online: https://reader.elsevier.com/reader/sd/pii/S0921509319309955?token=A3BC1920689C5B11B56D0A09575AB630D672AB22372 2688F9714331BCB53691D997EA1723A126C6289A5E7BD21D056E4&originRegion=us-east-1&originCreation=20210429171150 (accessed on 29 April 2021).

91. Chen, Z.; Zhao, D.; Liu, B.; Nian, G.; Li, X.; Yin, J.; Qu, S.; Yang, W. 3D Printing of Multifunctional Hydrogels. *Adv. Funct. Mater.* **2019**, *29*, 1900971. [CrossRef]

92. Bodkhe, S.; Ermanni, P. 3D Printing of Multifunctional Materials for Sensing and Actuation: Merging Piezoelectricity with Shape Memory. *Eur. Polym. J.* **2020**, *132*, 109738. [CrossRef]

93. Darabi, M.A.; Khosrozadeh, A.; Mbeleck, R.; Liu, Y.; Chang, Q.; Jiang, J.; Cai, J.; Wang, Q.; Luo, G.; Xing, M. Skin-Inspired Multifunctional Autonomic-Intrinsic Conductive Self-Healing Hydrogels with Pressure Sensitivity, Stretchability, and 3D Printability. *Adv. Mater.* **2017**, *29*, 1700533. [CrossRef]

94. Chen, Q.; Zhao, J.; Ren, J.; Rong, L.; Cao, P.-F.; Advincula, R.C. 3D Printed Multifunctional, Hyperelastic Silicone Rubber Foam. *Adv. Funct. Mater.* **2019**, *29*, 1900469. [CrossRef]

95. Tang, J.; Yin, Q.; Shi, M.; Yang, M.; Yang, H.; Sun, B.; Guo, B.; Wang, T. Programmable Shape Transformation of 3D Printed Magnetic Hydrogel Composite for Hyperthermia Cancer Therapy. *Extreme Mech. Lett.* **2021**, *46*, 101305. [CrossRef]

96. Ge, Q.; Chen, Z.; Cheng, J.; Zhang, B.; Zhang, Y.-F.; Li, H.; He, X.; Yuan, C.; Liu, J.; Magdassi, S.; et al. 3D Printing of Highly Stretchable Hydrogel with Diverse UV Curable Polymers. *Sci. Adv.* **2021**, *7*, eaba4261. [CrossRef] [PubMed]

97. Zhang, X.; Liu, Y.; Luo, C.; Zhai, C.; Li, Z.; Zhang, Y.; Yuan, T.; Dong, S.; Zhang, J.; Fan, W. Crosslinker-Free Silk/Decellularized Extracellular Matrix Porous Bioink for 3D Bioprinting-Based Cartilage Tissue Engineering. *Mater. Sci. Eng. C* **2021**, *118*, 111388. [CrossRef]

98. Butler, H.M.; Naseri, E.; MacDonald, D.S.; Andrew Tasker, R.; Ahmadi, A. Optimization of Starch- and Chitosan-Based Bio-Inks for 3D Bioprinting of Scaffolds for Neural Cell Growth. *Materialia* **2020**, *12*, 100737. [CrossRef]

99. Oliveira, E.P.; Malysz-Cymborska, I.; Golubczyk, D.; Kalkowski, L.; Kwiatkowska, J.; Reis, R.L.; Oliveira, J.M.; Walczak, P. Advances in Bioinks and in Vivo Imaging of Biomaterials for CNS Applications. *Acta Biomater.* **2019**, *95*, 60–72. [CrossRef]

100. Zadpoor, A.A.; Malda, J. Additive Manufacturing of Biomaterials, Tissues, and Organs. *Ann. Biomed. Eng.* **2017**, *45*, 1–11. [CrossRef]

101. Tytgat, L.; Van Damme, L.; Van Hoorick, J.; Declercq, H.; Thienpont, H.; Ottevaere, H.; Blondeel, P.; Dubruel, P.; Van Vlierberghe, S. Additive Manufacturing of Photo-Crosslinked Gelatin Scaffolds for Adipose Tissue Engineering. *Acta Biomater.* **2019**, *94*, 340–350. [CrossRef]

102. Zhu, W.; Qu, X.; Zhu, J.; Ma, X.; Patel, S.; Liu, J.; Wang, P.; Lai, C.S.E.; Gou, M.; Xu, Y.; et al. Direct 3D Bioprinting of Prevascularized Tissue Constructs with Complex Microarchitecture. *Biomaterials* **2017**, *124*, 106–115. [CrossRef] [PubMed]

103. Mollica, P.A.; Booth-Creech, E.N.; Reid, J.A.; Zamponi, M.; Sullivan, S.M.; Palmer, X.-L.; Sachs, P.C.; Bruno, R.D. 3D Bioprinted Mammary Organoids and Tumoroids in Human Mammary Derived ECM Hydrogels. *Acta Biomater.* **2019**, *95*, 201–213. [CrossRef] [PubMed]

104. Cetnar, A.D.; Tomov, M.L.; Ning, L.; Jing, B.; Theus, A.S.; Kumar, A.; Wijntjes, A.N.; Bhamidipati, S.R.; Do, K.P.; Mantalaris, A.; et al. Patient-Specific 3D Bioprinted Models of Developing Human Heart. *Adv. Healthc. Mater.* **2020**, *4*, 2001169. [CrossRef]

105. Chimene, D.; Kaunas, R.; Gaharwar, A.K. Hydrogel Bioink Reinforcement for Additive Manufacturing: A Focused Review of Emerging Strategies. *Adv. Mater.* **2020**, *32*, 1902026. [CrossRef] [PubMed]

106. Jose, R.R.; Rodriguez, M.J.; Dixon, T.A.; Omenetto, F.; Kaplan, D.L. Evolution of Bioinks and Additive Manufacturing Technologies for 3D Bioprinting. *ACS Biomater. Sci. Eng.* **2016**, *2*, 1662–1678. [CrossRef]

107. Xin, S.; Chimene, D.; Garza, J.E.; Gaharwar, A.K.; Alge, D.L. Clickable PEG Hydrogel Microspheres as Building Blocks for 3D Bioprinting. *Biomater. Sci.* **2019**, *7*, 1179–1187. [CrossRef]

108. Ling, Y.; An, T.; Yap, L.W.; Zhu, B.; Gong, S.; Cheng, W. Disruptive, Soft, Wearable Sensors. *Adv. Mater.* **2020**, *32*, 1904664. [CrossRef] [PubMed]

109. Jiang, Y.; Liu, Z.; Matsuhisa, N.; Qi, D.; Leow, W.R.; Yang, H.; Yu, J.; Chen, G.; Liu, Y.; Wan, C.; et al. Auxetic Mechanical Metamaterials to Enhance Sensitivity of Stretchable Strain Sensors. *Adv. Mater.* **2018**, *30*, 1706589. [CrossRef]

110. Nag, A.; Feng, S.; Mukhopadhyay, S.C.; Kosel, J.; Inglis, D. 3D Printed Mould-Based Graphite/PDMS Sensor for Low-Force Applications. *Sensors Actuators A Phys.* **2018**, *280*, 523–534. [CrossRef]
111. Asymmetric Structure Based Flexible Strain Sensor for Simultaneous Detection of Various Human Joint Motions | ACS Applied Electronic Materials. Available online: https://pubs.acs.org/doi/abs/10.1021/acsaelm.9b00386 (accessed on 21 October 2020).
112. Kisic, M.; Blaz, N.; Zivanov, L.; Damnjanovic, M. Elastomer Based Force Sensor Fabricated by 3D Additive Manufacturing. *AIP Adv.* **2020**, *10*, 015017. [CrossRef]
113. Multi-Material 3D Printing of a Soft Pressure Sensor | Elsevier Enhanced Reader. Available online: https://reader.elsevier.com/reader/sd/pii/S2214860419305238?token=B81C23A18C57D3A0C17FE0D40A7A57082C6CF8AE00D7C310530B9CBA394D7CA66DBD4BD558AC9EBE9154CA53647EEAFB (accessed on 10 January 2021).
114. Hohimer, C.J.; Petrossian, G.; Ameli, A.; Mo, C.; Pötschke, P. 3D Printed Conductive Thermoplastic Polyurethane/Carbon Nanotube Composites for Capacitive and Piezoresistive Sensing in Soft Pneumatic Actuators. *Addit. Manuf.* **2020**, *34*, 101281. [CrossRef]
115. Xiang, D.; Zhang, X.; Han, Z.; Zhang, Z.; Zhou, Z.; Harkin-Jones, E.; Zhang, J.; Luo, X.; Wang, P.; Zhao, C.; et al. 3D Printed High-Performance Flexible Strain Sensors Based on Carbon Nanotube and Graphene Nanoplatelet Filled Polymer Composites. *J. Mater. Sci.* **2020**, *55*, 15769–15786. [CrossRef]
116. 3D-Printed Highly Stable Flexible Strain Sensor Based on Silver-Coated-Glass Fiber-Filled Conductive Silicon Rubber | Elsevier Enhanced Reader. Available online: https://reader.elsevier.com/reader/sd/pii/S0264127520303221?token=79DD1D0239D01AACCB524D928DF87E8B65CCFA749FB69A6490AD6DCA72A49E6D4DE364D0F7B433EDA8DFAE14E90FA9FE (accessed on 12 October 2020).
117. Shi, G.; Lowe, S.E.; Teo, A.J.T.; Dinh, T.K.; Tan, S.H.; Qin, J.; Zhang, Y.; Zhong, Y.L.; Zhao, H. A Versatile PDMS Submicrobead/Graphene Oxide Nanocomposite Ink for the Direct Ink Writing of Wearable Micron-Scale Tactile Sensors. *Appl. Mater. Today* **2019**, *16*, 482–492. [CrossRef]
118. Abshirini, M.; Charara, M.; Liu, Y.; Saha, M.C.; Altan, M.C. *Additive Manufacturing of Polymer Nanocomposites with In-Situ Strain Sensing Capability*; American Society of Mechanical Engineer: New York, NY, USA, 2019.
119. Davoodi, E.; Fayazfar, H.; Liravi, F.; Jabari, E.; Toyserkani, E. Drop-on-Demand High-Speed 3D Printing of Flexible Milled Carbon Fiber/Silicone Composite Sensors for Wearable Biomonitoring Devices. *Addit. Manuf.* **2020**, *32*, 101016. [CrossRef]
120. Xiang, D.; Zhang, X.; Harkin-Jones, E.; Zhu, W.; Zhou, Z.; Shen, Y.; Li, Y.; Zhao, C.; Wang, P. Synergistic Effects of Hybrid Conductive Nanofillers on the Performance of 3D Printed Highly Elastic Strain Sensors. *Compos. Part Appl. Sci. Manuf.* **2020**, *129*, 105730. [CrossRef]
121. Taherkhani, B.; Azizkhani, M.B.; Kadkhodapour, J.; Anaraki, A.P.; Rastgordani, S. Highly Sensitive, Piezoresistive, Silicone/Carbon Fiber-Based Auxetic Sensor for Low Strain Values. *Sens. Actuators Phys.* **2020**, *305*, 111939. [CrossRef]
122. Yin, X.-Y.; Zhang, Y.; Cai, X.; Guo, Q.; Yang, J.; Wang, Z.L. 3D Printing of Ionic Conductors for High-Sensitivity Wearable Sensors. *Mater. Horiz.* **2019**, *6*, 767–780. [CrossRef]
123. Wang, Z.; Gao, W.; Zhang, Q.; Zheng, K.; Xu, J.; Xu, W.; Shang, E.; Jiang, J.; Zhang, J.; Liu, Y. 3D-Printed Graphene/Polydimethylsiloxane Composites for Stretchable and Strain-Insensitive Temperature Sensors. *ACS Appl. Mater. Interfaces* **2019**, *11*, 1344–1352. [CrossRef]
124. Li, X.; Yang, Y.; Xie, B.; Chu, M.; Sun, H.; Hao, S.; Chen, Y.; Chen, Y. 3D Printing of Flexible Liquid Sensor Based on Swelling Behavior of Hydrogel with Carbon Nanotubes. *Adv. Mater. Technol.* **2019**, *4*, 1800476. [CrossRef]
125. Kiziroglou, M.E.; Yeatman, E.M. Materials and techniques for energy harvesting. In *Functional Materials for Sustainable Energy Applications*; Elsevier: Amsterdam, The Netherlands, 2012; pp. 541–572. ISBN 978-0-85709-059-1.
126. Seol, M.-L.; Ivaškevičiūtė, R.; Ciappesoni, M.A.; Thompson, F.V.; Moon, D.-I.; Kim, S.J.; Kim, S.J.; Han, J.-W.; Meyyappan, M. All 3D Printed Energy Harvester for Autonomous and Sustainable Resource Utilization. *Nano Energy* **2018**, *52*, 271–278. [CrossRef]
127. Li, J.; Wong, W.-Y.; Tao, X. Recent Advances in Soft Functional Materials: Preparation, Functions and Applications. *Nanoscale* **2020**, *12*, 1281–1306. [CrossRef] [PubMed]
128. Ou, C.; Jing, Q.; Busolo, T.; Kar-Narayan, S. Chapter 13 - Manufacturing routes toward flexible and smart energy harvesters and sensors based on functional nanomaterials. In *Advances in Nanostructured Materials and Nanopatterning Technologies*; Guarino, V., Focarete, M.L., Pisignano, D., Eds.; Advanced Nanomaterials; Elsevier: Amsterdam, The Netherlands, 2020; pp. 381–437. ISBN 978-0-12-816865-3.
129. Zou, Y.; Tan, P.; Shi, B.; Ouyang, H.; Jiang, D.; Liu, Z.; Li, H.; Yu, M.; Wang, C.; Qu, X.; et al. A Bionic Stretchable Nanogenerator for Underwater Sensing and Energy Harvesting. *Nat. Commun.* **2019**, *10*, 2695. [CrossRef] [PubMed]
130. Chen, K.; Zhang, L.; Kuang, X.; Li, V.; Lei, M.; Kang, G.; Wang, Z.L.; Qi, H.J. Dynamic Photomask-Assisted Direct Ink Writing Multimaterial for Multilevel Triboelectric Nanogenerator. *Adv. Funct. Mater.* **2019**, *29*, 1903568. [CrossRef]
131. Tzounis, L.; Petousis, M.; Grammatikos, S.; Vidakis, N. 3D Printed Thermoelectric Polyurethane/Multiwalled Carbon Nanotube Nanocomposites: A Novel Approach towards the Fabrication of Flexible and Stretchable Organic Thermoelectrics. *Materials* **2020**, *13*, 2879. [CrossRef]
132. Yuan, X.; Gao, X.; Yang, J.; Shen, X.; Li, Z.; You, S.; Wang, Z.; Dong, S. The Large Piezoelectricity and High Power Density of a 3D-Printed Multilayer Copolymer in a Rugby Ball-Structured Mechanical Energy Harvester. *Energy Environ. Sci.* **2020**, *13*, 152–161. [CrossRef]
133. Keong, B.A.W.; Hua, R.Y.C. A Novel Fold-Based Design Approach toward Printable Soft Robotics Using Flexible 3D Printing Materials. *Adv. Mater. Technol.* **2018**, *3*, 1700172. [CrossRef]

134. Tognato, R.; Armiento, A.R.; Bonfrate, V.; Levato, R.; Malda, J.; Alini, M.; Eglin, D.; Giancane, G.; Serra, T. A Stimuli-Responsive Nanocomposite for 3D Anisotropic Cell-Guidance and Magnetic Soft Robotics. *Adv. Funct. Mater.* **2019**, *29*, 1804647. [CrossRef]
135. Hu, F.; Lyu, L.; He, Y. A 3D Printed Paper-Based Thermally Driven Soft Robotic Gripper Inspired by Cabbage. *Int. J. Precis. Eng. Manuf.* **2019**, *20*, 1915–1928. [CrossRef]
136. Vogt, D.M.; Becker, K.P.; Phillips, B.T.; Graule, M.A.; Rotjan, R.D.; Shank, T.M.; Cordes, E.E.; Wood, R.J.; Gruber, D.F. Shipboard Design and Fabrication of Custom 3D-Printed Soft Robotic Manipulators for the Investigation of Delicate Deep-Sea Organisms. *PLOS ONE* **2018**, *13*, e0200386. [CrossRef]
137. Neumann, W.; Pusch, T.P.; Siegfarth, M.; Schad, L.R.; Stallkamp, J.L. CT and MRI Compatibility of Flexible 3D-Printed Materials for Soft Actuators and Robots Used in Image-Guided Interventions. *Med. Phys.* **2019**, *46*, 5488–5498. [CrossRef]
138. Zolfagharian, A.; Kaynak, A.; Khoo, S.Y.; Kouzani, A.Z. Polyelectrolyte Soft Actuators: 3D Printed Chitosan and Cast Gelatin. *3D Print. Addit. Manuf.* **2018**, *5*, 138–150. [CrossRef]
139. Ang, B.W.K.; Yeow, C. Design and Characterization of a 3D Printed Soft Robotic Wrist Sleeve with 2 DoF for Stroke Rehabilitation. In Proceedings of the 2019 2nd IEEE International Conference on Soft Robotics (RoboSoft), Seoul, Korea, 14–18 April 2019; pp. 577–582.
140. Saed, M.O.; Ambulo, C.P.; Kim, H.; De, R.; Raval, V.; Searles, K.; Siddiqui, D.A.; Cue, J.M.O.; Stefan, M.C.; Shankar, M.R.; et al. Molecularly-Engineered, 4D-Printed Liquid Crystal Elastomer Actuators. *Adv. Funct. Mater.* **2019**, *29*, 1806412. [CrossRef]
141. Han, D.; Farino, C.; Yang, C.; Scott, T.; Browe, D.; Choi, W.; Freeman, J.W.; Lee, H. Soft Robotic Manipulation and Locomotion with a 3D Printed Electroactive Hydrogel. *ACS Appl. Mater. Interfaces* **2018**, *10*, 17512–17518. [CrossRef]
142. Roach, D.J.; Kuang, X.; Yuan, C.; Chen, K.; Qi, H.J. Novel Ink for Ambient Condition Printing of Liquid Crystal Elastomers for 4D Printing. *Smart Mater. Struct.* **2018**, *27*, 125011. [CrossRef]
143. Kim, T.-H.; Vanloo, J.; Kim, W.S. 3D Origami Sensing Robots for Cooperative Healthcare Monitoring. *Adv. Mater. Technol.* **2021**, *6*, 2000938. [CrossRef]
144. Kim, S.; Kim, S.; Majditehran, H.; Patel, D.K.; Majidi, C.; Bergbreiter, S. Electromechanical Characterization of 3D Printable Conductive Elastomer for Soft Robotics. In Proceedings of the 2020 3rd IEEE International Conference on Soft Robotics (RoboSoft), New Haven, CT, USA, 15 May–15 July 2020; pp. 318–324.
145. Manthrammel, M.A.; Yahia, I.S.; Shkir, M.; AlFaify, S.; Zahran, H.Y.; Ganesh, V.; Yakuphanoglu, F. Novel Design and Micro-electronic Analysis of Highly Stable Au/Indigo/n-Si Photodiode for Optoelectronic Applications. *Solid State Sci.* **2019**, *93*, 7–12. [CrossRef]
146. Mariappan, S.M.; Shkir, M.; Alshahrani, T.; Elangovan, V.; Algarni, H.; AlFaify, S. Insight on the Optoelectronics and Enhanced Dielectric Properties of Strontium Decorated PbI2 Nanosheets for Hot Carrier Solar Cell Applications. *J. Alloys Compd.* **2021**, *859*, 157762. [CrossRef]
147. Zhang, H.; Rogers, J.A. Recent Advances in Flexible Inorganic Light Emitting Diodes: From Materials Design to Integrated Optoelectronic Platforms. *Adv. Opt. Mater.* **2019**, *7*, 1800936. [CrossRef]
148. Park, S.H.; Su, R.; Jeong, J.; Guo, S.-Z.; Qiu, K.; Joung, D.; Meng, F.; McAlpine, M.C. 3D Printed Polymer Photodetectors. *Adv. Mater.* **2018**, *30*, 1803980. [CrossRef] [PubMed]
149. Black Phosphorus Ink Formulation for Inkjet Printing of Optoelectronics and Photonics | Nature Communications. Available online: https://www.nature.com/articles/s41467-017-00358-1 (accessed on 1 February 2021).
150. Allen, B.; Grant, P.S.; Isakov, D.; Stevens, C.J.; Wu, Y. 3D-Printed Optical Devices with Refractive Index Control for Microwave Applications. *TechConnect Briefs* **2018**, *4*, 92–95.
151. Loke, G.; Yuan, R.; Rein, M.; Khudiyev, T.; Jain, Y.; Joannopoulos, J.; Fink, Y. Structured Multimaterial Filaments for 3D Printing of Optoelectronics. *Nat. Commun.* **2019**, *10*, 4010. [CrossRef] [PubMed]

materials

Review

3D Printing of Fiber-Reinforced Plastic Composites Using Fused Deposition Modeling: A Status Review

Salman Pervaiz [1,*], Taimur Ali Qureshi [1], Ghanim Kashwani [2] and Sathish Kannan [3]

1 Department of Mechanical and Industrial Engineering, Rochester Institute of Technology, Dubai Campus, Dubai P.O. Box 341055, United Arab Emirates; taq3950@rit.edu
2 Engineering Division, New York University Abu Dhabi, Abu Dhabi P.O. Box 129188, United Arab Emirates; gak289@nyu.edu
3 Department of Mechanical Engineering, American University of Sharjah, Sharjah P.O. Box 26666, United Arab Emirates; skannan@aus.edu
* Correspondence: spervaiz@rit.edu

Abstract: Composite materials are a combination of two or more types of materials used to enhance the mechanical and structural properties of engineering products. When fibers are mixed in the polymeric matrix, the composite material is known as fiber-reinforced polymer (FRP). FRP materials are widely used in structural applications related to defense, automotive, aerospace, and sports-based industries. These materials are used in producing lightweight components with high tensile strength and rigidity. The fiber component in fiber-reinforced polymers provides the desired strength-to-weight ratio; however, the polymer portion costs less, and the process of making the matrix is quite straightforward. There is a high demand in industrial sectors, such as defense and military, aerospace, automotive, biomedical and sports, to manufacture these fiber-reinforced polymers using 3D printing and additive manufacturing technologies. FRP composites are used in diversified applications such as military vehicles, shelters, war fighting safety equipment, fighter aircrafts, naval ships, and submarine structures. Techniques to fabricate composite materials, degrade the weight-to-strength ratio and the tensile strength of the components, and they can play a critical role towards the service life of the components. Fused deposition modeling (FDM) is a technique for 3D printing that allows layered fabrication of parts using thermoplastic composites. Complex shape and geometry with enhanced mechanical properties can be obtained using this technique. This paper highlights the limitations in the development of FRPs and challenges associated with their mechanical properties. The future prospects of carbon fiber (CF) and polymeric matrixes are also mentioned in this study. The study also highlights different areas requiring further investigation in FDM-assisted 3D printing. The available literature on FRP composites is focused only on describing the properties of the product and the potential applications for it. It has been observed that scientific knowledge has gaps when it comes to predicting the performance of FRP composite parts fabricated under 3D printing (FDM) techniques. The mechanical properties of 3D-printed FRPs were studied so that a correlation between the 3D printing method could be established. This review paper will be helpful for researchers, scientists, manufacturers, etc., working in the area of FDM-assisted 3D printing of FRPs.

Keywords: FRP; 3D printing; defense; FDM

Citation: Pervaiz, S.; Qureshi, T.A.; Kashwani, G.; Kannan, S. 3D Printing of Fiber-Reinforced Plastic Composites Using Fused Deposition Modeling: A Status Review. *Materials* **2021**, *14*, 4520. https://doi.org/10.3390/ma14164520

Academic Editor: Tuhin Mukherjee

Received: 30 June 2021
Accepted: 7 August 2021
Published: 12 August 2021

Publisher's Note: MDPI stays neutral with regard to jurisdictional claims in published maps and institutional affiliations.

1. Introduction

The properties of materials play a significant role in manufacturing of equipment for various industries, such as defense, automobile, aerospace, healthcare, and many similar sectors, with demanding applications. Fiber reinforced polymer (FRP) composites are a combination of fibers and matrixes that can be either thermoplastic, elastomer, or thermoset. The fiber provides a strength-to-weight ratio, the polymer composites cost less, and the process of making their matrix is quite easy. In 1960, FRP composites were the major structural component of the aerospace sector. In 1990–2006, FRP composites were the low cost and flexible solution for many manufacturing processes [1].

FRP composites have a high strength-to-weight ratio, good anti-wear properties and improved anti-aging capacities compared to traditional metal materials. FRP composites are light weight and provide high performance in many industries. The matrix in FRP composites mostly consists of thermosetting and thermoplastics. The thermosetting plastics in FRP are of an epoxy and polyurethane type and that of thermoplastics are polypropylene (PP), polyamide (PA), and polyetheretherketone (PEEK). FRP composites are also characterized by different fiber materials such as glass, carbon, aramid, and Kevlar. FRPs are also divided into subtypes by fiber length parameter such as short fibers (0.2–0.4 mm), and long fibers (10–25 mm). The method of forming and processing long and short CFRPs is by extrusion or injection molding. For the other type of continuous fiber-reinforced plastic, the processing is performed by winding, molding, impregnating, and pultrusion. The construction of continuous fiber-reinforced plastic components is a lengthy and complex process, and complex structure achievement is difficult because the viscosity is high for infusion during wet-out [2,3].

Additive manufacturing (AM)/rapid prototyping is said to be a method of integrating materials to make objects from computer aided design (CAD) models in consecutive films [4]. AM manufacturing provides low-cost, versatile products; thus, over the past years, the use of this technology to design improved products has become a large trend. Either with monolithic structures or micrometer solutions, AM is proving to beneficial [5]. The literature on FRPs focuses only on straight processes describing the properties of the product and the potential applications for it. Due to the complex interfacial adhesion in FDM printing of fiber-polymer composites, the performance of the component varies significantly, and there is a need to better understand its performance. The mechanical properties of 3D-printed carbon fiber-reinforced polymers are created so that the correlation between the types of additive manufacturing methods can be understandable. The techniques of combination composite materials impact on the weight-to-strength ratio as well as the tensile strength of the components and can play a critical role towards the service life of the components. Fused filament fabrication (FFF) is also a technique for 3D printing; it also allows for layered fabrication of parts using thermoplastic composites. Complex shape and geometry with enhanced mechanical properties can be obtained using this technique.

The literature [6–12] has revealed that the strength of FDM-printed parts is highly dependent on the printing phenomena and quality of the bond formation. Weak strength is associated with insufficient bond strength between the layers. It is also observed that the formation of bonds between layers is based on three phases, namely, surface contact, neck growth, and molecular diffusion [6]. It has been revealed that the second phase of neck growth is very important and can play a vital role towards the strength of FDM-printed parts. The quality of the bonds between layers is dependent on the size of neck formation and is controlled by molecular diffusion that happens between the polymeric chains at the interface. Gurrala and Regalla [13] also studied the coalescence of filament towards the strength of FDM-printed parts. In the study, it was revealed, by scanning electron microscopy, that neck growth was not uniform throughout the process, and at some locations there was no neck formation at all. The reason attributed to this observation was linked with the localized non-uniform cooling rates and temperature variations at different locations. Sun et al. [14] provide an in-depth study on the coalescence mechanism between filament layers. The study mentions the importance of phase 2, which facilitates adhesion and formation of molecular diffusion. The study also revealed that phase 2 is dependent on the contact angle between two filaments. Bonding occurs between adjacent layers of the filament and successive layers of the filament, known as intra-layers and inter-layers, respectively, as shown in the Figure 1 [6].

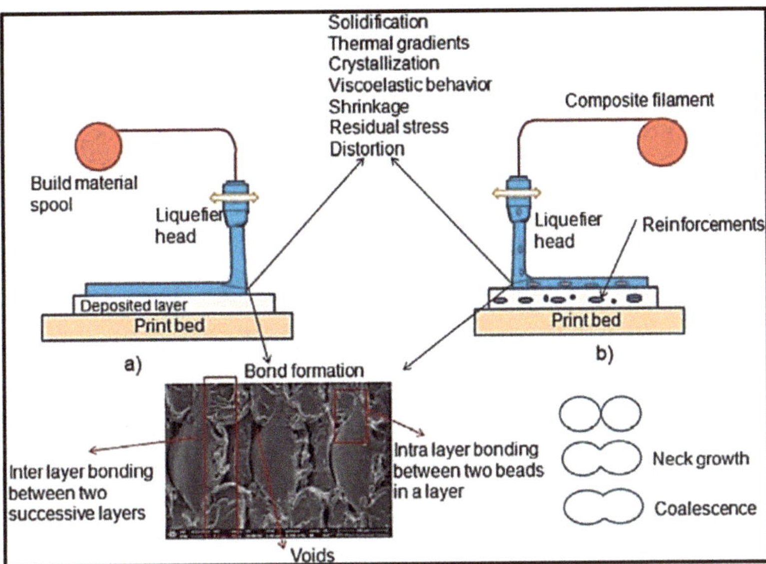

Figure 1. Schematic representation of the FDM process: (**a**) 3D printing of neat polymer; (**b**) 3D printing of polymer reinforced with particle fillers or short fibers [6] (reprinted with kind permission from Elsevier).

The review section of paper highlights the stated challenges in the development of carbon fiber-reinforced polymers and the challenges associated with its mechanical properties. The future prospects for the carbon fiber-reinforced polymers are also mentioned in this study, while it also puts a spotlight on areas requiring further investigation in rapid prototyping. The objective of this review paper was to corelate the mechanical properties of 3D-printed carbon fiber-reinforced polymers with respect to various 3D-printing techniques.

2. Industrial Significance of FRP Composites

FRP provides the desired strength and stiffness while being light weight at the same time. This lightweight property is controlled by using the weight of the matrix in the fiber-matrix material system. In addition to these properties, FRPs are capable of operating at higher temperatures, chemical inertness, and have the ability to provide better damping [15]. Due to the fact of these qualities, FRP composites are rapidly replacing conventional ferrous and non-ferrous metals and their alloys. It can also be observed that the global market growth for FRP composites is increasing rapid. As reflected in Figure 2, the US composite market (FRP) is forecasted to have a compound annual growth rate of 11.3% from 2017 to 2025 [16]. At the same time, industry is adopting 3D-printing technologies to print FRP composites. As shown in Figure 3, the 3D-printed composites market is forecasted to grow to 111 million USD between 2017 and 2022 [17].

As shown in Figure 4, a visible increase in market growth can be noticed regarding 3D-printed FRP composite products with respect to the aerospace and defense sectors. In the below subsections, these sectors and related FRP applications will be discussed in detail.

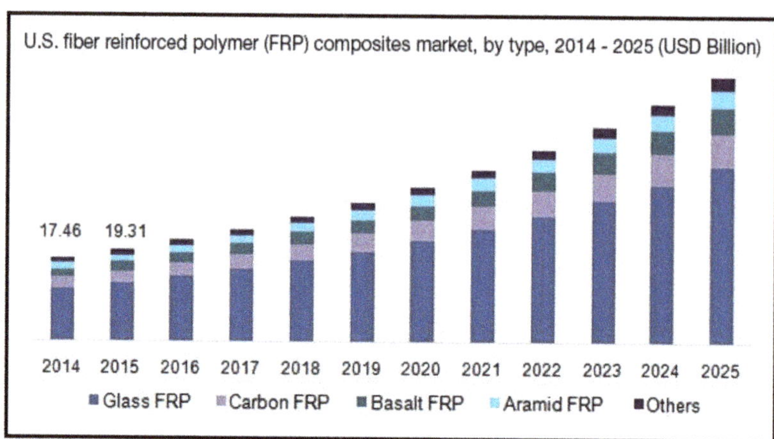

Figure 2. Forecast of the global growth of the FRP composite market, 2014–2025 [16].

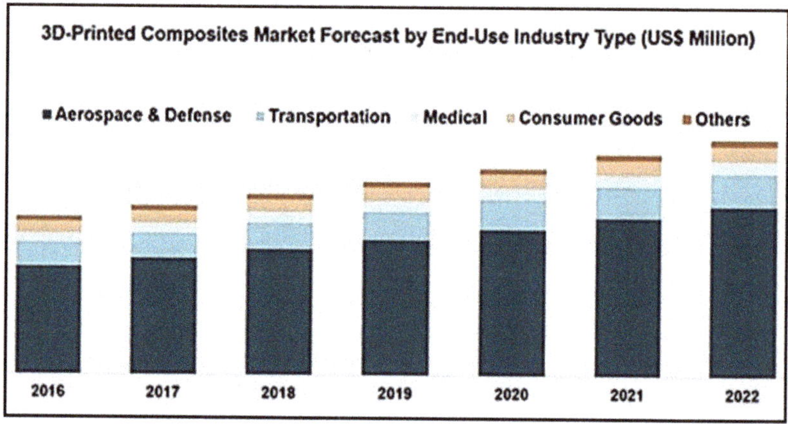

Figure 3. Forecast of the market growth of 3D-printed composites [17].

2.1. Defense and Military Sector

FRP composites are well suited for defense and military applications due to the fact of their high strength, light weight, corrosion resistance, and prototyping of complex geometries, etc. This class of materials was widely used in defense and military applications after World War II. They gained popularity over conventional metals and steels in the defense sector because of their anti-corrosiveness, fatigue resistance, and light weight. This was because structures experience excessive corrosion in salty seawater [18,19]. FRP composites are used in diverse applications such as military vehicles, shelters, war fighting safety equipment, fighter aircrafts, naval ships, and submarine structures. Figure 4 shows FRP composites in a submarine, an F-35 fighter jet, and a helicopter. These materials were favored due to the fact of their light weight and high strength, reliable performance, and ease in maintenance during service life. FRP composites are also utilized extensively in the protective clothing used by security enforcement bodies [20].

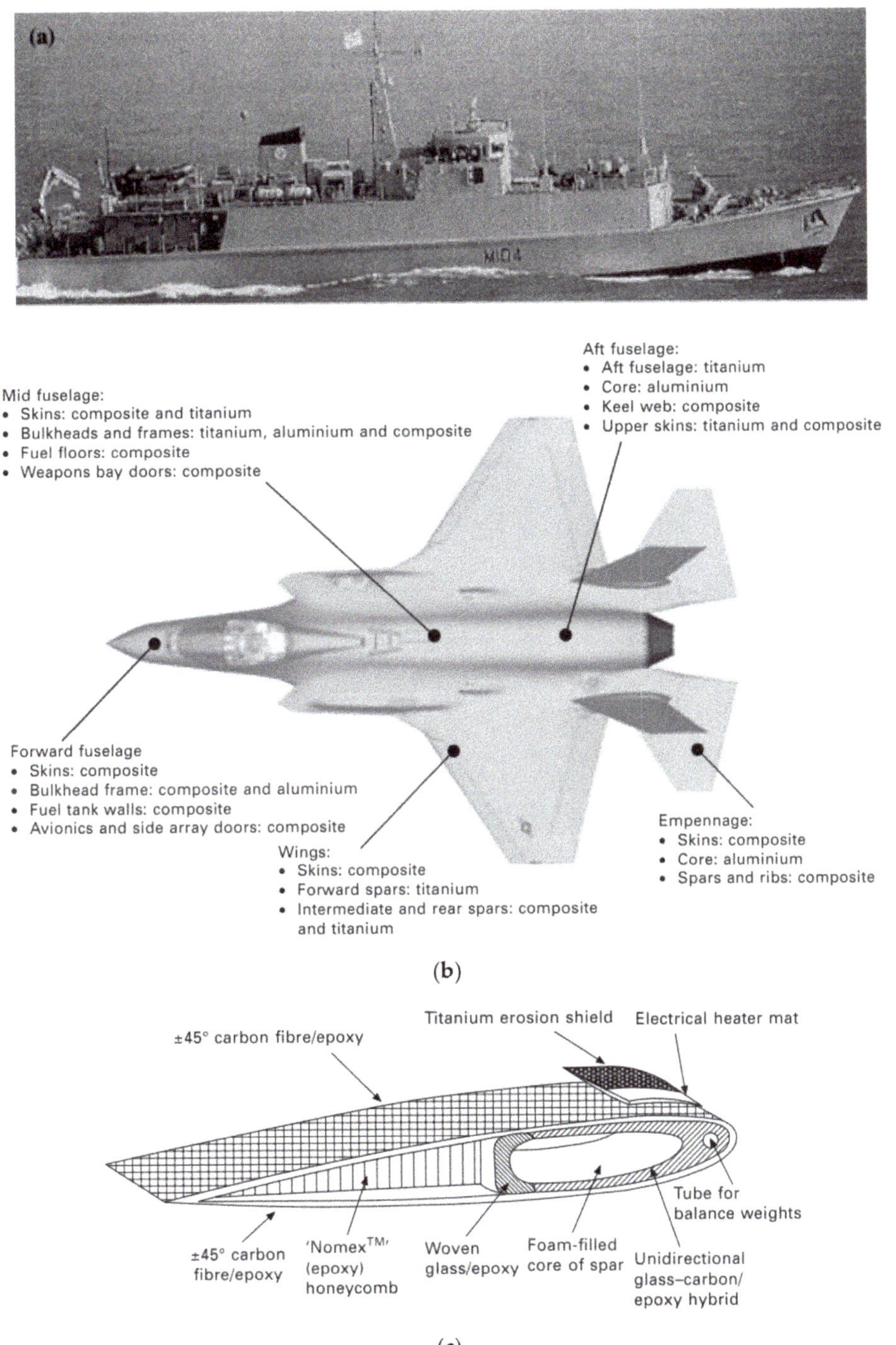

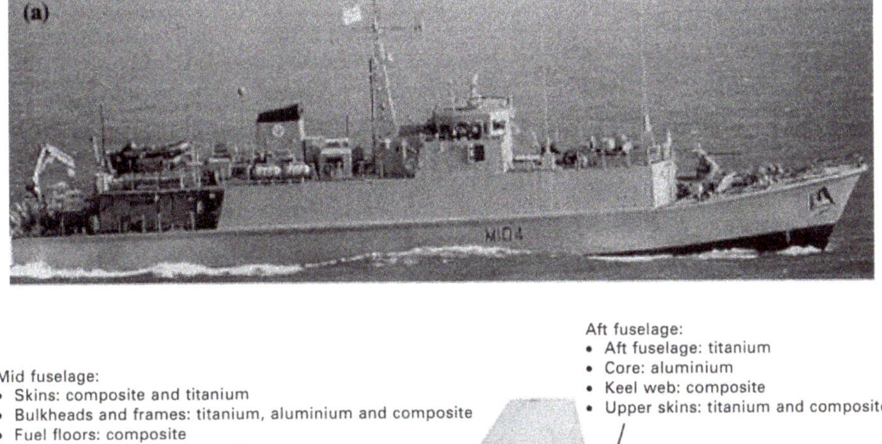

(**b**)

(**c**)

Figure 4. FRP composites in (**a**) Naval ships [19–22]; (**b**) an F-35 fighter jet [23]; (**c**) Helicopter blade [23] (reprinted with kind permission from Elsevier).

2.2. Aerospace Sector

FRP composites are widely used in the construction of passenger aircrafts. The Airbus A300 utilized CFRPs for spoilers, rudders, and airbrakes. Different famous FRP composites, such as CFRP, GFRP, and AFRP, were utilized in the Airbus A330/340 [15]. It is estimated that approximately 50% (by weight) of modern aircrafts are made of composites as shown below in the Boeing 787 Dream Liner in Figure 5. The Boeing 787 was the first air jetliner that utilized composite materials as leading structural materials in the airframe structure. The Boeing 787 carries 23 tons of composite materials. FRP composites are used in key body aircraft parts such as fuselage, upper and lower wing skins, radom, wing flaps, elevators, vertical fins, and horizontal stabilizer. Carbon-fiber epoxy is laid up with the help of robotic heads and fibers are reinforced in the desired directions to support maximum loads [21].

Figure 5. Material utilization in the Boeing 787 Dream Liner manufactured by HdH [24] (reprinted with kind permission from Elsevier).

2.3. Automotive Sector

The Automotive sector was first to introduce better engineering materials to replace conventional metals and alloys to reduce cost and weight. It is reported that 75% of fuel consumption is directly linked to the weight of automobiles [25]. In addition, the automotive sector is highly competitive, and a higher performance with an extended service life is desired from different automotive components. Composite materials are used heavily in the load bearing and structural parts, such as the body, chassis, hoods, brakes, and electronics, of an automobile [20]. In the past, engine parts in the automobile sector were made from cast iron; the drawback of using cast iron was the reduced fuel efficiency and the speed of the engine was also slow. Now, cast iron parts are being replaced by aluminum alloys. A simple material cannot provide all the properties required for an effective product, so combinations of two or more materials were introduced to obtain the desired properties; these are known as composite materials [25]. Figure 6 shows some automotive parts constructed out of composite materials.

2.4. Construction Sector

FRP composites are utilized extensively in the construction sector. They have successfully replaced conventional steels, previously used to produce reinforcement bars for concrete constructions. In the construction sector, FRP composites are popular because of their low weight, high strength, corrosion resistance, high-impact strength, electromagnetic transparency, low operational cost, and low thermal conductivity. However, with respect to the construction industry, they also have limitations such as high brittleness, low shear and bending strength, vulnerability to fire, and high initial investment [1]. FRP composites are widely used in bridge construction. The major limitation associated with FRP composite curing is linked with exothermic resin. It releases heat during the curing process, and thicker samples restrict this heat that, in some cases, result in instant combustion. Many of the bridge structures being used all over the world are pultruded profiles generated from FRP composites. Figure 7 shows different ways in which FRP composite materials are used in bridge structures.

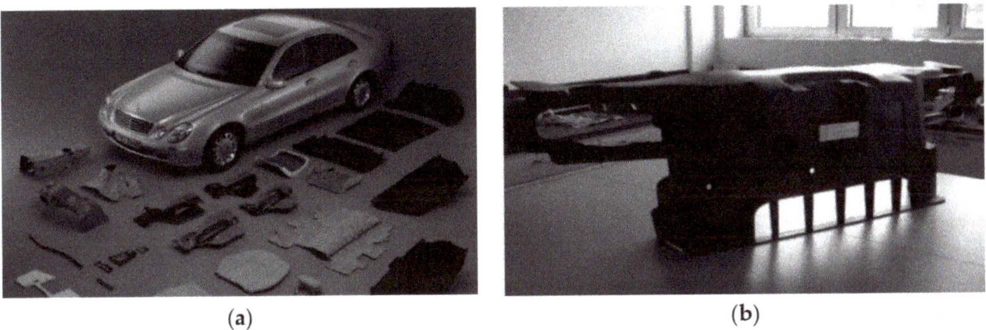

(a)   (b)

Figure 6. Natural composites applied in the automotive sector (**a**) Components of Mercedes-Benz E-Class (**b**) Ford Montagetrager front-end grill (available under a Creative Commons license [26]).

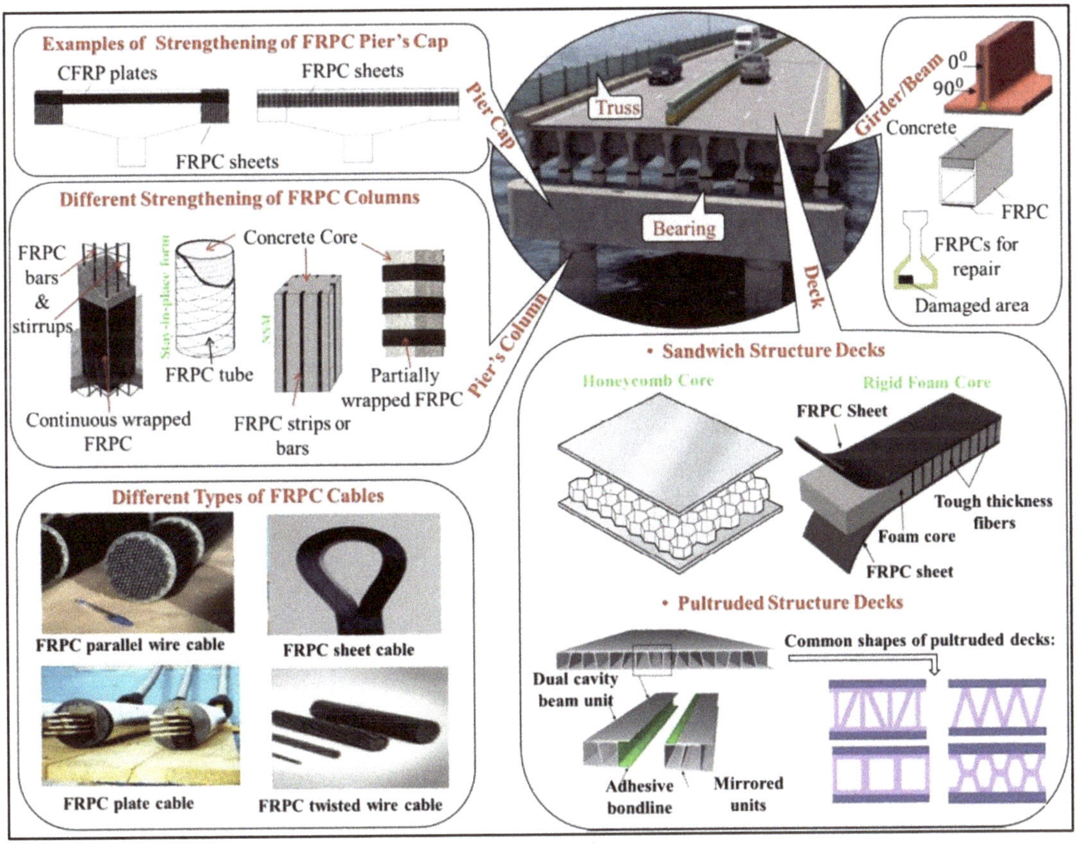

Figure 7. FRP utilization in bridge construction [27] (reprinted with kind permission from Elsevier).

3. Fiber-Matrix Material System

Composites are the materials that are made up of mixture of two or more distinct constituents. In order to be referred as composite materials, it is important that distinct constituents mix with each other at the macro level without any chemical interaction. In mixture form, the resulting product has properties significantly different from each constituent. Moreover, another important feature of composite materials is that one or more discontinuous constituent is embedded in a continuous constituent [28]. As per the nomenclature, the discontinuous constituents are called reinforcements and carry more strength and hardness, whereas the continuous constituent is referred to as the matrix and is generally weaker with respect to the reinforcements. Figure 8 represents schematic illustrations of the fiber-matrix material concept. Reinforcements can come in different geometric shapes and directly controls the strengthening. These reinforcements are classified as fiber or particle based. The resulting composites are referred to as fiber-reinforced composites or particulate-reinforced composites.

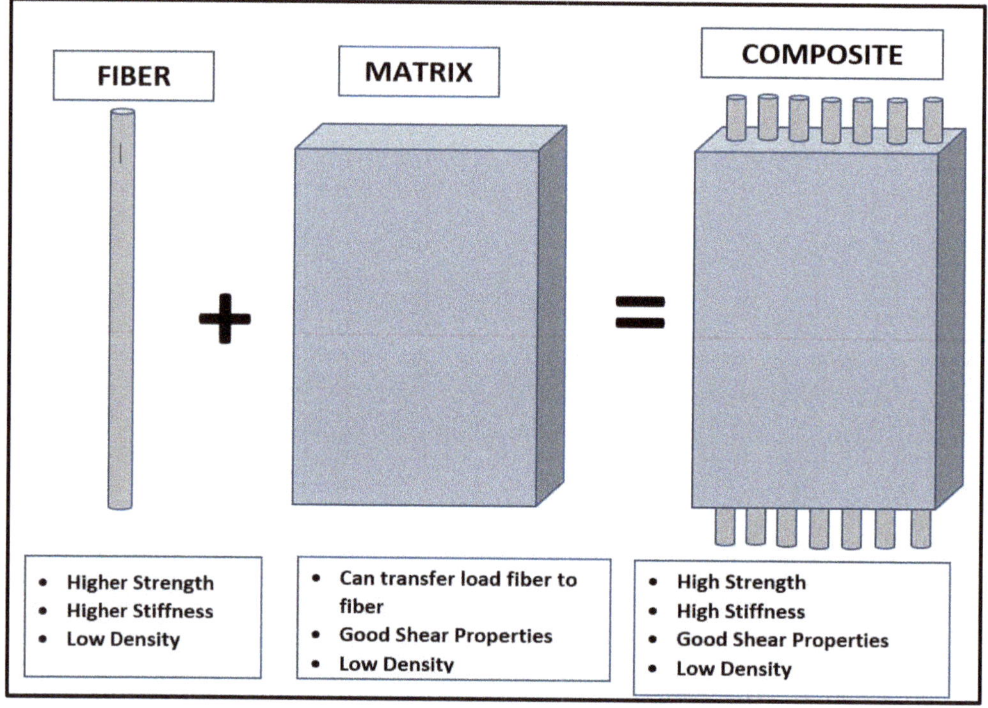

Figure 8. Schematic illustration of fiber-matrix mixture as composite material.

As fiber reinforcements, there are many contender materials that provide beneficial results. Some famous fiber materials are silica, alumina, aramid, carbon, and boron. These fibers possess high strength but low plasticity. Fibers are selected to make FRP composites based on their specific gravity, strength, and modulus. Higher values of specific modulus and strength points to the fiber material being light weight with high strength and stiffness. Desired properties can be tailormade by introducing the correct fiber material, their volume fraction, and their orientation.

Matrix material is weaker in strength and shows relatively more plasticity with respect to the fiber materials. Matrix materials should be chemically and thermally compatible with the fiber materials. Matrix materials bind fiber materials and perform load transfer mechanism between fibers. Matrix material is selected based on the resulting FRP composite operating temperature. Different types of materials such as polymers, metals, and ceramics can be used to serve the matrix materials. In the case of FRP composites, matrix materials are generally polymers. There are two major types of matrix materials, namely, thermosetting plastics and thermoplastics [29]. The most commonly used thermosetting resins are polyester resin, vinyl-ester resin, and epoxy resin. Table 1 represents the most commonly used matrix-fibers FRP composites. However, polymers as matrix material also contain several flaws such as low operating temperature, high thermal expansion coefficient, interaction with moisture, and interaction with radiation [30–34]. Figure 9 shows the properties of commonly used fiber materials.

Table 1. Different properties and industrial applications of matrix materials [30–34].

Sr. No.	Matrix Material	Properties	Major Industrial Sector
1	Polyether sulfone	Flame resistance	Automotive
2	Polyphenylene sulfide	Chemical and high temperature resistance	Electrical
3	Polysulfone	Low moisture absorption and high creep strength	Marine
4	Polyethylene (PE)	Corrosion resistance	Piping construction
5	Polypropylene (PP)	Chemical resistance	Packaging
6	Polylactic acid (PLA)	Biodegradable nature	Biomedical
7	Polyurethane (PU)	Wear resistance and waterproof	Structural
8	Natural Rubber	Low density	Automotive
9	Epoxy Resin	High strength	Automotive and aerospace
10	Polyester	Durable and resistance to water	Structural

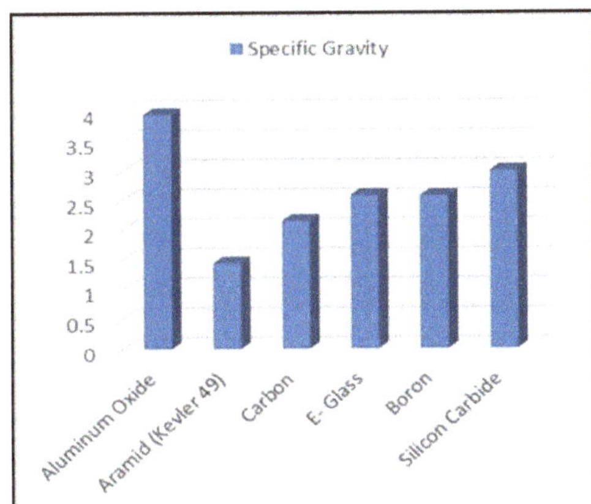

(a)

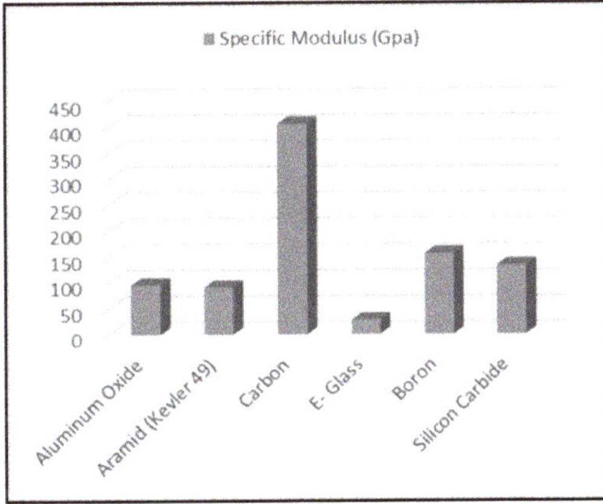

(b)

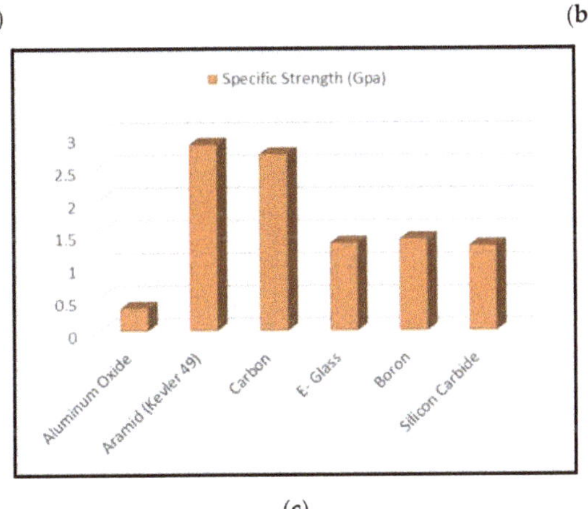

(c)

Figure 9. Comparison of different commonly used fiber materials: (**a**) specific gravity; (**b**) specific modulus (GPa); (**c**) specific strength (GPa) (based on data from [35]).

In order to bond matrix and fiber together, a bonding agent is utilized at the interface of both as shown in the Figure 10A [36]. To perform adequately, a composite material should be bonded properly at the interface region [37]. The interface has an important role, as it transfers a load from the matrix to the reinforcements. Better bonding at the interface results in higher stiffness and strength of the resulting composite [38].

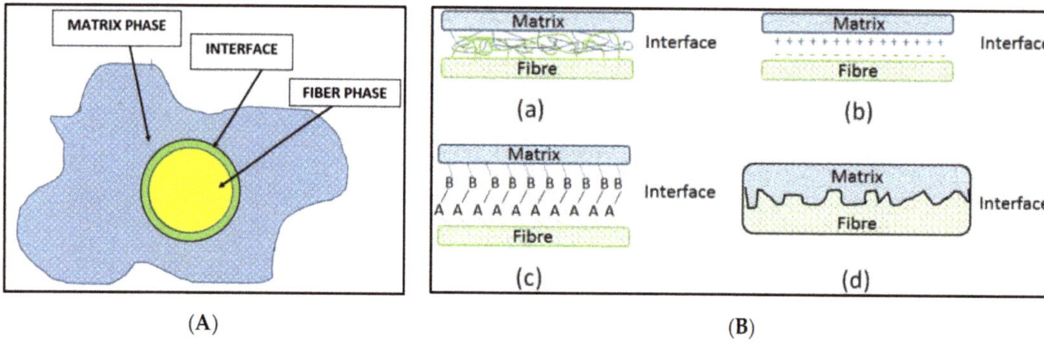

Figure 10. (**A**) Fiber-matrix composite material structure. (**B**) Fiber-matrix interfacial bonding mechanisms: (**a**) molecular entanglement following interdiffusion, (**b**) electrostatic adhesion, (**c**) chemical bonding, and (**d**) mechanical interlocking [39] (reprinted with kind permission from Elsevier).

Figure 10B shows the different types of interfacial bonding mechanisms between fiber and matrix. Interfacial adhesion between the fiber and matrix interface is the critical factor that significantly dictates the stress transfer between matrix and fibers. In the literature [40], different mechanisms for fiber-matrix bonding have been reported such as inter-diffusion, electro-static adhesion, chemical reaction, and mechanical interlocking as reported in the Figure 10B. These mechanisms contribute in combination to make the interfacial adhesion between layers. When fiber and matrix materials came into contact at the molecular level, a diffusion process occurs due to the Van der Waal interaction and hydrogen bonding. This process is controlled by the good wetting behavior between fiber and matrix materials and the degree of diffusion. The degree of diffusion is mainly dependent on the chemical compatibility of the fiber and matrix materials. In electrostatic adhesion mechanism, the bonding between fiber and matrix surfaces is the result of anion and cation formation at the surfaces that results in adhesion. In the chemical bonding mechanism, surface chemistry has a controlling influence. Mechanical interlocking is the result of penetration and locking of peaks and valleys between fiber and matrix materials.

Poor bonding at the interface provides lower strength and stiffness. Failure mechanisms are also characterized by fracture at the interface or away from interface, namely, adhesive and cohesive fractures, respectively.

4. 3D Printing of FRP Composites

It has been observed that conventional FRP composite fabrication methods are expensive, time consuming, and rigid to design modifications [41]. 3D printing is an additive manufacturing technique; in 3D printing, layer-by-layer components are fabricated instead of cutting or molding the material. 3D printing can produce complex and customized products with a short delivery time, lower production cost, and lower material consumption. Currently, 3D printing is used in applications, such as individual production, complex products manufacturing, on-demand manufacturing, new business models, new services, and decentralized products. 3D printing is without a doubt a technology that will rule the future and will represent new stage of smart manufacturing [42]. The global forecast for growth in additive manufacturing of composites is predicted to be a 10 billion USD overall opportunity as shown in Figure 11 [43]. The process of fabricating CFRP in a cost-effective way is under investigation. It is also predicted that 3D-printing technologies will bring positive changes in the cost, energy, and emissions throughout the life cycle of parts [44]. Another comparative study showed that cumulative energy demand was reduced by 41–64% using 3D-printed polymeric materials [45].

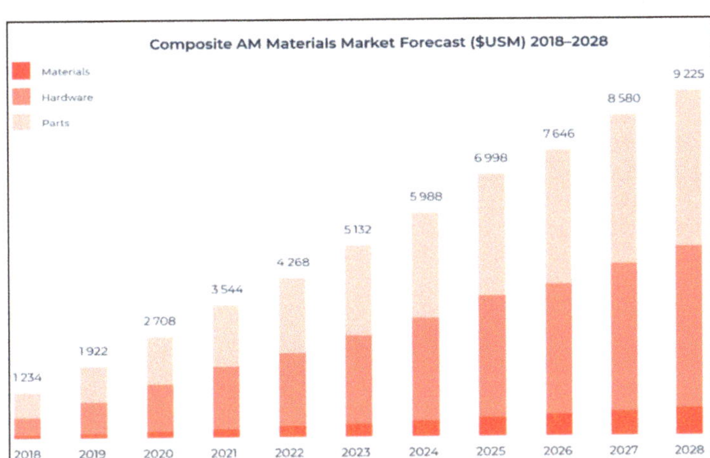

Figure 11. Growth in additive manufacturing of composites [43].

The most popular type of 3D-printing technologies is FDM. Other techniques that are used in 3D printing are selective laser sintering (SLS), laminated object manufacturing (LOM), PolyJet, digital light processing (DLP), selective deposition lamination (SDL), and electron beam melting (EBM). The materials used in FDM 3D printing are polymer filaments (PLA, TPU, ABS). The strength of 3D-printed materials is discussed in a later part of the manuscript.

Currently, the strength of 3D-printed components is not up to the industry's requirements, especially load-bearing parts and fully functional industry parts [46,47]. It is because of the lower end of the polymeric material system, commonly known as commodity thermoplastics, mainly used in 3D printing applications but have limited functionality based on load bearing capacity. Thus, if CFRPs and 3D printing are combined together, we can fabricate the best products that are light weight, high performance, complex in structure, and have good prospects in future industries [48,49]. Therefore, if we can combine CFRPs and 3D printing together and take advantage of both, we can produce light-weight and high-performance components more efficiently with more complex structures, which will have very good prospects in future manufacturing industries [50]. CFRPs have mechanical properties that can improve the life span of the products manufactured. These products are then used in the aerospace industry and small manufacturing sector so that light-weight products with durability can be manufactured [51,52].

4.1. Fused Deposition Modeling (FDM)

FDM is a 3D-printing method based on material extrusion. In this process, extruded heated material comes out of nozzle and adds, layer by layer, to result in a printed composite part [53]. Filaments are made up of thermal plastics. The FDM process consists of the deposition of two materials: one to build the actual component and the other with disposable structural support. Filament is available in the form of spools and fed to the extrusion head. The extrusion head has a temperature controlling mechanism that facilitates heating of filament and conversion into a semi-liquid state. Figure 12a represents the schematic illustration of the FDM process. A spool of filament is utilized to melt first in the nozzle and then print the material in a layer-by-layer sequence to generate required geometric feature. To print the FRP composites using the FDM printing method, there are two methodologies. Figure 12b shows the structural representation of the FRP printed product. It shows the nomenclature in terms of bead, lamina, and laminate and intra-bead and inter-bead voids.

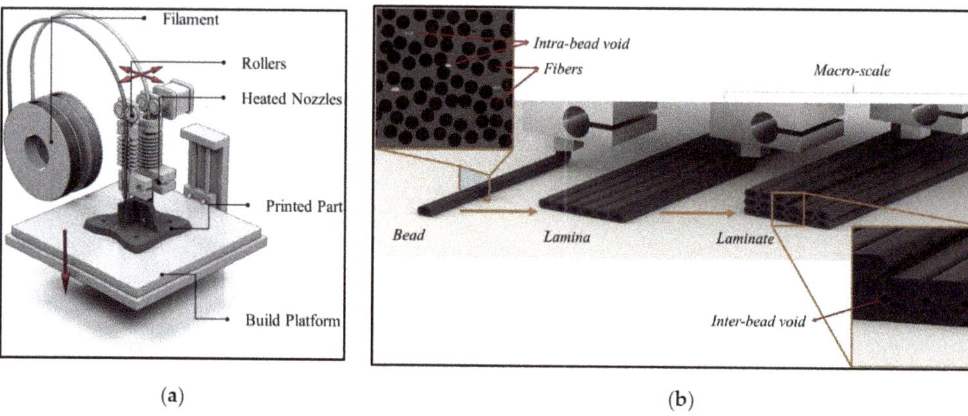

Figure 12. (**a**) Schematic illustration of conventional FDM Processes; (**b**) structural illustration of the FRP-printed part [54] (reprinted with kind permission from Elsevier).

4.2. Short Fiber-Polymer Composites Using FDM

In the first method, a filament is a mixture of short fibers and polymer and printing happens using the conventional FDM-based printing process. Figure 13 shows the schematic illustration of the material flow in the FDM printing of the FRPs. The figure shows that CF and plastic pellets are blended together in the blender and extruded in the form of filament. The prepared CF-polymer filament is then utilized in the conventional FDM printer for printing purpose.

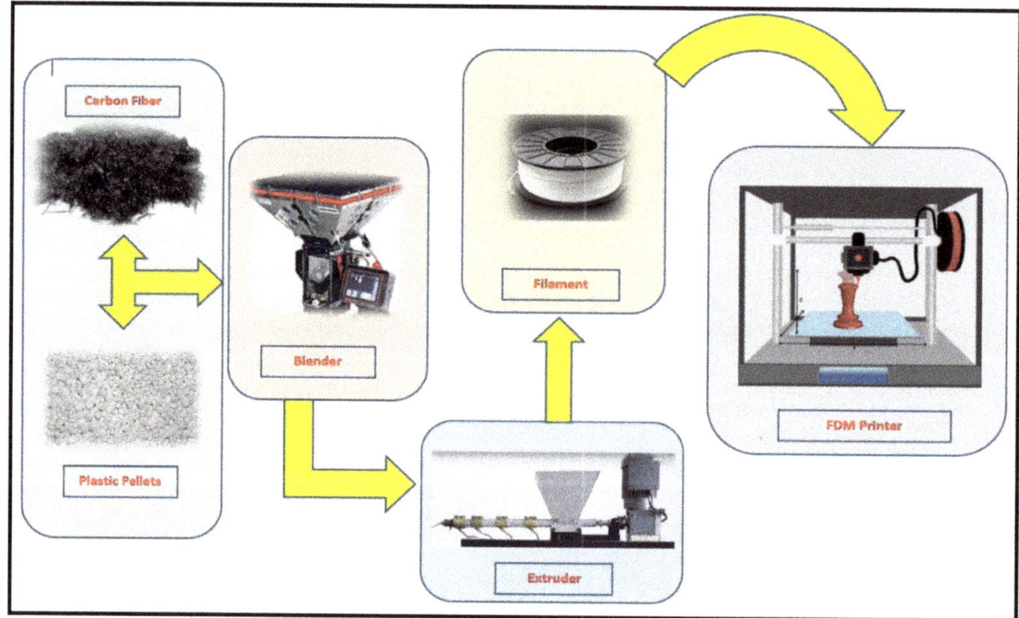

Figure 13. Schematic illustration of material flow in the first FRP printing strategy using a mixture of short fibers and polymer.

4.3. Continuous Fiber-Polymer Composites Using FDM

In the second approach, fiber is in a continuous form and mixed with polymeric matrix to print functional parts. The literature [55–57] reveals that there are three different options to incorporate continuous fiber in the matrix. These methods are differentiated in terms of timing and location of fiber-matrix mixing. In the first simple approach, prefabricated composite filaments are utilized to print the part using conventionally available FDM printers. It is referred to as a simple approach because it requires little change from the conventional FDM setup. In the second approach, the fiber and matrix are separated prior to reaching the print head, and they mix in the printing head that makes the mixing flexible. However, it comes with a challenging printing head setup. Air inclusion should be avoided by precise control during the fiber infiltration process [55]. Figure 14 represents the second type of printing arrangement. In the third case, fiber is deposited directly onto the polymeric component using separate mechanisms. Fiber impregnation is critical in this case, and inappropriate fiber deposition causes defects in the 3D-printed FRP composites. However, temperature-controlled post-processing can increase the strength significantly [58]. Table 2 shows different types of FDM-based FRP printing methods.

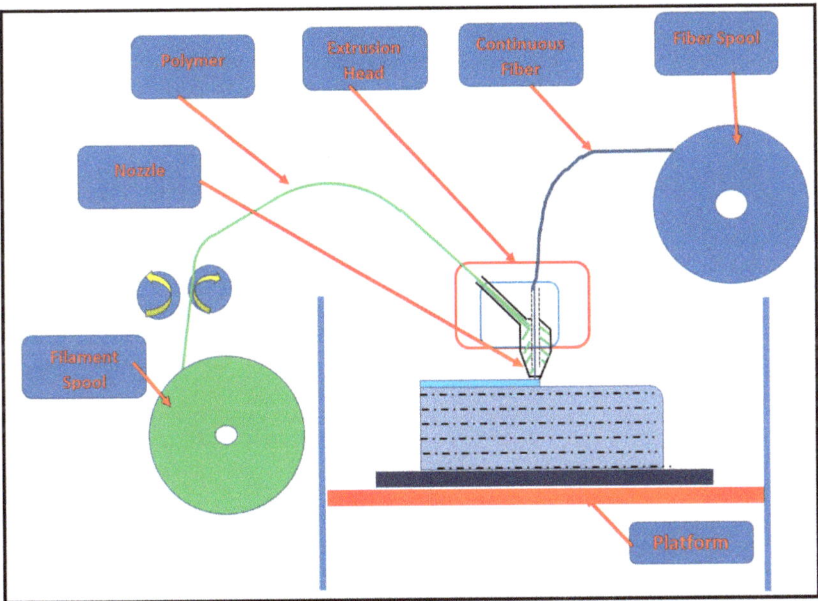

Figure 14. Schematic illustration of the second FRP printing strategy using continuous fibers and polymer (redrawn from [59]).

Table 2. 3D printing method.

		Pre-Embedded Composite Filament		Embedding in Print Head		Embedding on Component
Advantages	-	Improved handling while printing due to the prefabrication of filament.	-	Ability to print both pure plastic and plastic–fiber mixed parts.	-	Process is more versatile and different fibers and matrices materials can be added.
	-	Easy to incorporate in a conventional FDM printer.	-	Easy to vary fiber volume ratio.		
Disadvantages	-	Rigid in terms of material mixture, as it provides a fixed fiber–volume ratio.	-	Requires special printing head with precise control over mixing and air inclusion.	-	Adjustment of fiber orientation requires additional complex machine capabilities.

5. Mechanical Properties of Fiber-Reinforced Polymer Composites

The performance of any material is assessed from the mechanical properties. Mechanical character of material is judged by its yield strength, tensile strength, modulus of elasticity, and flexural strength. These properties are important in determining the performance of materials. Several studies were conducted to analyze the mechanical character of both short and long fiber 3D-printed FRP composites.

5.1. Elastic Modulus and Strength

Ning et al. [5] investigated if mixing carbon fiber in different content percentages and fiber lengths could improve the mechanical properties in comparison to pure thermoplastic, which was ABS plastic in the study. The study incorporated different fiber content, such as 3%, 5%, 7.5%, 10%, and 15%. Two different fiber lengths of 100 and 150 μm were investigated. The study revealed that highest tensile strength was observed for the fiber content between 5% and 7.5%. For higher fiber content, tensile strength was reduced by approximately 10%. Higher levels of porosity were observed for fiber contents of 10% and linked with low strength. The 3D printing of FRP composites showed that by increasing the carbon fiber content in the product, it resulted in large void areas, and these voids negatively impacted on the tensile strength of the material [5]. Tekinalp et al. [60] mixed chopped short carbon fibers in the ABS matrix. The study revealed that the tensile strength and modulus was improved significantly when compared with the conventional compression molded composite samples. It was observed that 3D printed samples showed 115% increase in tensile strength, and approximately 700% increase in the modulus. Karsli and Aytac [61] prepared FRP composite by mixing short carbon fiber in polyamide 6 (PA6) matrix. The study investigated the mechanical properties of the prepared FRP by considering the fiber length and fiber content as the main parameters. Increasing fiber content resulted in better strength, modulus, and hardness values at the expense of ductility. Zhong et al. [62] mixed short glass fibers in the ABS matrix and improved the strength significantly. Abeykoon et al. [63] investigated the performance of five commercially available printing materials such as polylactic acid (PLA), acrylonitrile butadiene styrene (ABS), carbon fiber-reinforced PLA (CFR-PLA), carbon fiber-reinforced ABS (CFR-ABS), and carbon nanotube-reinforced ABS (CNT-ABS). The study aimed to investigate the effects of infill pattern, infill density, and infill speeds. It was observed in the study that higher infill density increased the modulus. Out of the different infill patterns, linear provided the highest modulus as shown in Figure 15C,D. The best performance of linear pattern was attributed with the layer arrangement with lowest pores and higher bonding between layers. The strongest material among the five materials was CFR-PLA as shown in Figure 15A. A nozzle temperature of 215 °C was found to be appropriate for PLA as shown in Figure 15B. Table 3 shows the summary of few studies with short fiber reinforcement.

Table 3. Summary of the literature on short fiber reinforcement.

Source	Matrix	Reinforcement	Important Findings
Ning et al. [5]	ABS	Carbon fiber powder (100 μm, 150 μm)	- Tensile strength of 42 MPa was highest for 5% wt fiber and lowest 34 MPa for by 10% wt. - 100 μm fiber length specimen showed higher ductility and toughness with respect to 150 μm.
Tekinalp et al. [60]	ABS	Short carbon fiber (0.2–0.4 mm, after mixing: 0.26 mm)	- 3D printed composite samples showed 115% higher tensile strength and 700% higher modulus when compared with the conventional compression molded composites.

Table 3. Cont.

Source	Matrix	Reinforcement	Important Findings
Karsli and Aytac [61]	Polyamide 6	Short carbon fiber (0–50 µm)	- Increasing fiber content improved strength, modulus, and hardness. - However, increasing fiber content resulted in lower strain at break value.
Zhong et al. [62]	ABS	Short glass fiber	- Strength of ABS was improved significantly at the expense of low hand ability and poor flexibility.
Abeykoon et al. [63]	Polylactic acid (PLA), acrylonitrile butadiene styrene (ABS), carbon fiber-reinforced PLA (CFR-PLA), carbon fiber-reinforced ABS (CFR-ABS), and carbon nanotube-reinforced ABS (CNT-ABS)		- To obtain desirable results, the printing speed and nozzle temperature should match. - Higher modulus was obtained for higher infill density. - Linear pattern provided highest modulus due to the lower number of spaces in the sample. - CFR-PLA was found to be the strongest material.

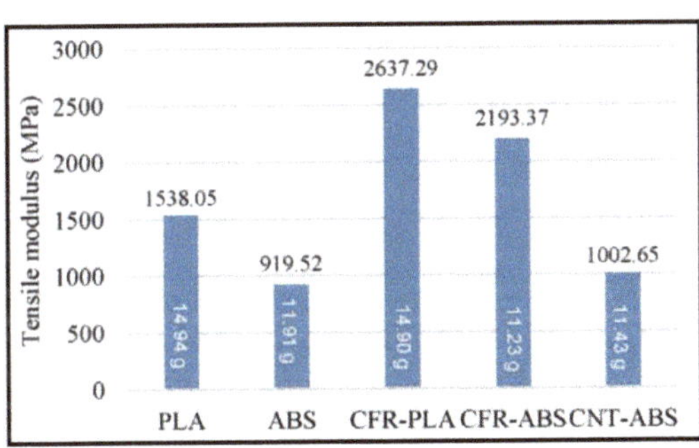

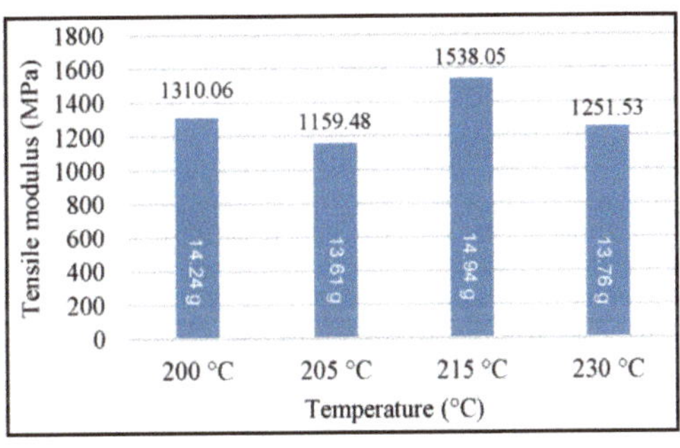

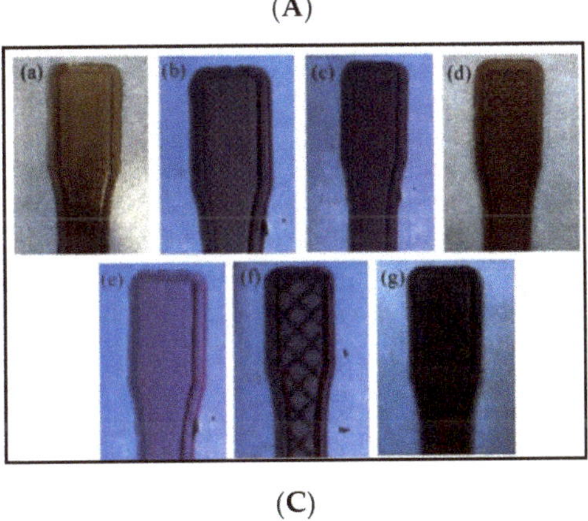

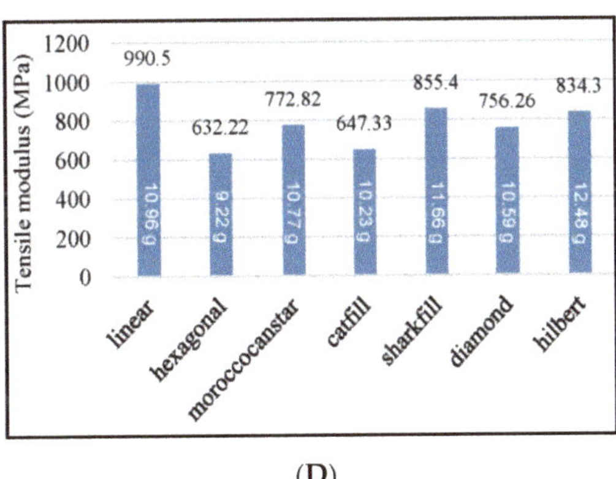

Figure 15. (**A**) Modulus of five different 3D printed materials. (**B**) Modulus of PLA at different nozzle temperatures. (**C**) PLA sample with infill patterns: (**a**) linear, (**b**) hexagonal, (**c**) moroccanstar, (**d**) catfill, (**e**) sharkfill, (**f**) diamond, and (**g**) Hilbert. (**D**) Modulus of PLA@50% infill density under different infill patterns (available under Creative Commons license [63]).

The weak bond between the polymer matrix and the carbon fiber highly impacts the mechanical properties of the material. In the same case of weak bonds, however, the tensile strength and flexural strength could be improved by surface treatment with methylene dichloride and PLA trimmings [4]. Several studies were conducted on continuous fiber reinforced printing as well. Li et al. [64] produced 3D printed FRP composites by using PLA matrix and continuous carbon fiber filament. It was observed that desirable interface bonding was achieved using 3D printing method, as a result tensile strength was improved by 13.8% and flexural strength was improved by 164%. Yang et al. [65] fabricated the composite sample using a 10% fiber part of continuous carbon fiber (CCF) and ABS polymer using an additive manufacturing technique. The samples made proved to have an improved flexural strength of 127 MPa and a tensile strength of 147 MPa. These results were close in nature to the injection moldering made CCF/ABS composites. Liao et al. [66] developed FRP composite using continuous carbon fiber in polyamide 12 (PA12) matrix. The study revealed that better performance was observed for the carbon content of 10% wt as shown in the Figure 16a–c. Heidari-Rarani et al. [57] investigated the 3D printing of continuous fiber-based PLA matrix. The study aimed to investigate the influence of printing temperature, printing speed, fiber tension, and fiber surface conditions on tensile and bending properties. In addition, the study explored the development and designing of a user-friendly extruder that can be used with conventional FDM printers. The study utilized embedding on the workpiece method of continuous fiber printing method. The experimental findings revealed that tensile and bending strengths were improved by 35% and 108% when PLA matrix was printed with the continuous carbon fiber. Table 4 provides a brief summary of some studies with continuous fiber FDM 3D printing method.

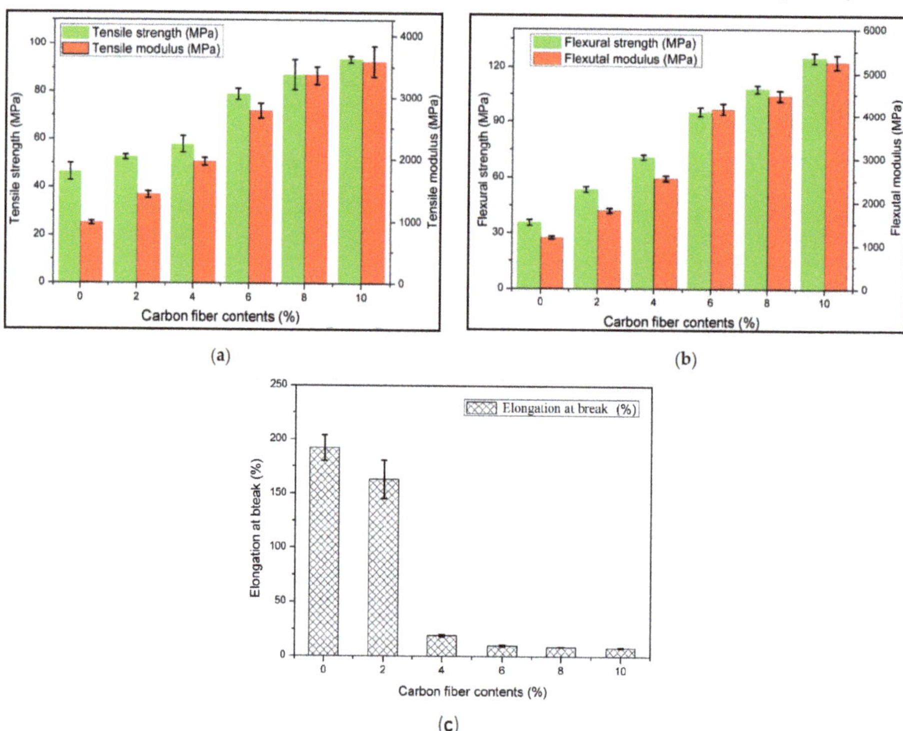

Figure 16. Effect of increasing fiber content % wt: (**a**) tensile modulus and strength; (**b**) flexural modulus and strength; (**c**) elongation at break (%) [66] (reprinted with kind permission from Elsevier).

Table 4. Table displaying the matrix, type of fiber reinforcement, and type of printing technique.

Source	Matrix	Reinforcement	Important Findings
Li et al. [64]	PLA	Continuous carbon fiber	- 3D printed samples showed desirable bonding at the interface and increased tensile and flexural strengths by 13.8% and 164%, respectively.
Yang et al. [65]	ABS	Continuous carbon fiber	- 10% wt continuous fiber in ABS increased the tensile and flexural strengths to 127 MPa and 147 MPa.
Liao et al. [66]	Polyamide	Continuous carbon fiber (6–7 μm)	- Adding 10% wt. fiber content to PA 12 increased tensile strength, flexural strength, and modulus without affecting behavior.
Heidari-Rarani et al. [57]	PLA	Continuous carbon fiber	- The tensile and bending strengths were improved by 35% and 108% when PLA matrix was printed with the continuous carbon fiber.

5.2. Fatigue Strength

Fatigue strength is considered an important criterion that contributes significantly towards the functionality of the 3D printed component. Shanmugam et al. [67] provided a detailed study to reveal the fatigue strength of 3D-printed polymeric material, 3D-printed polymeric composites, and 3D-printed cellular materials. The fatigue behavior of 3D-printed composite materials is very complex due to the anisotropic nature and layer by layer adhesion. Figure 17 represents the failure mechanism in the fiber-polymer composites in the form of three phases. The initial phase of failure is linked with the fiber matrix debonding in the regions where poor bonding is present. The reasons for poor bonding are reported to be fiber misalignment, matrix richness and poor surface conditions such as pores or voids. In the second phase, delamination occurs between fiber and matrix materials. In the final stage, crack propagation occurs at the fiber, and localized fractures occur [68]. Travieso-Rodriguez et al. [69] utilized the Taguchi design of experiments to study the fatigue performance of wood-PLA-based composite material. The study revealed the that 75% infill density, a 0.7 mm nozzle diameter, and a 0.4 mm layer height was the optimal combination. Printing velocity was found to have no significant influence on the performance. The lower endurance limit was found to be 17.9 MPa. Fatigue performance of the 3D-printed fiber-reinforced composites was highly dependent on the fiber volume fraction. Higher fatigue strength is achieved by higher volume fraction. Higher fiber orientation can result in poor fatigue performance. Higher fiber orientation can result in poor fatigue performance. It was also found that fatigue life was dependent on the stress ratio. A higher stress ratio provides a low fatigue life [68].

5.3. Creep Strength

Creep performance of 3D-printed FRP materials is an important parameter towards the reliable functionality of a product. It is rare in the literature for the creep performance of 3D-printed FRP materials to be investigated. Waseem et al. [70] performed a study where creep performance of 3D printed PLA was investigated using multiple response optimization. The study utilized different performance parameters such as infill pattern, layer height and infill percentages. Three levels of each parameter were investigated such as layer height which varied from 0.1–0.3 mm; infill patterns were linear; hexagonal, diamond, and infill percentages varied from 10–100%. The optimal condition of the hexagonal pattern, a 0.1 mm layer height, and 100% infill density was recommended in the study.

Zhang et al. [71] investigated the creep performance of 3D printed ABS materials. The study also investigated creep performance under different printing orientations. The study revealed that a 90-degree printing orientation provided the highest creep resistance. Mohammadizadeh et al. [72] investigated the tensile, fatigue and creep performance of 3D printed fiber polymer composites. The samples were prepared using nylon and fibers of Kevlar, carbon fiber, and fiber glass. Higher void formation was observed in the SEM observed for creep. Higher creep strains were observed when the temperature increased to the glass transition temperature due to the higher macromolecular mobility in the polymeric chains. Figure 18 the scanning electron micrographs of fractured samples in creep testing.

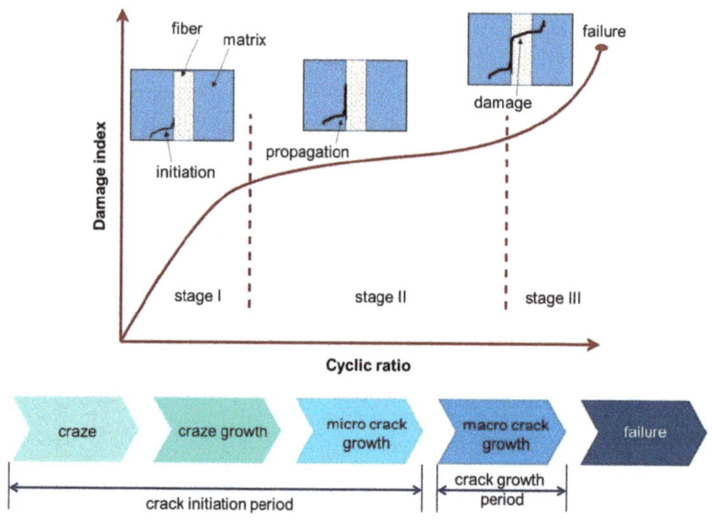

Figure 17. Phases of failure in fiber-polymer composites (available under Creative Commons license [67]).

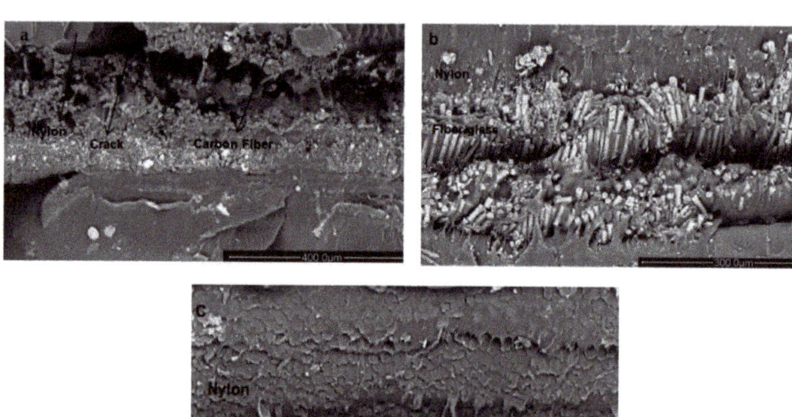

Figure 18. SEM images of creep fractured samples: (**a**) nylon-carbon fiber; (**b**) nylon-fiber glass; (**c**) nylon-Kelvar [72] (reprinted with kind permission from Elsevier).

6. Complexities in FRP Composite 3D Printing Using FDM

Generally, the traditional way of molding carbon fiber is a time taking process and costly as well. But integrating it with additive manufacturing helps reduce cost and save time as well, increasing efficiency in case of complex shapes especially [3,47,51–53,73]. For short fiber reinforced composites, carbon fibers (in short length segments) are mixed with other thermoplastics for printing. They are extruded to obtain a filament which can be used in 3D printer to manufacture various shapers and parts. Increasing the carbon fiber content increases the tensile strength and hardness but the effect of reinforcement for short fibers is less than long fibers. It was also observed that fibers have low wettability when combined with the thermoplastics resulting in poor behavior and also it makes fiber loading in filament problematic. The performance of 3D printed parts is linked with interlayer bonding between the consecutive layers. The melt dynamics of plastic are linked with the temperature and viscosity behavior during extrusion process [74,75]. The literature revealed that voids form during the printing of adjacent layers, and it has a controlling influence on the strength-related character of the printed FRP composite materials [14,76,77]. Different printing strategies can be adopted to decrease the void density in the printed samples as shown in the Figure 19. Reducing the interbead voids in the sample, increases the load bearing capacity of the printed sample. Figure 20A shows fractographs of neat ABS and CF-ABS under FDM and Conventional compression molded samples. It can be observed that fibers are bulging out showing poor interfacial adhesion between fiber and matrix. It can also be observed that larger pores were present in the FDM (ABS/CF) as compared to the compression molded sample. Figure 20B shows that as the CF is added to the neat ABS, the triangular channel between beads is reduced. It is associated with a reduction in die-swell and improvement in the thermal conductivity due to the CF.

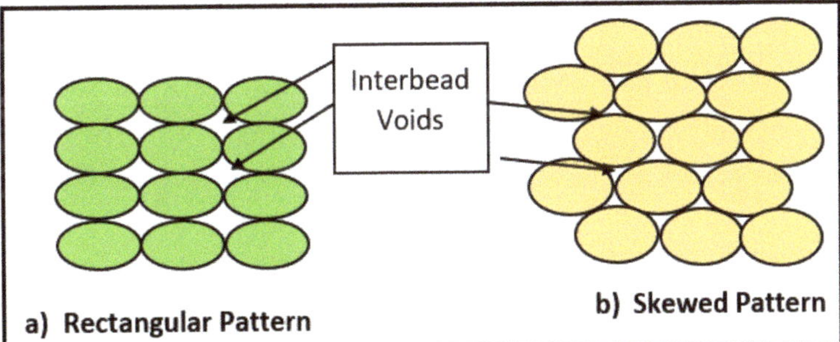

Figure 19. Schematic illustration of interbead void formation under different printing strategies (**a**) Rectangular pattern (**b**) Skewed pattern (redrawn from [78]).

The functionality of 3D printed FRP composites is strongly associated with the sintering of polymeric materials. Sintering in 3D printing is controlled through temperature and surface contact. Temperature is important because it governs the flow properties by influencing the surface tension and viscosity of the molten polymeric material. Heat distribution in the printed layers is controlled by the thermal conductivity and heat capacity [79,80]. The whole process becomes more complex due to the involvement of fiber content and dependency of material parameters with respect to the printing parameters. Figure 21 shows the influencing parameters to get optimal level of sintering between different polymeric layers.

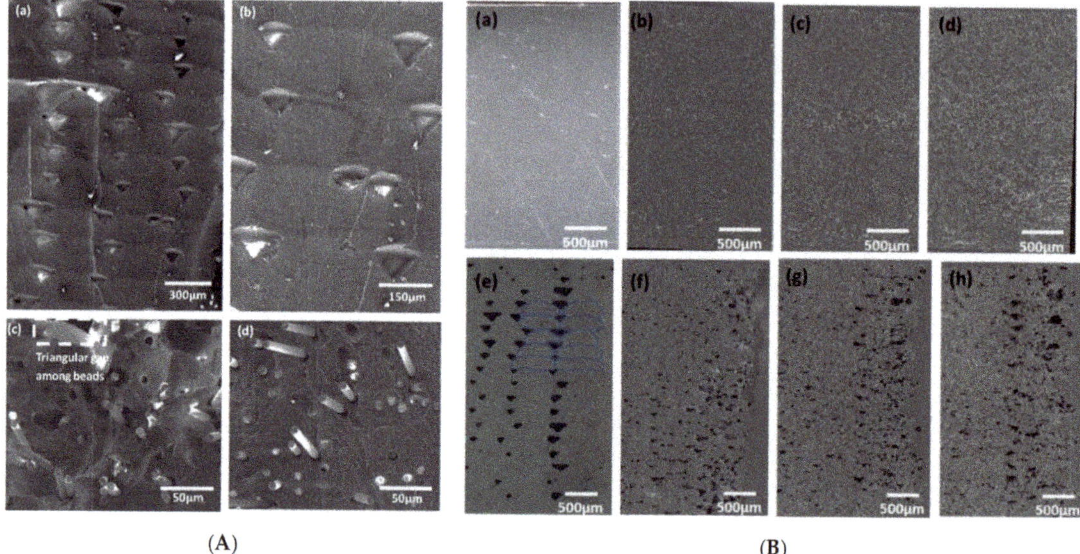

(A)   (B)

Figure 20. (**A**) Fractography of (**a,b**) conventional FDM-ABS (**c**) FDM printed with 10% wt. carbon fiber and (**d**) 10% wt. carbon fiber compression molded [60]. (**B**) Short fiber-based composites by compression molding (**a–d**), FDM-printed short fiber composites (**e–h**) with the fiber volume fraction changing from unfilled to 30 wt.% with a 10% increment [60] (reprinted with kind permission from Elsevier).

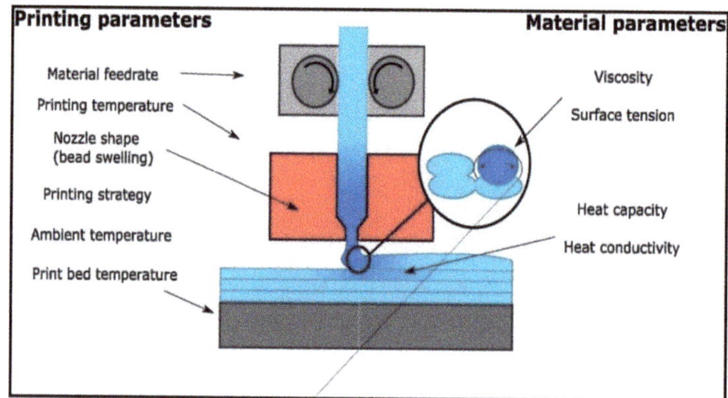

Figure 21. Influencing parameters to attain best possible sintering between different polymer layers (available under a Creative Commons license [81]).

It has also been observed that stiffness improves but strength is not improved significantly. This is due to the fiber pull out phenomenon that takes place before fiber fracture. In addition, increasing carbon fibers lead to larger areas of void which starts affecting the tensile strength negatively. Moreover, the resulting composite starts losing ductility and yield strength. Poor bonding between other materials and carbon fiber can significantly affect mechanical characteristics. Pure continuous carbon fiber when 3D printed has a better performance, but its major weaknesses are longer processing times and they cost more [37].

7. Industrial Developments to Print FRP Using FDM

According to a recent study, it is expected that the global market for 3D printing is projected to grow from USD 12.6 billion (in 2021) to USD 34.8 billion (by 2026) at a 22.5% compound annual growth rate (CAGR). 3D printing of composites is still in an emerging stage. But many industrial technologies, including defense, automotive, and aerospace, possess huge opportunities for 3D CFRP printing. This has many advantages such as reducing part manufacturing time and waste, achieving intricate geometries, and no expensive tooling is required. Currently, 3D printing is being used for the manufacturing of tools made of composites and the composite prototype parts. But the continuous advancement in 3D printing of FRP and the increasing interest of companies in additive manufacturing of composites will take the market to new heights. Many big names are at the forefront of using FRP-based 3D printing technology. For example, in 2017, Stratasys, an additive manufacturing company, launched a nylon-filled carbon fiber product for rapid proto-typing, composite tools, and high-end applications instead of using metals.

3D composite printing plays an imperative role in meeting customer needs by manufacturing various parts utilizing less time and reducing wastage. The market is segmented based on composite type (continuous fibers or discontinuous fibers), reinforcement type (e.g., carbon fiber or glass fiber) or the technology type (e.g., extrusion, powder bed fusion etc.). Also, carbon fiber reinforcement is preferable due to the wide variety of applications and advantages such as high strength, low weight and resistance to fatigue and corrosion. It is also high in demand in major industries including aerospace, automotive, defense and medical sectors due to the biocompatibility and light-weight parts for structural applications to improve the fuel efficiency and reducing carbon emissions. Based on regions, North America is the most dominant market for 3D printing of composites while the European market contributes as the 2nd largest. Some of the well-known names for the 3D printed composite parts in market include Stratasys Ltd., Cincinnati Incorporated, Arevo Labs, Mark Forged, 3D Systems Corporation, Inc., Graphite Additive Manufacturing Limited, and CRP Group. For example, the Figure 22 below shows a MarkOne printer with dual nozzles for nylon and fiber injection [81].

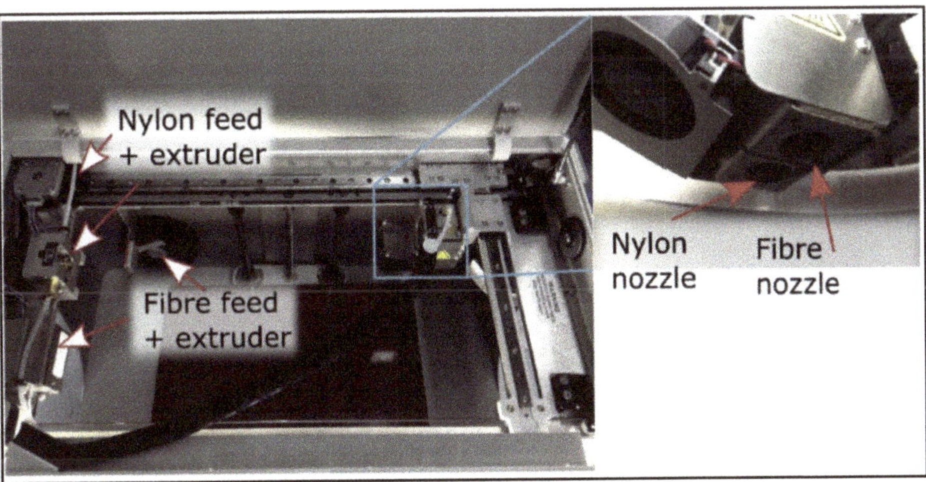

Figure 22. Dual nozzle system on MarkOne printer (available under a Creative Commons license [81]).

Additive manufacturing or 3D printing of polymer fiber composites, such as carbon fiber, is a vigorous manufacturing model. These composites provide flexibility in structure and enhanced mechanical properties. By reading the papers added in the review, one thing observed is that the FDM is the most commonly used additive manufacturing technique for the preparation of carbon fiber-reinforced polymers. Currently, FDM compatible thermoplastics are limited to amorphous polymers and polymers with low crystalline level acrylonitrile butadiene styrene (ABS) and polylactic acid (PLA) [17,31]. The study's focus is primarily on the mechanical properties of FRP composites fabricated using FDM techniques of 3D printing. The major usage of carbon fiber in products fabrication is because of the high-strength-to-weight-ratio and light weightiness of the material. The literate reviewed also reported the usage of varied thermoplastic polymers and short carbon fiber of 0.1 mm, this matrix is reinforced using 3D printing by using a slow extrusion process. Using short fibers proved to give improved strength of printed products. It has been reported that average fiber length reduces as the fiber loading increases during FDM. During the mixing of fibers with resin, the fiber length reduced drastically due to the following reasons such as contact of mixing instrument, resin contact, and contact with other fibers [35,36].

8. Data-Driven Based Machine Learning (ML) Approaches

There is a lot of attention given to artificial intelligence (AI) and machining learning (ML)-based data-driven approaches these days. ML approaches are based on recognizing the patterns from complex data. Goh et al. [82] provided a very detailed review of the ML approaches (supervised, semi-supervised, reinforced, and unsupervised) with respect to the 3D printing technologies. Charalampous et al. [83] conducted a study using regression-based machine learning approach to investigate the deviations between CAD model and the actual part. The study also discussed strategies to provide compensation related to the FDM process. Noriega et al. [84] conducted a study where an artificial neural network (ANN)-based algorithm was used to study the dimensional accuracy of FDM printed parts. The study revealed that 50% and 30% of dimensional errors were reduced for external and internal features using the proposed optimization approach. Vahabli and Rahmati [85] conducted a study using radial basis function neural networks (RBFNNs) to predict the surface finish of FDM printed parts. Optimization was performed using imperialistic competitive algorithm. The study revealed an error percentage of 7.11–3.64% for both models. Delli and Chang [86] provided a methodology to automatically monitor the 3D printed products using a machine learning approach, namely, support vector machine (SVM). Rayegani and Onwubolu [87] correlated the FDM process parameters with product strength using group method of data handling. The study developed a mathematical model by using the controlling parameters of orientation, raster angle, raster width and air gap. The results of this work were very practical and encouraging and can be easily implemented in the industry. Hooda et al. [88] utilized AI data-driven approach to reduce manufacturing time and cost of the product. The study discussed the deposition angle optimization with respect to the product geometry. Prediction accuracy of 94.57% was obtained by the proposed methodology. Figure 23 and Table 5 shows the parameters used in this work and the CAD models used to train the model. Yanamandra et al. [89] revealed an important application of reverse engineering of 3D printed composite part using imaging and machine learning assisted approach. The study analyzed the microstructure using the machine learning approach and even the tool was reconstructed. Figure 24a,b shows the CAD model and micro-level CT scans of layers. The approach revealed an error of only 0.33%. Jiang et al. [90] utilized back propagation neural network (BPNN) to study the parameters involved in unsupported bridge length for 3D printed sample component. It has been revealed that BPNN correctly provide the optimal longest distance between point that can hang unsupported.

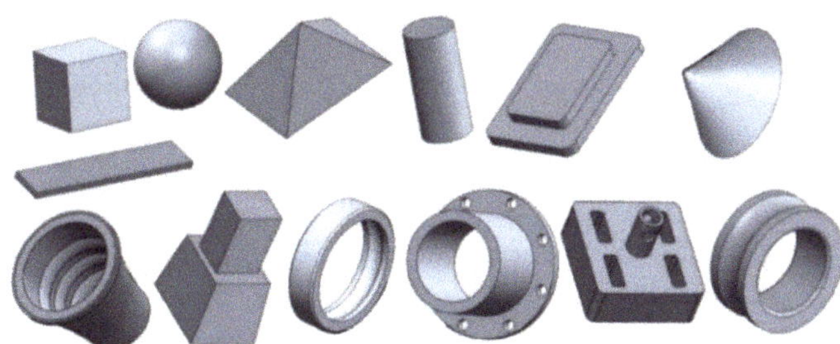

Figure 23. Data prepared to train a machine learning model [88] (reprinted with kind permission from Elsevier).

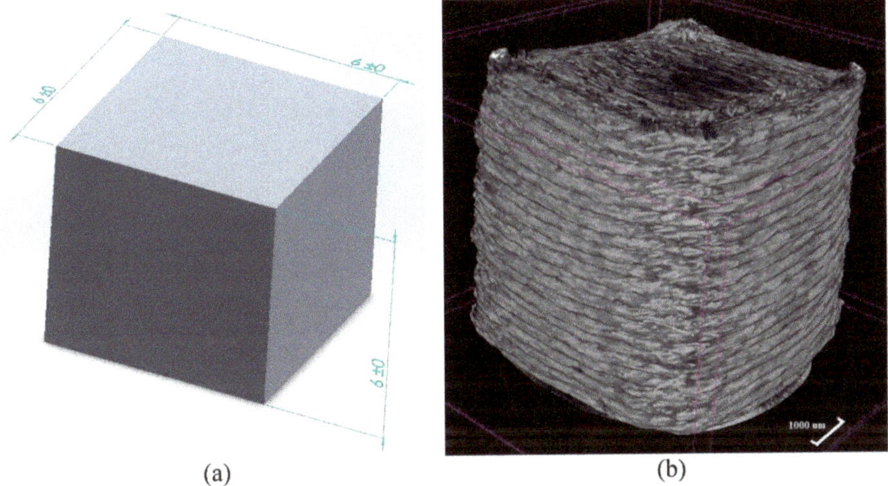

(a) (b)

Figure 24. (**a**) CAD model with dimensions. (**b**) Micro-level CT scan of the 3D printed part [89] (reprinted with kind permission from Elsevier).

Table 5. FDM parameters and parts based data [88].

Process Time (min)	Weight Material (g)	Length of Material Wire (m)	Orientation	Deposition Angle
77	6	2.43	Flat	0
107	7	2.57	Flat	15
138	5	2.07	Upright	60
158	6	2.22	Edge	15
146	6	2.21	Edge	30
84	6	2.26	Flat	75
88	6	2.31	Flat	90
76	6	2.43	Upright	0
110	7	2.55	Upright	15

9. Conclusions

After reviewing the literature, we can say that new and enhanced mechanical properties of materials with light-weight composition and greater flexibility can be achieved using various 3D printing techniques with carbon-fiber reinforced polymers. The 3D printing of carbon-fiber reinforced polymers is preferred to be performed using FDM technology. The material used in this type are separate carbon fiber filaments or saturated carbon fiber filaments. The FDM printer is modified to achieve co-extrusion, cutting, and fixed-shape properties. LOM type printers are using carbon fiber-impregnated films for product manufacturing; this technique is not fully developed yet. New technologies for automated fiber placement and laser tape-assisted winding are close in nature to additive manufacturing, and the concept of using these looks promising in applications of continuous carbon fiber-reinforced polymers. The continuous carbon fiber-reinforced thermoplastics by 3D printing is a novel concept due to the high efficiency and low-cost potential.

The continuous fiber placement for 3D-printed CFRP composites requires new algorithms so that accurate placement of the fiber can be made possible. There is a research gap that exists in understanding the long-term performance of CFRPs products fabrication using 3D printing technology. FRP materials are used in product fabrication to attain better strength. The material selection is the critical criteria for predicting the strength of the CFRPs in the long term. Moreover, the CFRP sheets encounter the problems of fiber rupture, micro-cracks in the structure, and the resin de-bounding with the passage of the time strength. For future research, studies should be conducted on the use of recycled carbon fibers, so that cheaper and less energy consuming products can be fabricated. These recycled carbon fibers will be helpful in reducing the environmental and financial impacts of additive manufacturing of CFRPs using virgin fibers. Another area for future research work is to study the physical and mechanical properties of carbon fiber reinforced polymers from different aspects, different than the ductile strength and the flexural properties. The results of these studies will be helpful in analyzing the potential of CFRPs fabricated using additive manufacturing and will open ways to new markets. One more field highlighted is to investigate the long-term usage impact and wear and tear in the structure of 3D-printed carbon fiber-reinforced polymers.

In order to enhance the performance of 3D printed FRP materials, there is a need improve the interfacial bonding between the fiber and matrix to improve the functional performance of the 3D printed materials. Limiting the porosities in the 3D printed FRP materials can also result in the improved performance. Parameters, such as fiber volume fraction, atmospheric environment, cooling rate, temperature of the nozzle, and printing speed. can be optimized to limit porosities.

Fatigue performance of the 3D-printed fiber-reinforced composites is highly dependent on the fiber volume fraction. Higher fatigue strength is achieved by higher volume fraction. Higher fiber orientation can result in poor fatigue performance. It was also found that fatigue life was dependent on the stress ratio. A higher stress ratio provides a low fatigue life. Higher creep strains were observed when the temperature increased to the glass transition temperature due to the higher macromolecular mobility in the polymeric chains.

Automated quality inspection of the 3D printed part is an emerging area these days. It has been seen that artificial intelligence-based machine learning approaches have good potential in this application. However, the majority of work is performed on conventional FDM printing for polymers only. There is a need to enhance this area for potential quality inspection of FDM-printed fiber-reinforced composites as well. Machine learning was also identified as a power tool to reengineer the microstructure of 3D-printed composite product. Machine learning based approaches can also be utilized efficiently to optimize the hanging lengths of the 3D printed product for optimal design solutions.

Author Contributions: Conceptualization, S.P. and S.K.; methodology, T.A.Q. and G.K.; formal analysis, T.A.Q.; investigation, T.A.Q.; data curation, S.P. and G.K.; writing—original draft preparation, S.P. and T.A.Q.; writing—review and editing, G.K. and S.K.; supervision, S.P. and S.K.; project administration, S.P.; funding acquisition, G.K. and S.P. All authors have read and agreed to the published version of the manuscript.

Funding: The APC was partly funded by New York University and Rochester Institute of Technology—Dubai.

Institutional Review Board Statement: Not applicable.

Informed Consent Statement: Not applicable.

Data Availability Statement: Not applicable.

Conflicts of Interest: The authors declare no conflict of interest.

References

1. Sonnenschein, R.; Gajdosova, K.; Holly, I. FRP Composites and their Using in the Construction of Bridges. *Procedia Eng.* **2016**, *161*, 477–482. [CrossRef]
2. Soutis, C. Carbon fiber reinforced plastics in aircraft construction. *Mater. Sci. Eng. A* **2005**, *412*, 171–176. [CrossRef]
3. Matsuzaki, R.; Ueda, M.; Namiki, M.; Jeong, T.-K.; Asahara, H.; Horiguchi, K.; Nakamura, T.; Todoroki, A.; Hirano, Y. Three-dimensional printing of continuous-fiber composites by in-nozzle impregnation. *Sci. Rep.* **2016**, *6*, 23058. [CrossRef]
4. Parandoush, P.; Lin, D. A review on additive manufacturing of polymer-fiber composites. *Compos. Struct.* **2017**, *182*, 36–53. [CrossRef]
5. Ning, F.; Cong, W.; Qiu, J.; Wei, J.; Wang, S. Additive manufacturing of carbon fiber reinforced thermoplastic composites using fused deposition modeling. *Compos. Part B Eng.* **2015**, *80*, 369–378. [CrossRef]
6. Penumakala, P.K.; Santo, J.; Thomas, A. A critical review on the fused deposition modeling of thermoplastic polymer composites. *Compos. Part B Eng.* **2020**, *201*, 108336. [CrossRef]
7. Yuan, S.; Li, S.; Zhu, J.; Tang, Y. Additive manufacturing of polymeric composites from material processing to structural design. *Compos. Part B Eng.* **2021**, *219*, 108903. [CrossRef]
8. Bakır, A.A.; Atik, R.; Özerinç, S. Mechanical properties of thermoplastic parts produced by fused deposition modeling: A review. *Rapid Prototyp. J.* **2021**, *27*, 537–561. [CrossRef]
9. Mustapha, B.K.; Metwalli, K.M. A review of fused deposition modelling for 3D printing of smart polymeric materials and composites. *Eur. Polym. J.* **2021**, *156*, 110591. [CrossRef]
10. Hu, C.; Qin, Q.H. Advances in fused deposition modeling of discontinuous fiber/polymer composites. *Curr. Opin. Solid State Mater. Sci.* **2020**, *24*, 100867. [CrossRef]
11. Shanmugam, V.; Rajendran, D.J.J.; Babu, K.; Rajendran, S.; Veerasimman, A.; Marimuthu, U.; Singh, S.; Das, O.; Neisiany, R.E.; Hedenqvist, M.S.; et al. The mechanical testing and performance analysis of polymer-fibre composites prepared through the additive manufacturing. *Polym. Test.* **2020**, *93*, 106925. [CrossRef]
12. Singh, S.; Ramakrishna, S.; Singh, R. Material issues in additive manufacturing: A review. *J. Manuf. Process.* **2017**, *25*, 185–200. [CrossRef]
13. Gurrala, P.K.; Regalla, S. Part strength evolution with bonding between filaments in fused deposition modelling: This paper studies how coalescence of filaments contributes to the strength of final FDM part. *Virtual Phys. Prototyp.* **2014**, *9*, 141–149. [CrossRef]
14. Sun, Q.; Rizvi, G.M.; Bellehumeur, C.T.; Gu, P. Effect of processing conditions on the bonding quality of FDM polymer filaments. *Rapid Prototyp. J.* **2008**, *14*, 72–80. [CrossRef]
15. Park, S.-J.; Kim, K.-S. Recent Uses of Carbon Fibers. In *Carbon Fibers*; Springer: Singapore, 2015; Volume 210.
16. Basalt.Today. Fiber Reinforced Polymer (FRP) Composites Market to Grow to 2025. 2017. Available online: https://basalt.today/2017/12/14093/ (accessed on 22 June 2021).
17. Basalt.Today. Global 3D Printed Composites Market to Reach $ 111 Million in 2022. 2017. Available online: https://basalt.today/2017/08/12140/ (accessed on 22 June 2021).
18. Critchfield, M.O.; Judy, T.D.; Kurzweil, A.D. Low-cost design and fabrication of composite ship structures. *Mar. Struct.* **1994**, *7*, 475–494. [CrossRef]
19. Rubino, F.; Nisticò, A.; Tucci, F.; Carlone, P. Marine application of fiber reinforced composites: A review. *J. Mar. Sci. Eng.* **2020**, *8*, 26. [CrossRef]
20. Carr, D.; Lewis, E.A. *Ballistic-Protective Clothing and Body Armour*; Woodhead Publishing Limited: Cambridge, UK, 2014.
21. Giurgiutiu, V. *Structural Health Monitoring of Aerospace Composites*; Academic Press: Cambridge, MA, USA, 2015; pp. 1–23.
22. Mouritz, A.P.; Gellert, E.; Burchill, P.; Challis, K. Review of advanced composite structures for naval ships and submarines. *Compos. Struct.* **2001**, *53*, 21–42. [CrossRef]
23. Mouritz, A.P. (Ed.) Ch 15 Fibre–polymer composites for aerospace structures and engines. In *Introduction to Aerospace Materials*; Woodhead Publishing Limited: Cambridge, UK, 2012; pp. 338–393.
24. Georgiadis, S.; Gunnion, A.J.; Thomson, R.S.; Cartwright, B.K. Bird-strike simulation for certification of the Boeing 787 composite moveable trailing edge. *Compos. Struct.* **2008**, *86*, 258–268. [CrossRef]
25. Friedrich, K.; Almajid, A.A. Manufacturing aspects of advanced polymer composites for automotive applications. *Appl. Compos. Mater.* **2013**, *20*, 107–128. [CrossRef]
26. Kurien, R.A.; Selvaraj, D.P.; Sekar, M.; Koshy, C. Green composite materials for green technology in the automotive industry. *IOP Conf. Ser. Mater. Sci. Eng.* **2020**, *872*, 012064. [CrossRef]
27. Ali, H.T.; Akrami, R.; Fotouhi, S.; Bodaghi, M.; Saeedifar, M.; Yusuf, M.; Fotouhi, M. Fiber reinforced polymer composites in bridge industry. *Structures* **2021**, *30*, 774–785. [CrossRef]
28. Agarwal, B.D.; Broutman, L.J.; Chandrashekhara, K. *Analysis and Performance of Fiber Composites*, 3rd ed.; Wiley: New Delhi, India, 2006.
29. Campbell, F.C. (Ed.) Chapter 1 Introduction to Composite Materials. In *Structural Composite Materials*; ASM International: Materials Park, OH, USA, 1987; Volume 262, pp. 6962–6964.
30. Rajak, D.K.; Pagar, D.D.; Menezes, P.L.; Linul, E. Fiber-Reinforced Polymer Composites. *Polymers* **2019**, *11*, 1667. [CrossRef] [PubMed]

31. Chukov, D.; Nematulloev, S.; Zadorozhnyy, M.; Tcherdyntsev, V.; Stepashkin, A.; Zherebtsov, D. Structure, mechanical and thermal properties of polyphenylene sulfide and polysulfone impregnated carbon fiber composites. *Polymers* **2019**, *11*, 684. [CrossRef]

32. Sherif, G.; Chukov, D.; Tcherdyntsev, V.; Torokhov, V. Effect of formation route on the mechanical properties of the polyethersulfone composites reinforced with glass fibers. *Polymers* **2019**, *11*, 1364. [CrossRef]

33. Zagho, M.M.; Hussein, E.A.; Elzatahry, A.A. Recent overviews in functional polymer composites for biomedical applications. *Polymers* **2018**, *10*, 739. [CrossRef] [PubMed]

34. Monteiro, S.; De Assis, F.S.; Ferreira, C.L.; Simonassi, N.T.; Weber, R.P.; Oliveira, M.S.; Colorado, H.A.; Pereira, A.C. Fique fabric: A promising reinforcement for polymer composites. *Polymers* **2018**, *10*, 246. [CrossRef]

35. Callister, W.; Rethwisch, D.G. *Fundamentals of Materials Science and Engineering*, 4th ed.; Wiley: Hoboken, NJ, USA, 2011.

36. Lilholt, H.; Sorensen, B.F. Interfaces between a fibre and its matrix. *IOP Conf. Ser. Mater. Sci. Eng.* **2017**, *219*, 012030. [CrossRef]

37. Zhu, L.; Li, Y. Equivalent moisture diffusion coefficient of fiber reinforced composites. *IOP Conf. Ser. Mater. Sci. Eng.* **2019**, *544*, 012019. [CrossRef]

38. Karataş, M.A.; Gökkaya, H. A review on machinability of carbon fiber reinforced polymer (CFRP) and glass fiber reinforced polymer (GFRP) composite materials. *Def. Technol.* **2018**, *14*, 318–326. [CrossRef]

39. Zhou, Y.; Fan, M.; Chen, L. Interface and bonding mechanisms of plant fibre composites: An overview. *Compos. Part B Eng.* **2016**, *101*, 31–45. [CrossRef]

40. Drzal, L.T.; Madhukar, M. Fibre-matrix adhesion and its relationship to composite mechanical properties. *J. Mater. Sci.* **1993**, *28*, 569–610. [CrossRef]

41. Tse, L.Y.L.; Kapila, S.; Barton, K. Contoured 3D Printing of Fiber Reinforced Polymers. In Proceedings of the Solid Freeform Fabrication Symposium: An Additive Manufacturing Conference, Austin, TX, USA, 8–10 August 2016.

42. Gebler, M.; Uiterkamp, A.J.M.S.; Visser, C. A global sustainability perspective on 3D printing technologies. *Energy Policy* **2014**, *74*, 158–167. [CrossRef]

43. 3D Printing Media Network, Composites. Available online: https://www.3dprintingmedia.network/composites-additive-manufacturing-market-2028/ (accessed on 22 June 2021).

44. Pearce, J.M.; Blair, C.M.; Laciak, K.J.; Andrews, R.; Nosrat, A.; Zelenika-Zovko, I. 3-D Printing of Open Source Appropriate Technologies for Self-Directed Sustainable Development. *J. Sustain. Dev.* **2010**, *3*, 17–29. [CrossRef]

45. Kreiger, M.; Pearce, J.M. Environmental life cycle analysis of distributed three-dimensional printing and conventional manufacturing of polymer products. *ACS Sustain. Chem. Eng.* **2013**, *1*, 1511–1519. [CrossRef]

46. Christ, S.; Schnabel, M.; Vorndran, E.; Groll, J.; Gbureck, U. Fiber reinforcement during 3D printing. *Mater. Lett.* **2015**, *139*, 165–168. [CrossRef]

47. Horn, T.J.; Harrysson, O.L.A. Overview of current additive manufacturing technologies and selected applications. *Sci. Prog.* **2012**, *95*, 255–282. [CrossRef] [PubMed]

48. Zhang, H.; Yang, D.; Sheng, Y. Performance-driven 3D printing of continuous curved carbon fibre reinforced polymer composites: A preliminary numerical study. *Compos. Part B Eng.* **2018**, *151*, 256–264. [CrossRef]

49. Spackman, C.C.; Frank, C.R.; Picha, K.C.; Samuel, J. 3D printing of fiber-reinforced soft composites: Process study and material characterization. *J. Manuf. Process.* **2016**, *23*, 296–305. [CrossRef]

50. Quan, Z.; Wu, A.; Keefe, M.; Qin, X.; Yu, J.; Suhr, J.; Byun, J.-H.; Kim, B.-S.; Chou, T.-W. Additive manufacturing of multi-directional preforms for composites: Opportunities and challenges. *Mater. Today* **2015**, *18*, 503–512. [CrossRef]

51. Huang, S.H.; Liu, P.; Mokasdar, A.; Hou, L. Additive manufacturing and its societal impact: A literature review. *Int. J. Adv. Manuf. Technol.* **2013**, *67*, 1191–1203. [CrossRef]

52. Yang, Y.; Boom, R.; Irion, B.; van Heerden, D.J.; Kuiper, P.; de Wit, H. Recycling of composite materials. *Chem. Eng. Process. Process. Intensif.* **2012**, *51*, 53–68. [CrossRef]

53. Wang, X.; Jiang, M.; Zhou, Z.; Gou, J.; Hui, D. 3D printing of polymer matrix composites: A review and prospective. *Compos. Part B Eng.* **2017**, *110*, 442–458. [CrossRef]

54. van de Werken, N.; Tekinalp, H.; Khanbolouki, P.; Ozcan, S.; Williams, A.; Tehrani, M. Additively manufactured carbon fiber-reinforced composites: State of the art and perspective. *Addit. Manuf.* **2020**, *31*, 100962. [CrossRef]

55. Prüß, H.; Vietor, T. Design for Fiber-Reinforced Additive Manufacturing. *J. Mech. Des. Trans. ASME* **2015**, *137*, 111409. [CrossRef]

56. Mohan, N.; Senthil, P.; Vinodh, S.; Jayanth, N. A review on composite materials and process parameters optimisation for the fused deposition modelling process. *Virtual Phys. Prototyp.* **2017**, *12*, 47–59. [CrossRef]

57. Heidari-Rarani, M.; Rafiee-Afarani, M.; Zahedi, A.M. Mechanical characterization of FDM 3D printing of continuous carbon fiber reinforced PLA composites. *Compos. Part B Eng.* **2019**, *175*, 107147. [CrossRef]

58. Mori, K.I.; Maeno, T.; Nakagawa, Y. Dieless forming of carbon fibre reinforced plastic parts using 3d printer. *Procedia Eng.* **2014**, *81*, 1595–1600. [CrossRef]

59. Tian, X.; Liu, T.; Yang, C.; Wang, Q.; Li, D. Interface and performance of 3D printed continuous carbon fiber reinforced PLA composites. *Compos. Part A Appl. Sci. Manuf.* **2016**, *88*, 198–205. [CrossRef]

60. Tekinalp, H.L.; Kunc, V.; Velez-Garcia, G.M.; Duty, C.E.; Love, L.; Naskar, A.K.; Blue, C.A.; Ozcan, S. Highly oriented carbon fiber-polymer composites via additive manufacturing. *Compos. Sci. Technol.* **2014**, *105*, 144–150. [CrossRef]

61. Karsli, N.G.; Aytac, A. Tensile and thermomechanical properties of short carbon fiber reinforced polyamide 6 composites. *Compos. Part B Eng.* **2013**, *51*, 270–275. [CrossRef]

62. Zhong, W.; Li, F.; Zhang, Z.; Song, L.; Li, Z. Short fiber reinforced composites for fused deposition modeling. *Mater. Sci. Eng. A* **2001**, *301*, 125–130. [CrossRef]

63. Abeykoon, C.; Sri-Amphorn, P.; Fernando, A. Optimization of fused deposition modeling parameters for improved PLA and ABS 3D printed structures. *Int. J. Lightweight Mater. Manuf.* **2020**, *3*, 284–297. [CrossRef]

64. Li, N.; Li, Y.; Liu, S. Rapid prototyping of continuous carbon fiber reinforced polylactic acid composites by 3D printing. *J. Mater. Process. Technol.* **2016**, *238*, 218–225. [CrossRef]

65. Yang, C.; Tian, X.; Liu, T.; Cao, Y.; Li, D. 3D printing for continuous fiber reinforced thermoplastic composites: Mechanism and performance. *Rapid Prototyp. J.* **2017**, *23*, 209–215. [CrossRef]

66. Liao, G.; Li, Z.; Cheng, Y.; Xu, D.; Zhu, D.; Jiang, S.; Guo, J.; Chen, X.; Xu, G.; Zhu, Y. Properties of oriented carbon fiber/polyamide 12 composite parts fabricated by fused deposition modeling. *Mater. Des.* **2018**, *139*, 283–292. [CrossRef]

67. Shanmugam, V.; Das, O.; Babu, K.; Marimuthu, U.; Veerasimman, A.; Johnson, D.J.; Neisiany, R.E.; Hedenqvist, M.S.; Ramakrishna, S.; Berto, F. Fatigue behaviour of FDM-3D printed polymers, polymeric composites and architected cellular materials. *Int. J. Fatigue* **2021**, *143*, 106007. [CrossRef]

68. Ansari, M.T.A.; Singh, K.K.; Azam, M.S. Fatigue damage analysis of fiber-reinforced polymer composites—A review. *J. Reinf. Plast. Compos.* **2018**, *37*, 636–654. [CrossRef]

69. Travieso-Rodriguez, J.A.; Zandi, M.D.; Jerez-Mesa, R.; Lluma-Fuentes, J. Fatigue behavior of PLA-wood composite manufactured by fused filament fabrication. *J. Mater. Res. Technol.* **2020**, *9*, 8507–8516. [CrossRef]

70. Waseem, M.; Salah, B.; Habib, T.; Saleem, W.; Abas, M.; Khan, R.; Ghani, U.; Siddiqi, M.U.R. Multi-response optimization of tensile creep behavior of PLA 3D printed parts using categorical response surface methodology. *Polymers* **2020**, *12*, 2692. [CrossRef]

71. Zhang, H.; Cai, L.; Golub, M.; Zhang, Y.; Yang, X.; Schlarman, K.; Zhang, J. Tensile, Creep, and Fatigue Behaviors of 3D-Printed Acrylonitrile Butadiene Styrene. *J. Mater. Eng. Perform.* **2018**, *27*, 57–62. [CrossRef]

72. Mohammadizadeh, M.; Imeri, A.; Fidan, I.; Elkelany, M. 3D printed fiber reinforced polymer composites—Structural analysis. *Compos. Part B Eng.* **2019**, *175*, 107112. [CrossRef]

73. Chethan, K.N.; Keni, L.G.; Padmaraj, N.H.; Dias, A.; Jain, R. Fabrication and Mechanical characterization of aluminium [6061] with conventionally prepared bamboocharcoal. *Mater. Today Proc.* **2018**, *5*, 3465–3475. [CrossRef]

74. Turner, B.N.; Gold, S.A. A review of melt extrusion additive manufacturing processes: II. Materials, dimensional accuracy, and surface roughness. *Rapid Prototyp. J.* **2015**, *21*, 250–261. [CrossRef]

75. Turner, B.N.; Strong, R.; Gold, S.A. A review of melt extrusion additive manufacturing processes: I. Process design and modeling. *Rapid Prototyp. J.* **2014**, *20*, 192–204. [CrossRef]

76. Yardimci, M.A.; Guceri, S.I.; Agarwala, M.; Danforth, S.C. Part quality prediction tools for fused deposition processing. In *1996 International Solid Freeform Fabrication Symposium*; The University of Texas in Austin: Austin, TX, USA, 1996; pp. 539–548.

77. Liu, Z.Q.; An, Q.; Jinyang, X.; Ming, C.; Shu, H. Wear performance of (nc-AlTiN)/(a-Si3N4) coating and (nc-AlCrN)/(aSi3N4) coating in high-speed machining of titanium alloys under dry and minimum quantity lubrication (MQL) conditions. *Wear* **2013**, *305*, 249–259. [CrossRef]

78. Rodriguez, J.F.; Thomas, J.P.; Renaud, J.E. Characterization of the mesostructure of fused-deposition acrylonitrile-butadiene-styrene materials. *Rapid Prototyp. J.* **2000**, *6*, 175–185. [CrossRef]

79. Hwang, S.; Reyes, E.I.; Moon, K.S.; Rumpf, R.C.; Kim, N.S. Thermo-mechanical Characterization of Metal/Polymer Composite Filaments and Printing Parameter Study for Fused Deposition Modeling in the 3D Printing Process. *J. Electron. Mater.* **2015**, *44*, 771–777. [CrossRef]

80. Christiyan, K.G.J.; Chandrasekhar, U.; Venkateswarlu, K. A study on the influence of process parameters on the Mechanical Properties of 3D printed ABS composite. *IOP Conf. Ser. Mater. Sci. Eng.* **2016**, *114*, 012109. [CrossRef]

81. Blok, L.G.; Longana, M.L.; Yu, H.; Woods, B.K.S. An investigation into 3D printing of fibre reinforced thermoplastic composites. *Addit. Manuf.* **2018**, *22*, 176–186. [CrossRef]

82. Goh, G.D.; Sing, S.L.; Yeong, W.Y. A Review on Machine Learning in 3D Printing: Applications, Potential, and Challenges. *Artif. Intell. Rev.* **2021**, *54*, 63–94. [CrossRef]

83. Charalampous, P.; Kostavelis, I.; Kontodina, T.; Tzovaras, D. Learning-based error modeling in FDM 3D printing process. *Rapid Prototyp. J.* **2021**, *27*, 507–517. [CrossRef]

84. Lao, W.; Li, M.; Wong, T.N.; Tan, M.J.; Tjahjowidodo, T. Improving surface finish quality in extrusion-based 3D concrete printing using machine learning-based extrudate geometry control. *Virtual Phys. Prototyp.* **2020**, *15*, 178–193. [CrossRef]

85. Vahabli, E.; Rahmati, S. Application of an RBF neural network for FDM parts' surface roughness prediction for enhancing surface quality. *Int. J. Precis. Eng. Manuf.* **2016**, *17*, 1589–1603. [CrossRef]

86. Delli, U.; Chang, S. Automated Process Monitoring in 3D Printing Using Supervised Machine Learning. *Procedia Manuf.* **2018**, *26*, 865–870. [CrossRef]

87. Rayegani, F.; Onwubolu, G.C. Fused deposition modelling (fdm) process parameter prediction and optimization using group method for data handling (gmdh) and differential evolution (de). *Int. J. Adv. Manuf. Technol.* **2014**, *73*, 509–519. [CrossRef]

88. Hooda, N.; Chohan, J.S.; Gupta, R.; Kumar, R. Deposition angle prediction of Fused Deposition Modeling process using ensemble machine learning. *ISA Trans.* **2021**. [CrossRef] [PubMed]

89. Yanamandra, K.; Chen, G.L.; Xu, X.; Mac, G.; Gupta, N. Reverse engineering of additive manufactured composite part by toolpath reconstruction using imaging and machine learning. *Compos. Sci. Technol.* **2020**, *198*, 108318. [CrossRef]
90. Jiang, J.; Hu, G.; Li, X.; Xu, X.; Zheng, P.; Stringer, J. Analysis and prediction of printable bridge length in fused deposition modelling based on back propagation neural network. *Virtual Phys. Prototyp.* **2019**, *14*, 253–266. [CrossRef]

materials

Perspective

3D Printed Sand Tools for Thermoforming Applications of Carbon Fiber Reinforced Composites—A Perspective

Daniel Günther [1], Patricia Erhard [1], Simon Schwab [1] and Iman Taha [1,2,*]

1   Fraunhofer Institute for Casting, Composite and Processing Technology, Am Technologiezentrum 2, 86159 Augsburg, Germany; daniel.guenther@igcv.fraunhofer.de (D.G.); patricia.erhard@igcv.fraunhofer.de (P.E.); simon.schwab@igcv.fraunhofer.de (S.S.)
2   Faculty of Engineering, Ain Shams University, El-Sarayat Street 1, Cairo 11517, Egypt
*   Correspondence: iman.taha@igcv.fraunhofer.de; Tel.: +49-(0)821-90678-252

Abstract: Tooling, especially for prototyping or small series, may prove to be very costly. Further, prototyping of fiber reinforced thermoplastic shell structures may rely on time-consuming manual efforts. This perspective paper discusses the idea of fabricating tools at reduced time and cost compared to conventional machining-based methods. The targeted tools are manufactured out of sand using the Binder Jetting process. These molds should fulfill the demands regarding flexural and compressive behavior while allowing for vacuum thermoforming of fiber reinforced thermoplastic sheets. The paper discusses the requirements and the challenges and presents a perspective study addressing this innovative idea. The authors present the idea for discussion in the additive manufacturing and FRP producing communities.

Keywords: binder jetting; sands; vacuum thermoforming; fiber reinforced composite

Citation: Günther, D.; Erhard, P.; Schwab, S.; Taha, I. 3D Printed Sand Tools for Thermoforming Applications of Carbon Fiber Reinforced Composites—A Perspective. *Materials* **2021**, *14*, 4639. https://doi.org/10.3390/ma14164639

Academic Editor: Tuhin Mukherjee

Received: 19 July 2021
Accepted: 13 August 2021
Published: 18 August 2021

Publisher's Note: MDPI stays neutral with regard to jurisdictional claims in published maps and institutional affiliations.

1. Introduction

A main industrial focus today lies in energy-efficient and resource-saving manufacturing. This is one way to meet relevant ecological and economic targets as well as to ensure the competitiveness of products in the long term. As a result, lightweight components of high mechanical performance are increasingly attracting attention. Thus, fiber reinforced polymers (FRP), especially involving carbon fibers, are currently in the spotlight of developments. Related manufacturing processes, equipment and tools must be adapted to the new material and functionality requirements. A significant amount of time in the optimization process of a component lies in the creation of initial prototypes. The production of small series often proves to be particularly cost-intensive. These are primarily manufactured in manual processes. Depending on part complexity, costly and time-consuming tool manufacturing might be necessary.

Additive manufacturing provides a good solution regarding the rapid fabrication of tools. Next to functional integration, additively manufactured molds can combine complexity and lightweight. As an example, additive manufacturing makes it possible to realize complex cooling channels within the tool without the need of parting the mold. Some additive manufacturing techniques, such as the Binder Jetting process have a high volume throughput compared to the established CNC machining, thus allowing for lower manufacturing time and energy consumption. Further, additive manufacturing contributes to the conservation of material resources, since it relies on generating the final structure layer by layer as opposed to the material removal concept in established machining processes. For processing FRPs, both tools for low temperature applications (such as hand layup or Resin Transfer Molding) as well as high temperature applications (such as autoclave curing or compression molding) are relevant. Generally, tools do not require high strength [1,2]. In high pressure processes, as in the case of autoclave curing, compression molding or vacuum-assisted processes, loads are primarily compressive. In fact, these loads are often more favorable for additively manufactured parts, as compared to tensile loads [2].

Several studies have been concerned with the fabrication of tools by additive manufacturing from different materials. Brotan et al. [3] fabricated innovative tools out of Marlok C1650 steel powder in a Powder Bed Fusion (PBF) process to realize complex gradient structures to increase the thermal fatigue resistance and allow for a defined thermal management of the tools. The target applications of these low weight tools are injection molding and die casting. Similarly, Fette et al. [4] compared common additive manufacturing techniques—Additive Layer Manufacturing (ALM), Electron Beam Melting (EBM) and Selective Laser Melting (SLM)—with conventional methods to manufacture metal tooling. The authors claim that the use of integrated heating channels close to the mold surface allows the even dissipation of heat.

Warden [5] studied the application of Fused Deposition Modeling (FDM) to manufacture tools for compression molding thermoplastic multiaxial prepreg systems. The author proposed a low-temperature curing cycle in order to reduce the thermal degradation of the FDM tooling. Further, Bere et al. [6] processed Carbon Fiber Reinforced Polymers (CFRP) through vacuum bagging and oven curing using an FDM polylactic acid (PLA) versus acrylonitrile butadiene styrene (ABS) mold. The PLA mold was treated with a layer of epoxy that contains aluminum powder to enhance the bonding with a subsequently applied polyester gel coat, which further acts as a release agent and facilitates demolding. Hassen et al. [7] proposed the use of the extrusion-based Big Area Additive Manufacturing (BAAM) process to fabricate CFRP tools for further molding purposes. The main advantage lies in the increased throughput (~16,400 cm^3/h) and the possibility to flexibly process material from granule form [8,9].

Further attempts have been made to reduce tooling costs for the thermoforming of non-reinforced plastics by additive manufacturing. Laser-sintered metal parts can be excluded due to their high costs. Although manufacturing costs can be reduced to as low as 14% compared to conventional tooling methods (milling), although this is offset by the high metal powder costs of 167% [10]. For the medical field, additive manufacturing of molds from calcium sulfate and gypsum powders was successfully demonstrated and even implemented for thermoforming of plastic structures [11,12]. In addition, Junk et al. [10] reported the application of the inkjet technology for thermoforming mold fabrication out of polymer gypsum. This was tested for a case study, in which a fairing made of ABS for an unmanned aerial vehicle was produced. Among other things, this study aimed to integrate vacuum channels in the manufacturing step without the need for additional post-processing (e.g., drilling). The study also explained the specific technical challenges, such as the demolding of undercuts or the separation of the tool. Results showed that the tool had sufficient strength for the subsequent thermoforming process. Another advantage of additive manufacturing is the design flexibility of both the overall geometry and the internal structure to influence strength and other physical properties for optimized vacuum guidance [11,13].

Little work has been carried out so far to elicit the advantages of using silica sand for tooling purposes. In this perspective, the authors claim that sand structures may prove to be efficient for tooling in general and for vacuum thermoforming processes in specific. Sand structures are thought to be cost-effective. Rough calculations evidenced a tooling cost of EUR 600 to 1000 for an aluminum or steel omega shaped tool compared to less than EUR 100 for a sand tool; the outer tool dimensions are 300 mm × 260 mm × 60 mm. In addition, their low thermal conductivity is favorable for the thermoforming process. First, maintaining the tool at constant temperature allows uniform manufacturing conditions for all parts that are to be formed. Second, when forming thermoplastics, the sheets need to be heated above their glass transition (for amorphous polymers) and slightly below their melting point (for semi-crystalline polymers). Within the process, it is necessary to ensure that forming is completed before the sheet temperature falls below these levels. Achieving that through low heat conduction through the tool may assist in allowing an energy-efficient thermoforming process. In the case of thermoforming, vacuum channels may be directly printed, eliminating the necessity of subsequent drilling steps. In addition, bonding of individual particles allows a porous structure, which might be advantageous for vacuum drawing.

This manuscript presents a novel idea and gives an insight regarding current and planned research activities targeting the use of sand molds for vacuum thermoforming of FRPs. The targeted manufacturing process for the mold is Binder Jetting. Further, the approach for mold modification for the anticipated application is presented. In addition, the challenges regarding the vacuum thermoforming of FRPs are discussed and a systematic procedure is followed to address them.

2. Thermoforming of Fiber Reinforced Composites

One of the established methods in the automotive sector to fabricate FRP thin-walled structures is the Resin Transfer Molding (RTM) process. This involves the layup of multi-axial textiles into stacks, followed by draping into a 3-dimensional shape by matched tool pressing and finally resin infusion. This process is mostly limited to thermosets due to their low viscosity (below 1000 mPa·s) compared to thermoplastics (beyond 100,000 mPa·s) and is able to fulfill the demands for good surface quality. A drawback lies in the long cycle times needed for part curing (standard RTM processes may require several hours or days for curing; enhanced RTM processes 2–10 min) [14], thus limiting the use of thermosets for mass production. In contrast, the use of thermoplastic matrices makes short cycle times (60–140 s) [15] more feasible. Thermoplastics further have the advantages of providing improved toughness behavior and allow for recyclability. For fiber reinforced thermo-plastics the thermoforming process is applicable. The process, as illustrated in Figure 1, implies forming a flat laminate into a 3D geometry under the influence of temperature. For heating IR lamps or convection ovens are often used [16]. The transfer time to the press has to be kept short in order to avoid extensive laminate cooling, which would prohibit adequate forming. For 3D forming of FRPs, both matched-die forming as well as deformable die forming is applicable. During shaping, the laminate is held in place using a support frame or a blank holder. This is crucial in order to keep the laminate under tension and accordingly prevent the creation of wrinkles. Further, the mold is kept closed during deformation to avoid heat loss and to maintain high pressures (10–40 bar), targeting proper compaction and full consolidation [17]. The final part is then left to cool within the mold, until it reaches a temperature below the glass transition temperature of the matrix.

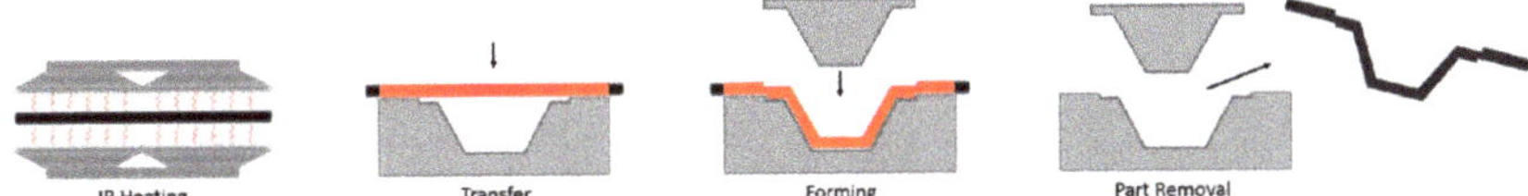

Figure 1. Schematic illustration of the thermoforming process.

Schug [17] summarized the main influencing parameters on the forming behavior and part quality. The author claims that the tool temperature has a major impact on the cooling time and morphology of the material. Further, the forming speed should be selected in such a way that allows a compromise between timely shaping and shear thickening response of the thermoplastic melt. The press load, where a high pressure favors consolidation on the one hand but may lead to dry spots and matrix flow on the other hand, mainly governs the surface quality. Further typical defects are fiber undulations, gap formation, out-of-plane wrinkles and folds [18,19]. Insufficient contact between mold surface and laminate at the end of the forming process may be the reason for rough surface areas. In the case of extensive tension of the laminate within the support frame, a fiber breakage and thus drastic losses in mechanical performance can be expected.

In the case of single and double diaphragm forming, pneumatic pressure is applied. The function of the diaphragm is to transfer the mechanical loads to the laminate and keeping a biaxial tension for wrinkle and undulation prevention. In single diaphragm forming, the laminate is placed over the mold having the diaphragm on top. The vacuum is applied and the laminate is forced against the tooling to take its shape. In double diaphragm forming the laminate is sandwiched between two flexible membranes and compacted by vacuum. Further vacuum is applied to press the sandwiched stack against the mold. This technique allows a greater degree of double curvature than single diaphragm forming [20]. For more intricate geometries and smaller fillets the hydroforming can be adopted. Here, a fluid pressure is applied to a rubber diaphragm to force the laminate stack to the mold. Hydroforming and diaphragm forming eliminate the need of mutual conforming tools and hence, the tooling cost is lowered [20].

Harrison et al. [21] investigated the use of springs instead of friction-based blank-holders to induce in-plane tension in the laminate in order to prevent the formation of wrinkles. This technique was evaluated as being flexible, easy to use and naturally facilitating heat transfer into the laminate prior to forming. However, the authors report that the focused application of tension might lead to in-plane buckling and localized zones of high shear.

The use of vacuum thermoforming for pure plastic sheets is a widespread process. In 2018, the market size for vacuum forming was estimated at USD 11.69 billion [22]. Various plastics (e.g., PA, PE, PP, PC or PET) are often applied. The basic process (Figure 2a) implies forming a sheet into a 3D geometry by heating it to a formable state, pressing it against a mold and holding it in that position until the sheet is cooled below the glass transition temperature. Finally, the part is trimmed to final shape [23]. In most cases, the processing temperatures are below the melting temperature and above the glass transition temperature (e.g., around 240–250 °C for PA6) [24,25]. Heating is mostly carried out by radiation, for example by using IR-heaters. Accordingly, bringing the plastic sheet to a viscoelastic state depends on the polymer's ability to absorb the long infrared wavelength energy. Plastics with low crystallinity are said to have better formability [25] since the crystals hinder the flow behavior of the viscous fluid.

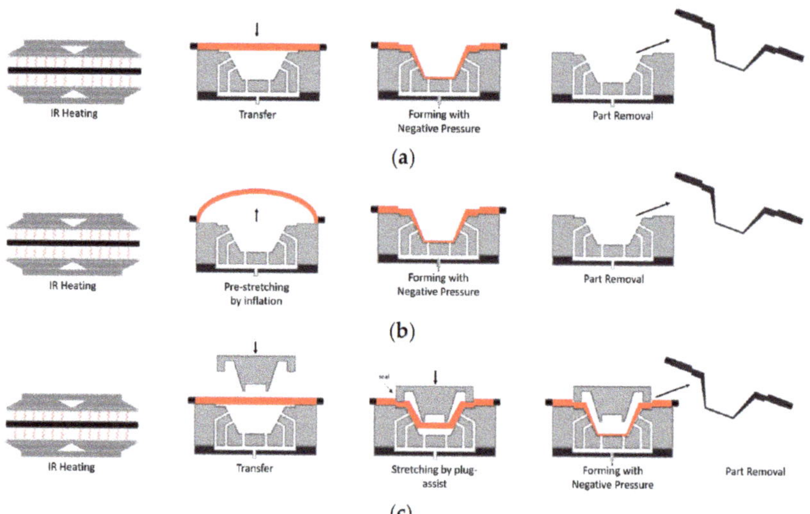

Figure 2. (**a**) Basic vacuum thermoforming process, (**b**) vacuum thermoforming with pre-stretching via inflation of a bubble, (**c**) vacuum thermoforming using a plug-assist.

Despite its high process flexibility and good suitability for prototype and series production, thermoforming is associated with some challenges. When using single-sided tools, the component thickness can vary significantly, due to varying stretching according to the geometry. Sheet areas first touching the mold are hindered in their biaxial motion by friction and thus remain thicker than other areas that continue to witness stretching. This is also associated with faster cooling by heat conduction from the touching area to the mold. This uneven deformation and cooling behavior can lead to residual stresses in the final component [23]. As a remedy, a pre-stretching step may be introduced after the heating step. In that case a clamping tool fixes the edges of the sheet and air pressure is applied to inflate the sheet into a bubble (Figure 2b) [26]. In the next step, forming is achieved by vacuum pressure. Nevertheless, the thickness distribution in the final components largely depends on the viscoelastic properties of the polymer materials.

In plug-assisted (Figure 2c) thermoforming, the heated sheet is pre-stretched through a mechanical plug. Next, actual forming is completed through the application of vacuum pressure [27]. Chen et al. [28] observed that thickness at the sidewalls increased with increasing mold temperature, preheating temperature, plug depth and holding time, but decreased with increasing plug speed. At a thermoforming temperature above the glass transition the sheet material becomes sticky and thus unable to slip over the plug surface due to increased friction [27,29,30].

3. Binder Jetting of Sand Tools

Conventional tool manufacturing relies on subtractive (machining) process starting from a material block. Such tools are often metallic (e.g., tool steel type AISI-SAE 1045 [31]) or in some cases made from polymers (e.g., RAKU®Tool [32]) (RAMPF Tooling Solutions GmbH & Co. KG, Grafenberg, Germany) and foams. Considering material costs and machining time, the fabrication of tools may prove to be extensively time consuming and costly [31].

In contrast to the subtractive manufacturing, additive manufacturing builds the structure in a layer-by-layer technique, applying material only there where it is needed. Figure 3 illustrates the Binder Jetting process according to ASTM [33]. Here, the following process steps are repeated until the desired component is created: A build platform is lowered by a layer thickness of ranging between 10 μm to 400 μm. The hereby-created free space is then filled with powdered material using a re-coater. In the third step, a binder is selectively deposited using an inkjet print head to bond the individual particles together. This creates a bond within the layer and with the layer below. In the case of Binder Jetting of sand, sand grains are used as the building material.

Figure 3. Schematic illustration of the steps characterizing the Binder Jetting process: First—lower platform, Second—recoat layer, Third—jet binder.

This setup makes Binder Jetting systems particularly easy to scale in terms of performance. The number of nozzles correlates with the overall performance [34]. Likewise, several layers can be applied simultaneously. This feature cannot be achieved with laser-based systems using beam deflection. Such unique arrangement leads to extraordinary build rates and, consequently, to special cost-effectiveness. Binder Jetting currently achieves a maximum build-up rate of approximately 400 L/h (Datasheet Exerial™ 3D) (ExOne, Gersthofen, Germany) printer and costs significantly less than 10 €/l. The basis of the Binder Jetting can be various particle materials. Plastic particles, metal powders or inorganic materials such as sand or ceramics can be used. The particles are adapted to the layer thickness to be achieved, and the spectrum of average particle sizes ranges from d_{50} = 20 μm to d_{50} = 400 μm. Silica sand is particularly cost-effective for use as tools. Here, the raw material costs are often less than 100 €/t.

The components manufactured by Binder Jetting can be used either directly or with post-treatment as a tool. Post-treatment usually increases the strength and wear resistance of the product. Used directly, sand binder systems achieve a maximum tensile strength of about 10 MPa (identified by preliminary examinations and further referred to as the base material). The strength can be significantly increased, for example, by infiltration with an epoxy resin. The sand or the underlying particle material can be sintered or an impression can be made with a higher-strength material (e.g., with a polymer cement) when even higher strengths are required.

The strength of the base material is affected by different aspects related to material and process, such as the strength of a sand grain, the density of the sand, the binder adhesion and cohesion as well as the amount of binder. The sand grains are packed more or less densely during the coating process, depending on the technique [35]. This results in a volume ratio of air to sand of about 50%. Depending on the binder system, approximately 2–5 wt% binder penetrates into the loose sand during the deposition of the binder. The capillary action causes liquid to accumulate at the contact points of the sand grains (Figure 4), which solidify by drying or polymerization, depending on the binder system. The cavity that now remains can be filled by infiltration with another material system, thus raising the strength due to the higher amount of binder compared to the base material.

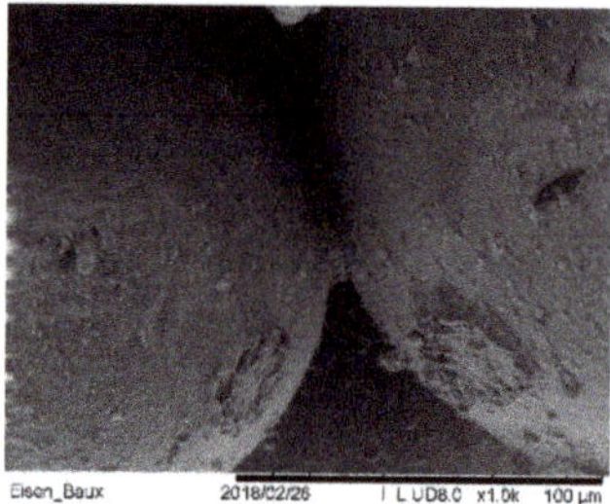

Figure 4. SEM-Micrograph of the connection between two grains.

The properties of printed sand molds can be used especially in applications where low strength is of secondary importance or even helpful. For example, a casting core needs to be as strong as necessary for safe handling but, after the casting process, has to be easily removed out of the cavity. Also relevant here are cores for hollow structures that have to be removed after the lamination process (washouts) [36].

Tools for casting purposes are also manufactured using tools fabricated by Binder Jetting. Very different casting materials can be used in this process. For example, concrete can be molded to produce structures in the construction industry (Figure 5) [37,38]. The molds must be pre-treated for casting by sealing the surface to prevent the ingress of mixing water. After the concrete has set, the mold is opened and removed from the structural member. Depending on whether the structure is undercut, the molds can be used once or several times. This is also the case, with laminating molds where the mold is used only for a few impressions in prototyping applications.

Sand molds are further widely spread in the foundry industry for the production of metal castings. Here, individual parts with small and large dimensions are often realized with printed molds. In particular, parts that conventionally require the storage of large molds over long periods can be produced economically using Binder Jetting [39]. On the other hand, large series in engine construction can nowadays also be realized with sand molds from 3D printers [40].

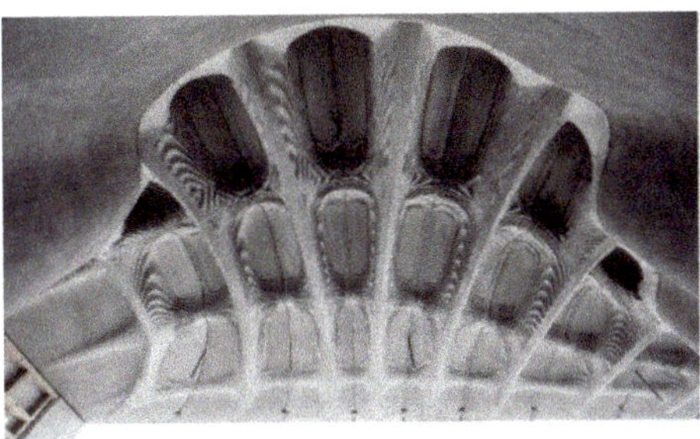

Figure 5. Concrete part formed on a Binder-Jetted tool [38].

Various sand types have been qualified for the usage in Binder Jetting of sand molds and cores [41]. Silica sand is most commonly used in foundries to make molds (amongst other methods also through additive manufacturing), due to its cost effectiveness. The sand morphology and particle size (Figure 6) distribution are known to strongly affect the resulting packing density [35] and surface properties [42]. The packing density influences both mechanical strength and permeability. In general, by choosing a sand with a smaller d_{50} medium grain size than a reference one, a higher casting surface quality can be expected [43]. However, this grain size implies low permeability to air and gases. The shape of the sand grain also has a decisive influence. Sharp edged grains, on the one hand, have the least contact with each other in a compacted structure and thus make the sand highly permeable to gases. On the other hand, they cannot be packed to the optimum extent during Binder Jetting and structures made from them have low strength [44].

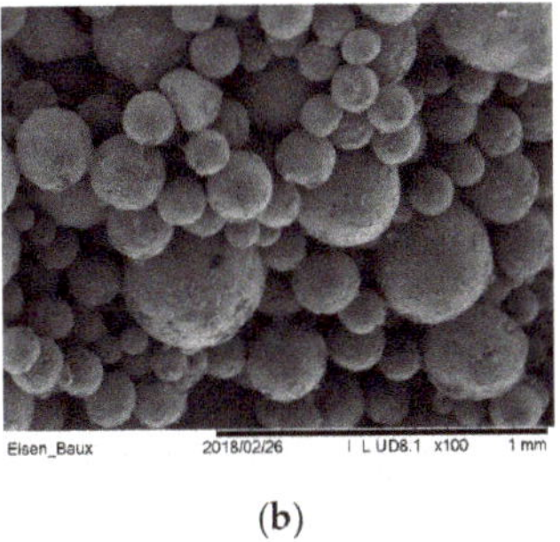

(**a**) (**b**)

Figure 6. SEM-Micrographs of different qualities of sand suitable for the Binder Jetting of tools: (**a**) Silica Sand type GS14 RP by Strobel Quarzsand GmbH Freihung, Germany, and (**b**) Bauxitsand by Hüttenes Albertus GmbH Düsseldorf, Germany.

Current binder systems in combination with sand as a molding material are inexpensive epoxy, furan or phenolic resin–based systems that bring sufficient thermal stability. Environmentally friendly inorganic binders are nowadays the subject of research and are preferred within this study [43,44].

For 3D printing of tools, the mechanical properties of the 3D-printed part prior to infiltration might be of minor importance. Thus, a sufficient strength for secure handling during the infiltration process is expected to be acceptable considering the overall process. The permeability, however, is essential to ensure a sufficient infiltration depth and thus to provide the basis for infiltrated sand tools of enhanced mechanical properties. Sand qualities that will be investigated within this study are silica sands obtained from natural reservoirs of varying powder size distributions as well as artificial sands produced by sintering. In contrast to natural sands, artificially produced sands typically show regular, round shapes and narrow particle size distributions.

4. Sand Tools for Vacuum Forming of FRP

The main target of the investigations is to develop an innovative process chain for the rapid and cost-effective production of large-area thermoplastic-based shell structures made of fiber reinforced plastics—especially during the product development phase, or where small part numbers are needed. In the solution approach, the advantages of additively manufactured sand molds are combined with the possibilities of a fast thermoforming process. This results in synergy effects that significantly advance prototype and small series production. In the following the main requirements regarding sand tools for vacuum thermoforming applications are given:

- Compression strength exceeding 20 MPa in order to stand the specific process loads;
- Acceptable wear resistance to withstand at least ten forming cycles;
- Surface finish with arithmetic mean roughness $R_a < 20$ µm to minimize indentations on visible surfaces;
- Porous structure to ensure a safe application of vacuum;
- Temperature resistance up to 280 °C;
- Cost-effective in terms of engineering and production comparable with state-of-the-art tools.

In order to address the aforementioned requirements, a segmented approach is applied (Figure 7). In the first stage, the focus is laid on the sand mold, where materials and process parameters will be defined. After 3D printing by Binder Jetting, a mockup tool is further treated to fulfill the requirements regarding mechanical properties and surface quality. In parallel, the feasibility of vacuum thermoforming is studied, addressing the main challenges such as the applicable clamping mechanism for FRP laminate to ensure vacuum pressure on the one hand and draping of the laminate under tension on the other hand.

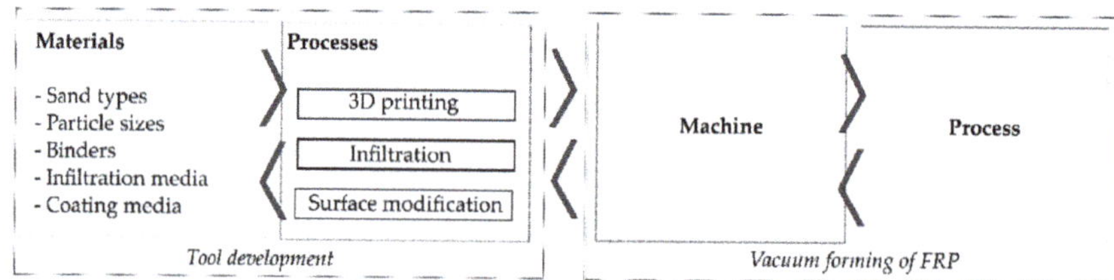

Figure 7. Investigation approach targeting the fabrication of sand tools for vacuum forming of FRP.

4.1. Initial Approaches and Results in Binder Jetting of Sand Tools

For preliminary tests, the silica sand GS14 (Strobel Quarzsand GmbH, Freihung, Germany) in combination with an inorganic binder system (IOB by voxeljet AG, Friedberg, Germany) was selected. The 3D printing tests were carried out using increased binder weight contents (150%, 200% and 300% with respect to conventionally used binder amount) in order to investigate the feasible limits and their effects on the mechanical properties considered relevant for sand tool fabrication for vacuum forming of FRP. Further, during 3D printing, the binder content is to be increased globally on the one hand and selectively varied locally on the one hand. Local variation (further referred to as the skin-core setup) implies the fabrication of a skin layer of 5 mm thickness with high binder content relative to the core segment. This allows for superior mechanical properties at the mold surface. In contrast, in the case of global binder application at high binder content, the entire sand structure is printed at the aforementioned increased values. Two sets of specimens were produced: bars with a cross-section of 22.4 × 22.4 mm for 3-point flexure tests and cylinders D50 × 50 mm for compression tests and density evaluation. Five specimens were tested for every parameter set. Figures 8–10 show the results of the preliminary investigations, indicating that the increase of binder content can be directly correlated with an increase in density, compressive strength and bending strength for globally increased binder contents

(a) and skin–core settings (b). The specimens' densities printed using the skin-core strategy were found to be higher than those of the respective specimens with overall increased binder contents (Figure 8). This effect was attributed to the fact that the printed binder of the underlying layer and its residual moisture impedes the deposition of new powder and its compressibility. The compressive strengths of the specimens was significantly reduced by the skin–core setting (Figure 9), while the bending strength slightly increased (Figure 10). This was related to the maximum bending stress, which increases indirectly proportional to the geometrical moment of inertia of the cross-section that is smaller for hollow rectangular cross-sections compared to filled rectangular areas.

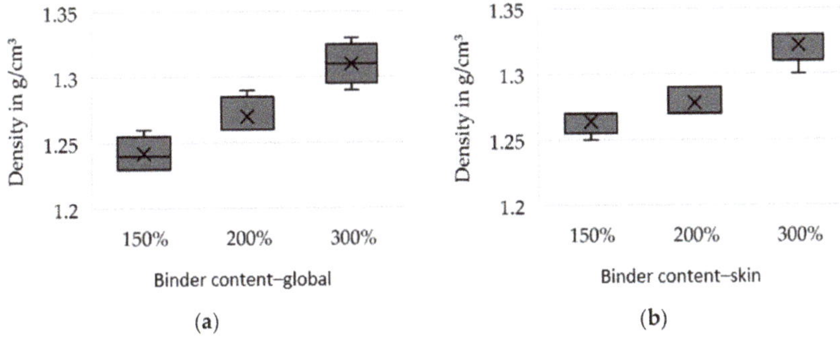

Figure 8. Density of specimens produced at varying binder contents. The binder content was increased (a) globally and (b) only in an outer skin of 5 mm.

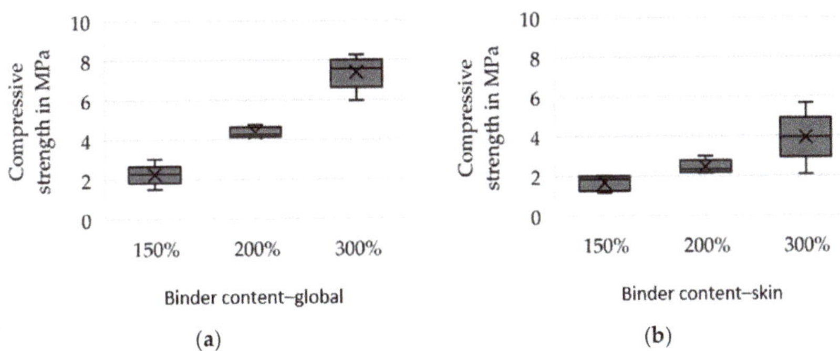

Figure 9. Compressive strength of specimens produced at varying binder contents. The binder content was increased (a) globally and (b) only in an outer skin of 5 mm.

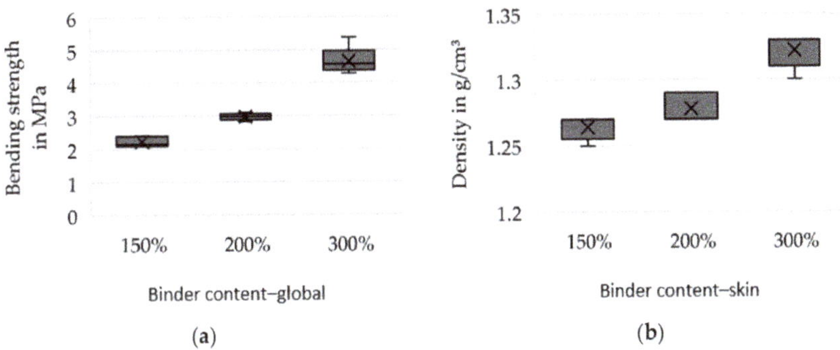

Figure 10. 3-point bending strength of specimens produced at varying binder contents. The binder content was increased (a) globally and (b) only in an outer skin of 5 mm.

Especially the dimensional accuracy in building direction decreased significantly with increasing binder contents. While the specimens' dimensions (measured by a Vernier caliper) in building direction roughly comply with targeted dimensions for moderately increased binder contents (mean deviations of +0.83% for 150% binder content and +1.56% for 200% binder content), the dimensions for 300% binder content deviated extremely from the targeted values (mean deviation of +3.91). The upper application limit of binder content was therefore identified to be 200% for the observed sand–binder combination. Especially the print strategy (e.g., the local density of the jetted binder), were found to play an important role and will be further investigated.

4.2. Initial Approaches and Results in Infiltration of Sand Tools

The requirements regarding the mechanical properties cannot be achieved with the basic material system. A conflict of targets exists regarding strength, penetrability and final permeability. This conflict can only be solved by considering the process, i.e., including printing and infiltration. Therefore, the 3D-printed sand structures were subsequently impregnated with various resins. In a preliminary test, first promising results were gained by infiltrating two D15 × 20 mm specimens. The samples were Binder-Jetted using GS14 sand and furan resin (VX-2C by voxeljet AG, Friedberg, Germany). The Epoxy resin (IH16 by Ebalta Kunststoff GmbH, Rothenburg ob der Tauber, Germany) was applied as the infiltration medium. Infiltration was carried out manually at room temperature using a brush. Figure 11a shows two infiltrated specimens: at the top of the figure, a radial cut shows the full impregnation of the sand specimen; at the bottom, the second specimen, previously tested for compression strength, is shown. The fully infiltrated specimen reached a compressive strength of ~70 MPa, thus exceeding the targeted value of 20 MPa. However, when altering the specimen dimensions to D50 × 50 mm, it was observed that only a comparably limited penetration depth could be achieved. Figure 11b demonstrates that such samples witnessed a spalling effect during compression testing. This was accompanied by a reduced mean compression strength of 5.9 MPa at a standard deviation of 0.3 MPa. When doubling the amount of infiltration medium for another five specimens of D50 × 50 mm, spalling could still be observed, while the compression strength increased to a mean compression strength of 12.1 MPa at a standard deviation of 3.1 MPa. These first experiments were intended to show the potential and challenges associated with the infiltration of 3D-printed sand structures. Further examinations with a larger quantity of specimens including a broader variety of materials are planned for future investigations.

(a)

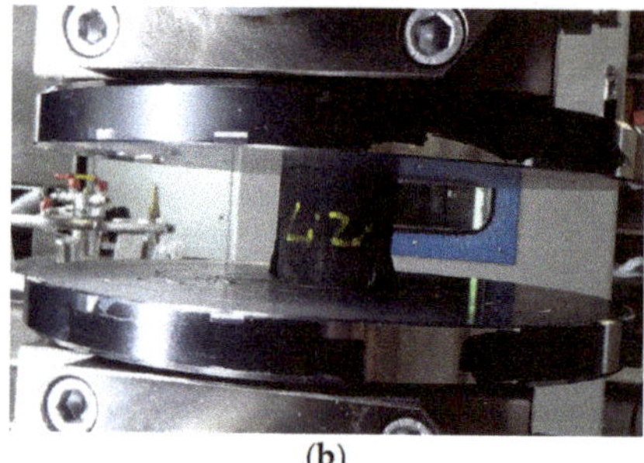
(b)

Figure 11. The 3D-printed specimens infiltrated with epoxy resin: (**a**) fully infiltrated specimens of dimensions D15 × 20 mm, (**b**) spalling effect during compression testing for specimens of dimensions D50 × 50 mm.

A rough evaluation of the surface quality achievable by grinding the epoxy-infiltrated structure showed R_a of 7 µm, fulfilling the requirements stated above. However, hand-grinding using sand 120-grit paper shows a low level of automation. Thus, alternative methods are also scope of the future investigations.

4.3. Intended Future Investigations Regarding Sand Tool Fabrication

The results of the first investigations showed that the requirements regarding the tools can in principle be met. The preliminary tests have to be refined in order to prove the feasibility of the production of suitable tools.

The planned study implies the evaluation of the mechanical performance, dimensional accuracy, density, permeability and surface quality of the sand structures with respect to the targeted requirements. Table 1 presents the applied test methods for evaluation.

Table 1. Materials testing methods for further investigations.

Material Testing Method	Standard
3-point flexure test	VDG P71
Compression test	DIN EN ISO 126
Dimensional accuracy	DIN 862
Permeability test	VDG P41
Density determination	DIN 862, DIN 8128-1
Roughness measurement	ISO 4288

Various sands are to be considered for the future study in order to investigate the effect of particle size and shape on density, strength, roughness, penetrability and permeability. Three different particle size distributions of natural silica sand will be investigated (GS14, GS19 and GS25 by Strobel Quarzsand GmbH, Freihung, Germany 3D) as well as one synthetic sand composed of aluminium silicate (Cerabeads ES650 by Hüttenes-Albertus Chemische Werke GmbH, Düsseldorf, Germany). Printing will be carried out using two different binder systems: a furan resin (VX-2C by voxeljet AG, Friedberg, Germany) that is the most frequently used for Binder Jetting of sand molds and an inorganic binder system (IOB by voxeljet AG, Friedberg, Germany), which is gaining increased attention due to its low emissions during casting. Further, a graded structure that allows for a controlled permeability for subsequent impregnation and the final vacuum forming purposes may be applicable in sand tools. Therefore, the binder content and the printing strategy will be further investigated in accordance with the preliminary tests. In addition, the lower limit regarding binder content will be identified. This is expected to result in lower densities, which may enhance the penetrability of the 3D-printed sand structure by the infiltration media. Figure 12 sums up the planned investigations with the various material systems and 3D printing parameters.

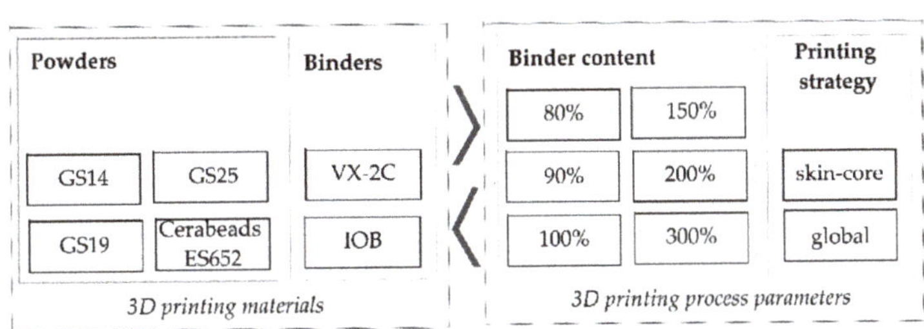

Figure 12. Planned variations regarding 3D printing of sand tools.

Further investigations target the enhancement of the penetration depth using alternative resins of higher penetrating capacity in addition to tailoring the permeability of the 3D-printed structure by locally altering the binder content. Figure 13 shows the suggested specimen geometry designed for investigating the depth of penetration. The sample is cylindrical with a central hole making an indentation of 5 ml volume. Accordingly, a pre-metered amount of infiltration media is to be applied into the indentation, which will allow a comparability of the impregnation efficiency with varying infiltration media at varying temperatures and printing material systems (involving powder and binder) and parameters. The penetration depth will be examined microscopically across the cross-section.

Figure 13. Suggested specimen geometry for the determination of the penetration depth.

Further investigations will include surface modifications, as smooth surfaces will not only enhance the surface quality of the FRP part but also are expected to reduce forces when demolding. Investigations will concern known application methods like grinding and coating of the infiltrated tools as well as the modification of the impregnation media.

4.4. Thermoforming of Organosheets

The implementation of vacuum thermoforming for FRPs requires a step-by-step approach to overcome the first obvious challenges. These mainly lie in the lack of ductility of the fibers in contrast to the surrounding thermoplastic matrix. Heating the organosheet above the melting point of the matrix allows the matrix to flow and thus to be formed. During this stage, the matrix is stretched within the blank holder or the support frame. In the case of endless fibers, these are tensed but remain unable to stretch. This leads to the withdrawal of the organosheet from the support frame or excessive folds. Accordingly, a flexible holder is necessary; this should allow the sheet to flow and feed new material necessary to envelope the 3D mold geometry.

A further challenge lies in the load available for deformation. It can be expected that the vacuum pressure alone is insufficient for deforming FRP sheets. Although the main target of the study is to keep the process as simple as possible, it is kept in mind that an application of a diaphragm might prove to be inevitable. Other than in the case of conventional thermoforming with unreinforced polymer sheets, it is assumed that heat must be applied to both surfaces of the sheet in order to achieve a homogeneous melt of the matrix throughout the thickness. This implies that the setup of a thermoforming unit must be adapted to allow heating both sheet surfaces in addition to the mold surface in order to avoid rapid cooling of the sheet during deformation. Figure 14 shows the envisaged machine setup. An IR-heating unit is moved into the thermoforming unit, to simultaneously heat both sides of the sheet and the mold surface. The heating unit is then wheeled out and the thermoforming process takes place.

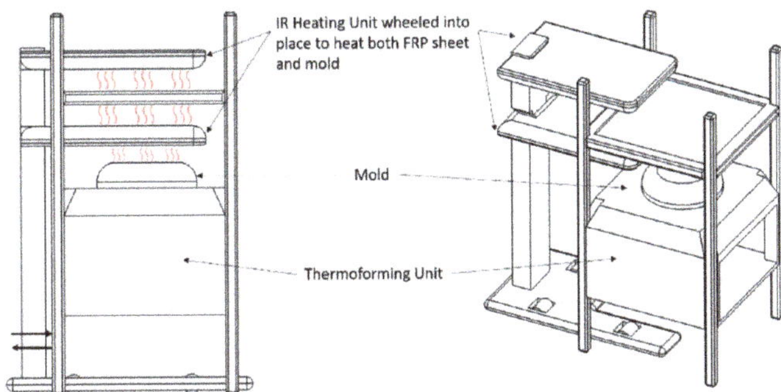

Figure 14. Preliminary concept of the heating and thermoforming unit.

In order to tackle these challenges, first, the focus was set on the development of the sand mold for vacuum thermoforming applications in general. Hence, the first tools developed in the above-mentioned stage were tested for their feasibility for pure thermoplastic sheets. In that stage, an ABS fender was vacuum thermoformed. The sand mold, illustrated in Figure 15a, is binder-jetted as a hollow structure in order to save weight (final weight 30 kg) and equipped with vacuum channels. The mold was further impregnated using an epoxy resin to increase the stability of the surface layer. For improved surface finish, the mold was ground using sandpaper (grit size 120). The process proved to be feasible and the surface quality, as presented in Figure 15b,c, was found acceptable.

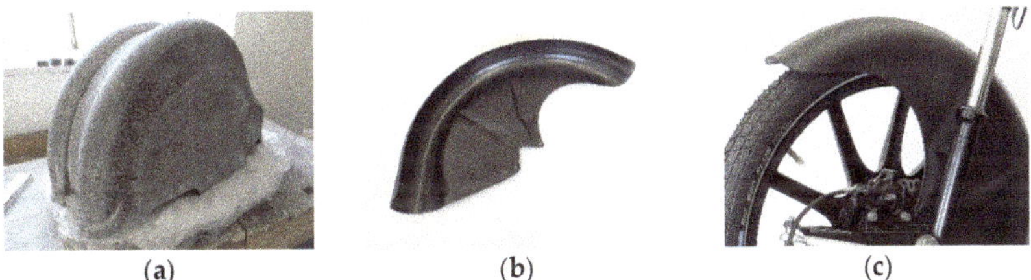

(a)	(b)	(c)

Figure 15. (**a**) Binder-jetted sand mold for thermoforming thermoplastic fenders, (**b**) thermoformed fender made from ABS, (**c**) application of the ABS fender.

In the next phase, the processing ability of FRPs is studied using metallic molds equipped with conventional vacuum channels. Here, issues like vacuum pressure, heating temperature, heating time and setup are examined. The difficulty level is increased by first considering randomly oriented short fiber reinforced sheets. Then, the fiber length and the orientation are altered step by step. When considering endless fibers, the behavior of woven textiles will be observed in contrast to multi-axial fabrics. These experiments are to be carried out on different levels of geometrical complexities. Figure 16 shows the envisaged geometries of the mold.

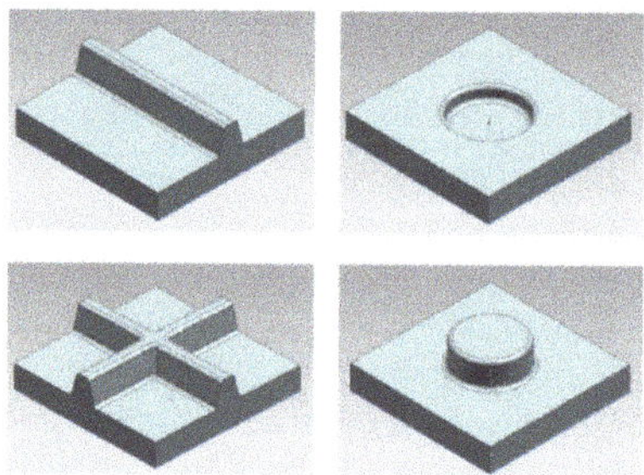

Figure 16. Targeted designs of mockup molds to tackle the challenges in various complexity levels.

As mentioned above, the support frame has to fulfill several functions and resembles one of the key factors for successful thermoforming. First, the frame has to seal the forming area to ensure vacuum build-up. Further, it should act as a blank holder that has to provide enough degrees of freedom to allow for feeding sufficient sheet material to compensate the desired 3D shape. Only if both features are provided by the clamping mechanism a good molding process and part quality can be expected.

In order to be able to identify the necessary clamping force, preliminary evaluation of the frictional forces between the above-mentioned materials and the anticipated frame materials were conducted. The investigations were performed using a self-constructed test rig, as depicted in Figure 17. The rig consists of a carriage for material A and a further fixture for material B. Material A is cut to the dimensions 180 mm × 50 mm and is clamped at both ends into the carriage. Material B is prepared to the dimensions 110 mm × 50 mm. The test setup involves sandwiching two layers of material A between two single layers of material B on each side (Figure 17 magnification). Hence, a contact area of 50 mm × 50 mm is created. The normal (press) force is controlled by weights. For the measurements, the test rig is mounted into a universal testing machine. The pulling rope is attached to the crosshead and translates the vertical movement into a horizontal movement between material A and B through the idler pulleys. When the carriage is moved back and forth, a friction between A and B can be measured.

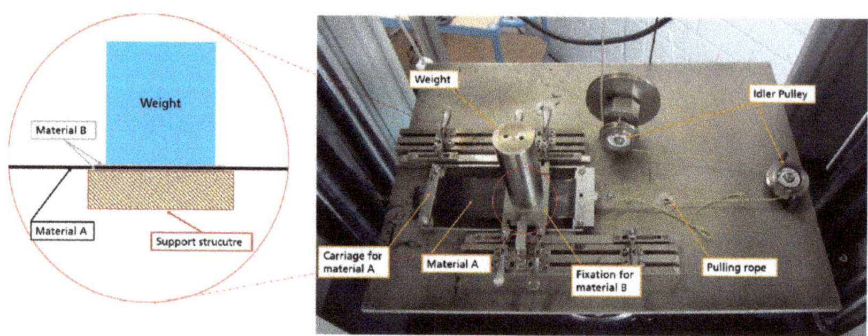

Figure 17. Test rig for measurement of frictional forces between FRP sheet and clamp materials.

The candidate materials for the support frame in direct contact with the sheets were selected to be polytetrafluoroethylene (PTFE) and aluminum. Regarding the anticipated sheet materials, a unidirectional carbon fiber and another short-recycled carbon fiber reinforced polyamide tape were investigated. The material combinations examined are summarized in Table 2. For this test campaign, the universal test machine ZwickRoell Z100 in combination with a KMD 5 KN load cell was used. For creating a specific press force between materials A and B, a 1 kg mass was applied on the materials. The crosshead was moved with a constant speed of 100 mm/min and stopped after a displacement of

80 mm. All tests were performed at room temperature. The load-displacement data were recorded for three replicates, where only material A was replaced by a new one after each examination. To calculate the friction coefficients, two forces were evaluated from the load-displacement curves. The static and dynamic coefficients of friction were calculated according to Equation (1), where μ_{PT} is the coefficient of friction, F is the recorded force, F_N is the applied normal force, m is the mass and g the gravitational acceleration.

Table 2. Overview of the material combinations, the measured forces and calculated friction values.

Material Combination		Average Max.	Average Sliding	μ_{PTs} (Static	μ_{PTd} (Dynamic
Material A	Material B	Force [N]	Force [N]	Friction)	Friction)
UD CF-PA	Teflon	4.78	3.06	0.243	0.156
UD CF-PA	Aluminium	6.91	5.08	0.352	0.259
UD CF-PA	UD CF-PA	4.68	3.7	0.238	0.189
Teflon	Aluminium	8.71	6.3	0.444	0.321
rCF_PA	Teflon	5.89	4.35	0.300	0.222
rCF_PA	Aluminium	12.37	7.02	0.624	0.358

In the case of the static friction coefficient μ_{PTs}, the force F is identified at the beginning of motion as the maximum force value between 0 and 10 mm displacement, whereas for the dynamic friction coefficient μ_{PTd} F is given by the average force value, between 10 and 60 mm displacement. The calculated values are presented in Table 2.

$$\mu_{PT} = \frac{F}{2 * F_N}; \; F_N = m * g = 1\,kg * 9.81\,m/s^2 = 9.81\,N \tag{1}$$

A simple model for the calculation of the clamping force was developed based on the coefficients of friction μ_{PT}, the achievable vacuum pressure and the required forming geometry. The model is presented in more detail in the Appendix A. Through the known pressure difference Δp (between vacuum and ambient pressure) and the surface area A of the sheet, the resulting force F_{vac} can be calculated. With this force and the assumption that the tape behaves like a rope (inelastic material, unable to transfer moments), the clamping force F_s can be estimated. Accordingly, Figure 18 gives a simplified estimation of the forming process at its initial state, where the pliable sheet is not yet in contact with any of the mold surfaces except the top surface (indicated by the plate in Figure 18).

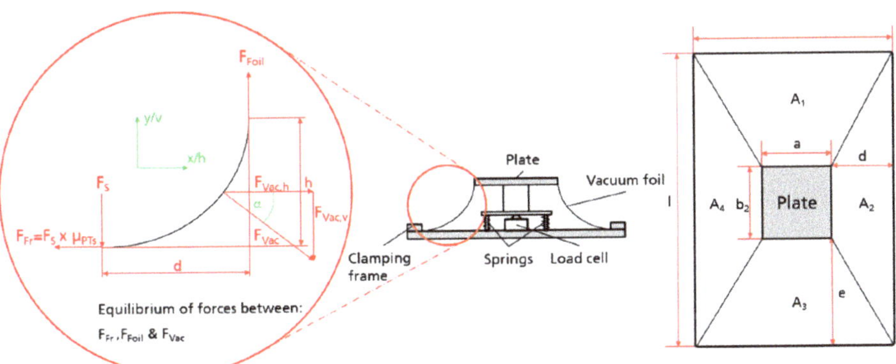

Figure 18. Side view of the test stand and acting forces and top view of the test stand.

The test rig is further illustrated in Figure 19. Knowing the inclination angles and dimensions of the test rig, the surface area of a clamped foil can be calculated. Further, the vacuum force F_{vac} applied to the foil can be determined and can be decomposed in its horizontal and vertical force components, for each area segments (A_1–A_4 in Figure 18). Based on the assumption that the tape behaves like a rope and that the tape does not transport any momentum in addition to the boundary condition, that $F_s \times \mu_{PTs} < F_{Foil}$ (where F_{Foil} is the tangential tensile force in the foil), the parameters for all area segments can be calculated.

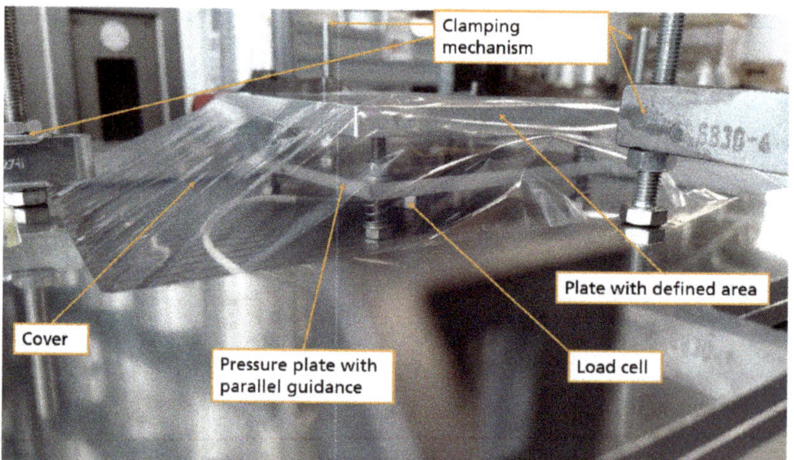

Figure 19. Test setup for the validation of the developed model.

To validate the model and calculations, a test (Figure 19) was performed, using the aforementioned setup, while integrating a load cell in an exact copy of the described model (Figure 18). A PA6 foil (commonly used in vacuum infusion processes) was used. A vacuum cleaner (Wieland IS-46h) was used to create the desired pressure difference of 270 mbar. After assembling the test stand and clamping the foil in the frame the load cell is tared, and the vacuum was applied. A nominal force of 0.91 kN was generated and was further used to validate the model. The first trials indicated that with the vacuum cleaner and the test stand the expected pressure difference of 270 mbar could not be achieved. However, the pressure difference was not measured, and it remains unclear what the absolute achieved vacuum pressure would lead to, regarding the pressure difference between ambient and vacuum. This difference is vital for the calculations. Hence, in the next step, a pressure sensor is integrated into the test stand to raise the necessary data for validation.

Once the theoretical model is validated, it can be applied to construct the support frame and to define the necessary clamping forces. These aspects will further be considered for a prototype vacuum thermoforming stand. The products, produced with the prototype machine, are to be evaluated regarding their dimensional accuracy, the fiber distribution and local fiber volume content.

First experimental trials have proved the above-mentioned challenges. Trials with endless fiber reinforced thermoplastic sheets (Figure 20a) confirmed the demand for a new clamping concept and a two-sided heating system with short transportation routes from the heating unit to the forming unit. Further, the trials have shown, that the short fiber material can be challenging, as it tends to loft (Figure 20b) and thus generate voids in the process, which in turn negatively affect the generation of the required vacuum pressure.

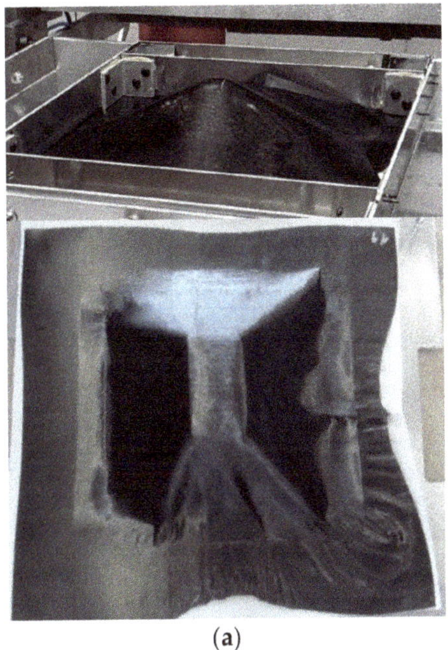 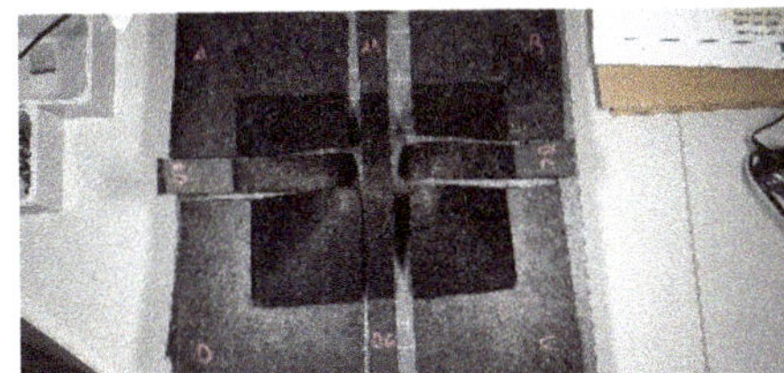 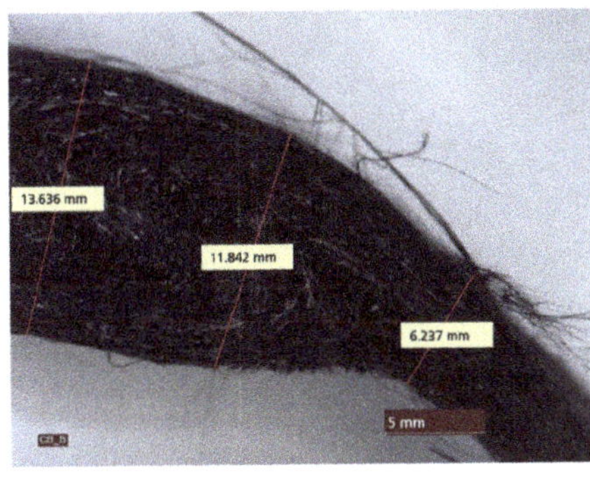

(a) (b)

Figure 20. (**a**) Vacuum thermoforming of a multiaxial endless CFRP sheet under non-optimal conditions. (**b**) Vacuum thermoformed short CFRP sheet and resulting lofting effect.

4.5. Summary of Technical Challenges

Preliminary investigations showed that 3D-printed sand specimens meet the requirements for the application of FRP vacuum forming tools regarding strength and surface quality on laboratory scale. It was shown that it is possible to customize density and strength by adapting the 3D printing parameters and through post-processing by infiltration, while varying impregnation parameters and media.

Technical challenges in developing 3D-printed sand tools for vacuum forming of FRP derive in particular from increased mechanical and thermal loads when forming FRP sheets. Process related mechanical peak loads are expected to occur during forming and ejecting of the formed part resulting in multi-axial stresses on the tool. Thus, a geometry-dependent stability of the tool material is needed. However, a residual permeability is expected to be advantageous for the vacuum-forming process. Especially the homogeneity of the material properties within the tool, strongly influenced by the infiltration process, are expected to play an important role for the long-term stability of the tools.

Moreover, temperature increases in the tool arise along with an increased number of production cycles as the sheets solidify in direct contact with the tool surface. However, using 3D printing offers extensive possibilities to overcome constraints by including enhanced functionalities, such as undercut cooling and vacuum channels.

Regarding the vacuum forming of FRPs, the main challenges lie in the construction of a suitable sheet holder that allows material feed to compensate its draping over the mold and that provides enough tension to prohibit wrinkles in the sheet. In addition, rapid cooling against the mold has to be prevented up to the point, where the geometry is accurately mapped. Further, the forming process itself, the dynamic process and the reproducibility of the material feeding behavior will be the key challenges for successful vacuum forming.

In view of the transfer to an industrial application, the central key challenge is defining an economically, sustainable and preferably automatable process for the production of tools of consistent quality.

Author Contributions: The individual contributions of the authors are as follows. Conceptualization, I.T. and D.G.; methodology, P.E. and S.S.; investigation, P.E. and S.S.; Formal analysis, P.E. and S.S.; validation, P.E. and S.S.; writing—original draft preparation, I.T., D.G., P.E. and S.S.; writing—review and editing, I.T.; supervision, D.G. and I.T.; project administration, I.T.; funding acquisition, I.T. All authors have read and agreed to the published version of the manuscript.

Funding: This research was funded by the Bavarian Research Foundation, grant number 1400-19.

Institutional Review Board Statement: Not applicable.

Informed Consent Statement: Not applicable.

Data Availability Statement: Not applicable.

Acknowledgments: The authors acknowledge the support of the partners within the scope of the aforementioned funded project: voxeljet AG (Binder Jetting Process), BBG GmbH (construction of sheet holder and machine), Miedl Kunststoff & Design GmbH (vacuum forming expertise), Gierl DCP GmbH (alternative mold materials) and solar velomobil pedilio (defition of product requirements).

Conflicts of Interest: The authors declare no conflict of interest.

Appendix A

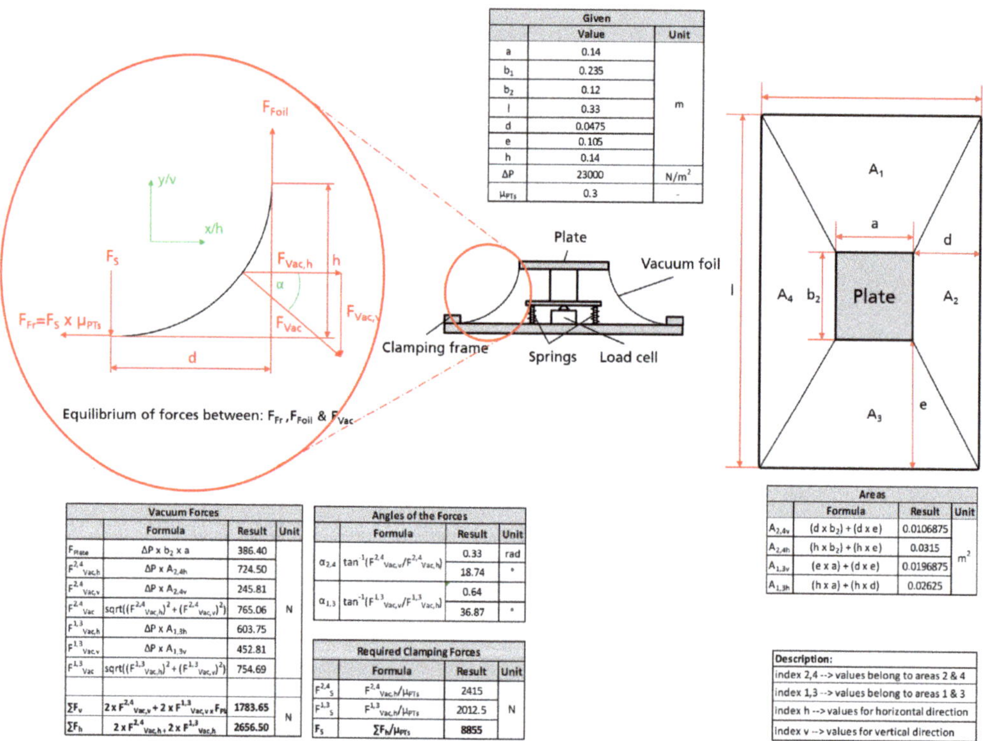

Given

	Value	Unit
a	0.14	
b_1	0.235	
b_2	0.12	
l	0.33	m
d	0.0475	
e	0.105	
h	0.14	
ΔP	23000	N/m^2
μ_{PTs}	0.3	-

Vacuum Forces

	Formula	Result	Unit
F_{Plate}	$\Delta P \times b_2 \times a$	386.40	
$F^{2,4}_{Vac,h}$	$\Delta P \times A_{2,4h}$	724.50	
$F^{2,4}_{Vac,v}$	$\Delta P \times A_{2,4v}$	245.81	
$F^{2,4}_{Vac}$	$sqrt((F^{2,4}_{Vac,h})^2 + (F^{2,4}_{Vac,v})^2)$	765.06	N
$F^{1,3}_{Vac,h}$	$\Delta P \times A_{1,3h}$	603.75	
$F^{1,3}_{Vac,v}$	$\Delta P \times A_{1,3v}$	452.81	
$F^{1,3}_{Vac}$	$sqrt((F^{1,3}_{Vac,h})^2 + (F^{1,3}_{Vac,v})^2)$	754.69	
ΣF_v	$2 \times F^{2,4}_{Vac,v} + 2 \times F^{1,3}_{Vac,v} \times F_{Pl}$	1783.65	N
ΣF_h	$2 \times F^{2,4}_{Vac,h} + 2 \times F^{1,3}_{Vac,h}$	2656.50	

Angles of the Forces

	Formula	Result	Unit
$\alpha_{2,4}$	$\tan^{-1}(F^{2,4}_{Vac,v}/F^{2,4}_{Vac,h})$	0.33	rad
		18.74	°
$\alpha_{1,3}$	$\tan^{-1}(F^{1,3}_{Vac,v}/F^{1,3}_{Vac,h})$	0.64	
		36.87	°

Required Clamping Forces

	Formula	Result	Unit
$F^{2,4}_S$	$F^{2,4}_{Vac,h}/\mu_{PTs}$	2415	
$F^{1,3}_S$	$F^{1,3}_{Vac,h}/\mu_{PTs}$	2012.5	N
F_S	$\Sigma F_h/\mu_{PTs}$	8855	

Areas

	Formula	Result	Unit
$A_{2,4v}$	$(d \times b_2) + (d \times e)$	0.0106875	
$A_{2,4h}$	$(h \times b_2) + (h \times e)$	0.0315	m^2
$A_{1,3v}$	$(e \times a) + (d \times e)$	0.0196875	
$A_{1,3h}$	$(h \times a) + (h \times d)$	0.02625	

Description:
index 2,4 --> values belong to areas 2 & 4
index 1,3 --> values belong to areas 1 & 3
index h --> values for horizontal direction
index v --> values for vertical direction

Figure A1. Model for the Calculation of the Necessary Clamping Force.

References

1. Francis, F.L. Solid Processes. In *Materials Processing–A Unified Approach to Processing of Metals, Ceramics and Polymers*, 1st ed.; Academic Press: Cambridge, MA, USA, 2016; Chapter 4; pp. 251–342.
2. Brenken, B. Extrusion deposition additive manufacturing of fiber reinforced semi-crystalline polymers. Ph.D. Thesis, Purude University, West Lafayette, IN, USA, 2017.
3. Brøtan, V.; Berg, O.Å.; Sørby, K. Additive Manufacturing for Enhanced Performance of Molds. *Procedia CIRP* **2016**, *54*, 186–190. [CrossRef]
4. Fette, M.; Sander, P.; Wulfsberg, J.P.; Zierk, H.; Herrmann, A.S.; Stoess, N. Optimized and Cost-efficient Compression Molds Manufactured by Selective Laser Melting for the Production of Thermoset Fiber Reinforced Plastic Aircraft Components. *Procedia CIRP* **2015**, *35*, 25–30. [CrossRef]
5. Warden, G.F. Development of an Additive Manufacturing Compression Molding Process for Low Cost In-House Prototyping. Bachelor's Thesis, California Polytechnic State University, San Luis Obispo, CA, USA, 2018.
6. Bere, P.; Neamtu, C.; Udroiu, R. Novel Method for the Manufacture of Complex CFRP Parts Using FDM-based Molds. *Polymers* **2020**, *12*, 2220. [CrossRef]
7. Hassen, A.A.; Lindahl, J.; Post, B.; Love, L.; Kunc, V. Additive manufacturing of composite tooling using high temperature thermoplastic materials. In Proceedings of the SAMPE Conference, Long Beach, CA, USA, 23–26 May 2016.
8. Yeole, P.; Herring, C.; Hassen, A.; Kunc, V.; Stratton, R.; Vaidya, U. Improve durability and surface quality of additively manufactured molds using carbon fiber prepreg. *Polym. Compos.* **2021**, *42*, 2101–2111. [CrossRef]
9. Hassen, A.A.; Springfield, R.; Lindhal, J.; Post, B.K.; Love, L.; Chad, D.; Uday, V.; Pipes, R.B.; Kunc, V. The Durability of Large-Scale Additive Manufacturing Composite Molds. In Proceedings of the CAMX Conference, Anaheim, CA, USA, 26–29 September 2016.
10. Junk, S.; Schrock, S.; Schröder, W. Additive tooling for thermoforming a cowling of an UAV using binder jetting. In Proceedings of the 22nd International Esaform Conference On Material Forming (Esaform 2019), Vitoria-Gasteiz, Spain, 8–10 May 2019; AIP Publishing: College Park, MD, USA, 2019; Volume 2113, p. 150001.
11. Chimento, J.; Highsmith, M.J.; Crane, N. 3D printed tooling for thermoforming of medical devices. *Rapid Prototyp. J.* **2011**, *17*, 387–392. [CrossRef]
12. Modi, Y.K. Calcium sulphate based 3D printed tooling for vacuum forming of medical devices: An experimental evaluation. *Mater. Technol.* **2018**, *33*, 642–650. [CrossRef]
13. Bhandari, S. Feasibility of Using 3D Printed Molds for Thermoforming Thermoplastic Composites. Master's Thesis, University of Maine, Orono, ME, USA, 2017.
14. Blanchard, P.J.; Rudd, C.D. Cycle time reduction in resin transfer moulding by phased cytalyst injection. *Compos. Sci. Technol.* **1996**, *56*, 123–133. [CrossRef]
15. Jamil, M.S.; Khalid, R.; Zulqarnain, A.; Salman, M. Improving Thermoform Productivity: Case of Design-of-Experiment. *J. Qual. Technol. Manag.* **2018**, *XV*, 87–106.
16. Henning, F.; Moeller, E. *Handbuch Leichtbau: Methoden, Werkstoffe, Fertigung*; Hanser: Munich, Germany, 2011.
17. Schug, A. Unidirectional Fibre Reinforced Thermoplastic Composites: A Forming Study. Ph.D. Thesis, Technical University Munich, Munich, Germany, 2020.
18. Sherwood, J.A.; Fetfatsidis, K.A.; Gorczyca, J.L.; Berger, L. Fabric thermostamping in polymer matrix composites. In *Manufacturing Techniques for Polymer Matrix Composites (PMCs)*; Elsevier: Amsterdam, The Netherlands, 2012; pp. 139–181.
19. Abdin, Y.; Taha, I.; Elsabbagh, A.; Ebeid, S. Description of draping behaviour of woven fabrics over single curvatures by image processing and simulation techniques. *Compos. Part B Eng.* **2013**, *45*, 792–799. [CrossRef]
20. Mills, A.R.; Cornero, L.; Bouquerel, L.; Macura, N.; Cruz, B.A. Materials and Process Development for High Rate Manufacturing of a Carbon Fibre Composite Automotive Floor. In Proceedings of the SAMPE Europe Conference, Stuttgart, Germany, 14–16 November 2017.
21. Harrison, P.; Gomes, R.; Curado-Correia, N. Press forming a 0/90 cross-ply advanced thermoplastic composite using the double-dome benchmark geometry. *Compos. Part A Appl. Sci. Manuf.* **2013**, *54*, 56–69. [CrossRef]
22. Thermoformed Plastics Market Size, Share. Industry Report, 2019-URL. Available online: https://www.grandviewresearch.com/industry-analysis/thermoformed-plastics-market (accessed on 2 August 2021).
23. Throne, J. Thermoforming. In *Applied Plastics Engineering Handbook*, 2nd ed.; Plastics Design Library: William Andrew, Applied Science Publishers: Norwich, NY, USA, 2017; pp. 345–375.
24. Schwarzmann, P. *Thermoforming: A Practical Guide*, 2nd ed; Hanser: Munich, Germany, 2018.
25. Morris, B. Thermoforming, Orientation, and Shrink. In *The Science and Technology of Flexible Packaging: Multilayer Films from Resin and Process to End Use Plastics Design Library*; Elsevier Inc.: Amsterdam, The Netherlands, 2017; pp. 401–433.
26. Cha, J.; Song, H.Y.; Hyun, K.; Go, J.S. Rheological measurement of the nonlinear viscoelasticity of the ABS polymer and numerical simulation of thermoforming process. *Int. J. Adv. Manuf. Technol.* **2020**, *107*, 2449–2464. [CrossRef]
27. O'Connor, C.; Martin, P.; Sweeney, J.; Menary, G.; Caton-Rose, P.; Spencer, P. Simulation of the plug-assisted thermoforming of polypropylene using a large strain thermally coupled constitutive model. *J. Mater. Process. Technol.* **2013**, *213*, 1588–1600. [CrossRef]
28. Chen, S.-C.; Huang, S.-T.; Lin, M.-C.; Chien, R.-D. Study on the thermoforming of PC films used for in-mold decoration. *Int. Commun. Heat Mass Transf.* **2008**, *35*, 967–973. [CrossRef]

29. Collins, P.; Harkin-Jones, E.M.A.; Martin, P.J. The Role of Tool/Sheet Contact in Plug-assisted Thermoforming. *Int. Polym. Process.* **2002**, *17*, 361–369. [CrossRef]

30. Morales, R.A.; Candal, M.V.; Santana, O.O.; Gordillo, A.; Salazar, R. Effect of the thermoforming process variables on the sheet friction coefficient. *Mater. Des.* **2014**, *53*, 1097–1103. [CrossRef]

31. Schützer, K.; Helleno, A.L.; Pereira, S.C. The influence of the manufacturing strategy on the production of molds and dies. *J. Mater. Process. Technol.* **2006**, *179*, 172–177. [CrossRef]

32. Sukhorukov, D.; Kreshchik, A.; Sharshin, V.; Sukhorukova, E. Recycling of polymer materials for foundry patterns. *IOP Conf. Series Mater. Sci. Eng.* **2020**, *919*, 062037. [CrossRef]

33. ASTM. ASTM F2792-12a-Standard Terminology for Additive Manufacturing Technologies. *Rapid Manuf. Assoc.* **2013**, 10–12. [CrossRef]

34. Gokuldoss, P.K.; Kolla, S.; Eckert, J. Additive Manufacturing Processes: Selective Laser Melting, Electron Beam Melting and Binder Jetting—Selection Guidelines. *Materials.* **2017**, *10*, 672. [CrossRef] [PubMed]

35. Mostafaei, A.; Elliott, A.M.; Barnes, J.E.; Li, F.; Tan, W.; Cramer, C.L.; Nandwana, P.; Chmielus, M. Binder jet 3D printing—Process parameters, materials, properties, modeling, and challenges. *Prog. Mater. Sci.* **2021**, *119*, 100707. [CrossRef]

36. Türka, D.A.; Triebea, L.; Meboldta, M. Combining Additive Manufacturing with Advanced Composites for Highly Inte-grated Robotic Structures. *Procedia. CIRP* **2016**, *50*, 402–407. [CrossRef]

37. Teizer, J.; Blickle, A.; King, T.; Leitzbach, O.; Guenther, D.; Mattern, H.; König, M. BIM for 3D Printing in Construction. In *Building Information Modeling*; Springer Science and Business Media LLC: Berlin, Germany, 2018; pp. 421–446.

38. Jipa, A.; Meibodi, M.A.; Giesecke, R.; Shammas, D.; Leschok, M.; Bernhard, M.; Dillenburger, B. 3D-Printed Formwork for Prefabricated Concrete Slabs. In Proceedings of the 1st International Conference on 3D Construction Printing (3DcP), Melbourne, Australia, 26–28 November 2018.

39. Sama, S.R.; Badamo, T.; Manogharan, G. Case Studies on Integrating 3D Sand-Printing Technology into the Production Portfolio of a Sand-Casting Foundry. *Int. J. Met.* **2020**, *14*, 12–24. [CrossRef]

40. Stegmaier, G. Zu 90 Prozent kein BMW-Motor. Auto-Motor-und-Sport Zeitung. Available online: https://www.auto-motor-und-sport.de/tech-zukunft/bmw-m3-motor-2020-sechszylinder-biturbo/ (accessed on 16 July 2021).

41. Recknagel, U.; Dahlmann, M. Spezialsande—Formgrundstoffe für die moderne Kern- und Formherstellung. *Giesserei. Rund-Schau.* **2009**, *56*, 9–17.

42. Günther, D.; Mögele, F. Additive Manufacturing of Casting Tools Using Powder-Binder- Jetting Technology. In *Igor V. Shishkovsky. New Trends in 3D Printing*; IntechOpen: London, UK, 2016; pp. 53–86.

43. Rao, P.N. Forming & Welding, Foundry. In *Manufacturing Technology*, 2nd ed.; Tata McGraw-Hill Pub.: New Delhi, India, 1998.

44. Wang, X. Thermal, Physical, and Mechanical Properties of Raw Sands and Sand Cores for Aluminum Casting. Master's Thesis, The University of Leoben, Leoben, Austria, 2014.

MDPI

St. Alban-Anlage 66

4052 Basel

Switzerland

Tel. +41 61 683 77 34

Fax +41 61 302 89 18

www.mdpi.com

Materials Editorial Office

E-mail: materials@mdpi.com

www.mdpi.com/journal/materials

9 783036 525853